Mobile Location Services
The Definitive Guide

Andrew Jagoe

PRENTICE
HALL
PTR

PRENTICE HALL
PROFESSIONAL TECHNICAL REFERENCE
UPPER SADDLE RIVER, NJ 07458
WWW.PHPTR.COM

ISBN 0-13-008456-5

90000

9 790130 084568

Library of Congress Cataloging-in-Publication Data

Jagoe, Andrew.
   Mobile location services : the definitive guide / Andrew Jagoe.
     p. cm.
   Includes bibliographical references and index.
   ISBN 0-13-008456-5
   1. Mobile computing. 2. Mobile communication systems. 3. Geographic information
systems. 4. Automatic tracking. I. Title.

QA76.59 .J34 2002
004.165--dc21

2002075284

Editorial/production supervision: *Jane Bonnell*
Composition: *Vanessa Moore*
Cover design director: *Jerry Votta*
Cover design: *Nina Scuderi*
Manufacturing buyer: *Maura Zaldivar*
Publisher: *Bernard M. Goodwin*
Editorial assistant: *Michelle Vincenti*
Marketing manager: *Dan DePasquale*

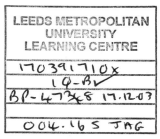

ISBN 0-13-008456-5

Pearson Education LTD.
Pearson Education Australia PTY, Limited
Pearson Education Singapore, Pte. Ltd.
Pearson Education North Asia Ltd.
Pearson Education Canada, Ltd.
Pearson Educación de Mexico, S.A. de C.V.
Pearson Education—Japan
Pearson Education Malaysia, Pte. Ltd.

About Prentice Hall Professional Technical Reference

With origins reaching back to the industry's first computer science publishing program in the 1960s, Prentice Hall Professional Technical Reference (PH PTR) has developed into the leading provider of technical books in the world today. Formally launched as its own imprint in 1986, our editors now publish over 200 books annually, authored by leaders in the fields of computing, engineering, and business.

Our roots are firmly planted in the soil that gave rise to the technological revolution. Our bookshelf contains many of the industry's computing and engineering classics: Kernighan and Ritchie's *C Programming Language,* Nemeth's *UNIX System Administration Handbook,* Horstmann's *Core Java,* and Johnson's *High-Speed Digital Design.*

PH PTR acknowledges its auspicious beginnings while it looks to the future for inspiration. We continue to evolve and break new ground in publishing by providing today's professionals with tomorrow's solutions.

PRENTICE
HALL
PTR

Contents

Foreword   **xi**

Preface   **xv**

Acknowledgments   **xvii**

Part 1   Introduction   1

Chapter 1   Mobile Location Services   1

The GIS and Wireless Industries Meet   2

Market Drivers   4

    Auto Manufacturers and Mobile Operators   4

    Government Mandate   6

New Opportunities and Challenges for Application Developers   7

Chapter 2   Building a Mobile Location Services Solution   9

Wireless and Mobile Network Basics   9

    Mobile Location Services Deployment Environment   9

    How Does a Wireless Network Work?   9

The Mobile Location Services Industry   14

Solution Components: What Makes a Mobile Location Solution?   15

Part 2   The Mobile Location Server   17

Chapter 3 The Application Server   17
What Is an Application Server?   18
Why Is an Application Server Important?   18
  Intellectual Property Protection   19
  Reuse   19
  Manageability   19
  Network Communications Security   19
  Performance   19
J2EE Application Server   20
  Why Java on the Server?   20
  What Is a J2EE Application Server?   21
  Application Server Architecture   22
  Additional Server Sizing Resources   28
Endnotes   28

Chapter 4 Spatial Analysis   29
Sample Spatial Analysis Server   29
  Hybrid Database Model   31
  Relational Database Model   31
  Object-Oriented Database Model   31
Digital Maps   32
  What Is a Digital Map?   32
  How Are Digital Maps Created and Maintained?   33
Geographical Data Types   34
  Points   34
  Nodes   35
  Lines and Arcs   35
  Polygons   36
  Linked Attributes   36
  Computer Storage of Geographic Data Structures   36
Projections and Coordinate Systems   37
Geocoding   39
  Technical Definition of Geocoding   39
  How Does Geocoding Work?   40

What Makes Geocoding So Difficult? 42
Why Is Geocoding Important to Mobile Location Services? 43
Reverse Geocoding 44
Routing 45
Routing Problems Defined 45
What Impacts Route Calculation Performance? 49
Map Image Generation 50
People Are Visual 50
Maps Aid Decision Making 50
Maps Consumers Can Use 50
Raster Maps 52
Vector Maps 55
POI Searches 56
Search Types 56
POI Databases 57
Routing With POIs 60
POI Sources 60
Real-Time Attribute Editing 61
Traffic Applications 61
Other Map Updates 61
Endnotes 61

Chapter 5 Mobile Positioning **63**
Cell of Origin 65
GPS 66
How GPS Works 66
A-GPS 67
Augmenting GPS With Dead Reckoning 68
Enhanced-Observed Time Difference 68
Angle of Arrival 69
Time Difference of Arrival 69
Location Pattern Matching 72
Map Matching 72
Positioning Accuracy and Speed Requirements
    for Mobile Location Services Applications 73

Chapter 6 Authentication and Security   **75**

Special Concerns in Wireless Networks   76

    GSM and Security   77

    Wireless Transport Layer Security (WTLS): Security for WAP   77

Endnotes   80

Chapter 7 Personalization and Profiling   **81**

What Is Personalization and Why Is It Important?   81

A Personalization and Profiling System   83

    Generation and Control of Location Information   84

    Integration With Customer Relationship Management   84

    P3P   84

    Microsoft .NET and Passport   87

Privacy Issues   88

    Industry Self-Regulation Efforts   89

Endnotes   90

Chapter 8 Billing   **91**

Technology and Business Models   92

Roaming   93

Who Will Pay?   94

Endnotes   95

Chapter 9 Mobile Commerce   **97**

What Is M-Commerce?   97

Mobile Electronic Transactions Standard   99

Part 3   The Mobile Location Client   **101**

Chapter 10 Client Platforms and Protocols   **101**

Platforms   102

    Palm OS   102

    Microsoft Windows CE   103

    Symbian OS   105

    Sample LBS Client Devices: The Car Dashboard   106

Client Protocols and Languages   109

    XML   109

    WAP and WML   112

    Voice and VoiceXML   113

    Java and J2ME   115

    BREW   116

Internationalization and Localization   117

    Localization: The Voice System Challenge in Europe   118

    Localization: The Map Data Challenge   118

Endnotes   118

Part 4   Mobile Location Service Technology   119

Chapter 11   Mobile Location Service Applications   119

Navigation and Real-Time Traffic   119

    Traffic   120

Emergency Assistance   123

Concierge and Travel Services   127

Location-Based Advertising and Marketing   129

Location-Based Billing   130

Endnotes   130

Part 5   Advanced Topics   131

Chapter 12   Digital Map Databases   131

"Best of Breed" Maps   131

    When Is a Best of Breed Map Required?   131

    Example Merge: Kivera Combined NavTech/GDT Solution   136

Endnotes   139

Appendix A   Abbreviations   141

Appendix B   Geography Markup Language   145

Appendix C LIF Mobile Location Protocol   **219**

Appendix D Platform for Privacy Preferences (P3P)   **313**

Appendix E Internet Resources   **435**

References   **441**

Index   **445**

Foreword

Mobile location services, the subject of Andrew Jagoe's excellent new book, are poised to be the leading technology market of the first decade of the new century. Just as the synergy of computing and data communications produced a whirlwind of development, investment, and innovation in the 1990s, the rapid evolution and combination of wireless technologies, location determination, and enterprise computing—globally linked through the Internet—are generating the next great wave of creative growth in technology markets.

Jagoe has written the first book that explains and dissects all of the previously discrete technologies that are driving the mobile location services revolution. Through our work at Autodesk Location Services, we have learned that the effective deployment of scalable and reliable mobile location services requires knowledge of a broad, and previously orthogonal, array of hard and soft technologies. This comprehensive text is the first that presents the reader with the latest trends and standards in the entire portfolio of critical technologies: location determination, mapping software and databases, wireless and mobile networks, application software languages and interfaces, and mobile applications and platforms. Engineers and business people with expertise in just one of the aforementioned component technologies will learn how their piece fits into a unified mobile location environment.

The author draws extensively from a growing body of successful mobile location deployments, concentrated in the area of automotive telematics, that are proving a forerunner for a broader, telephone handset-based mobile location services market. The more than 1 million auto-based telematics terminals installed by the end of 2001 are ample testimony of the opportunities and attractiveness of the mobile location services market. This large and growing installed base of subscribers also provides multiple implementation examples, which are incorporated into the text for the benefit of those readers developing new mobile location applications.

I only wish that this comprehensive book was available to our staff as we developed some of the earliest automotive telematics services. In partnership with Fiat, Autodesk Location Services deployed a car-based information system that offers drivers real-time traffic reporting and driving directions, detailed point-of-interest information, and native-language concierge services across Europe. The development team had to integrate global positioning system and global system for mobile communication technologies; database interfaces to leading information vendors (like Michelin's famous restaurant guide); mapping and geocoding software; an administrative, billing, and provisioning platform (Autodesk's LocationLogic); and Fiat's multilingual call center. Experts in various fields needed a common technological baseline to communicate with their development colleagues in other disciplines. System architects and marketing and business managers required a technically rich overview to assess the multiple and interlocking requirements and constraints of a complex mobile location solution. This book provides an excellent and detailed tutorial on all of the necessary and relevant technologies for the mobile application developer.

What we learned through our early experiences developing and deploying the Fiat telematics service was the need to integrate the contributions of multiple content providers and application developers. An attractive and comprehensive solution for the subscriber required that unique value be aggregated from many diverse data sources and applications. It is clear that mobile location services will only thrive if there is an extensive application developer community that applies their creative talents to the problems and the needs of mobile subscribers. The market will develop rapidly and organically if robust and competitive independent software vendors are empowered to address the mobile location market. At Autodesk, we have witnessed the benefits that a large, active, and creative development community has had on the success and proliferation of the geographic information software market, an important component of mobile location services. It is my hope that this book will help enlarge the mobile location application developer community by bringing knowledge of the various mobile location disciplines to a wide audience.

What types of mobile location services are likely to prove successful, and therefore worthy of the application developer's time and effort? Our early experience at Autodesk Location Services points to three critical attributes that successful services incorporate: localization, personalization, and actionability. Localization and personalization provide a context and relevancy, a reasonable set of constraints on the set of possible options that the subscriber can select from the interface. These choice limits, which are determined by both the physical location of the user and his or her personal preferences and data, enable mobile location services to employ a concise and efficient interface. The limitless selection options and endless surfing of the wired Internet is impractical and unattractive for mobile users and, in the case of automotive telematics, downright dangerous.

However, personalization and localization are not enough. Our experience clearly indicates that successful services combine these context elements with actionability—the ability of the subscriber to immediately employ or act on the information provided, whether that means accessing an alternate driving route to avoid traffic, changing a restaurant reservation because of delay, or buying a movie ticket at the nearest theater for the next showing.

Delivering these three elements to the subscriber requires a delicate and seamless interplay of database, Internet, location determination, and wireless services. It also requires the coordination and partnership of system operators, application service providers, m-commerce merchants, handset manufacturers, and platform providers. Mastering and merging the requirements and perspectives of these different players, and their technologies, is a central theme of this book.

For application developers that rise to the challenge of mastering the diverse mobile location landscape, the financial benefits will be considerable, in my view. With 20/20 hindsight, mobile location services can learn from the missteps of the early, fixed-line Internet application developers. Although this technical book does not advocate specific business models, it does address the critical need for secure billing and high-volume transaction accounting—the basis for profitable service delivery, and a process at which mobile system operators excel.

I expect *Mobile Location Services: The Definitive Guide* to inspire a new generation of engineers and help propel a new wave of independent software vendors. Like Mead on VLSI design, or Hennessy and Patterson on computer architecture, Jagoe's book on mobile location services will be viewed as a technology classic and a key reference work. I continue to consult it on a daily basis.

Joe Astroth, Ph.D.
Executive Vice President, Autodesk Location Services

Preface

The mobility provided by the proliferation of wireless devices presents a new class of opportunities and problems for application developers. Mobile applications must be much smarter than desktop applications. As a mobile user, you want personalized applications that instantly know where you are, what things are around you, and how you get to them. Application developers face many challenges in building these applications, including how to pinpoint a user's location and then retrieve relevant spatial data from a digital map database sometimes as large as 20 gigabytes with the sub-second response times necessary for a mobile user.

A mobile location service platform requires knowledge of three highly complex technologies: mobile telephony, geographic information systems, and enterprise (or perhaps, more precisely, Internet) application development. The idea for this book came from my experience working with a number of companies developing mobile location service platforms and from the observation that organizations developing mobile location service solutions typically had significant experience in only one or two of the necessary technology areas. Good mobile location service solutions must:

- Seamlessly interface with the mobile operator's network and consider bandwidth, security, and mobile device form factor constraints.
- Provide scalable and high-speed spatial data retrieval.
- Leverage the paradigms of the most successful Web architectures to promote user adoption and provide for massive scalability.

The primary goal of this book is to provide a high-level overview of how all the pieces of a mobile location service application fit together. A second goal of the book is to provide a basic overview of how each underlying technology component in a mobile location service architecture works and a framework for evaluating commercial mobile location service platforms.

Who Is This Book For?

This book is for people who want to understand what mobile location services are and what type of mobile location service applications are being built today. This book is also for those who want to grasp the underlying mobile location service platform technology for evaluation purposes and who want to understand the technical architecture that must be employed to support a carrier-class mobile location service application. For the technically oriented, the book focuses on the many technical challenges that mobile location service applications face. Although some technology industry background is assumed, this book is not a programming manual and is accessible to readers with even the most basic programming and systems engineering background.

Acknowledgments

I have benefited from the insights and help of Alex Tschobokdji, Babak Fouladi, Brent Hoberman, Mark Winberry, Victoria Gray, Jet Broekman, Simon Buckingham, Hans Kammann, and Larry Williams. Thanks also go to Dave Wood, Eric Carlson, and all the other great people at Telcontar, Inc.

Special thanks go to everyone at Autodesk Location Services for their contributions. I especially benefited from the insights and support of Mike McGill, Brian McDonough, Tara Cooper, and Spencer Horowitz. Joe Astroth has been an exceptional supporter, and I believe Joe will lead and inspire a new class of software developers to build mobile location services applications that we cannot imagine today.

This book would not have been possible without the contributions of everyone at Kivera. My thanks to Clay Collier and Michael Fisher for seeing a great opportunity to help application developers. Rajiv Synghal's insights on mobile location services are without compare, and I always enjoy his input and criticism. I have benefited from the insights of Mark Strassman and Carlo Cardilli, and my thanks also go to John Gibb, Bill Patow, Noelle Murata, Michael Schwener, and Igor Grinkin for the first-class support they gave me. Kivera has an excellent engineering team, and I have benefited from the insights of Jerry Smythe, Matt Hartfield, Rajat Ahuja, Ashwini Verma, Vishwas Goel, and Tom Singer. The one person who has contributed more than any other to this book and to my understanding of mobile location services is Brian Shenson.

My thanks to my friends Jim Dunn, Richard Tung, Allen Rea, Mark Abraham, Erick Hendricks, and Jason Harris for all their support. The greatest thanks of all go to my parents, Michael and Susan Jagoe, who are the best parents imaginable, and to my sister, Christy Jagoe.

PART 1

1

Mobile Location Services

From mainframes to minicomputers to PCs, computer hardware has long been bulky—so bulky that not only was it not mobile, but it was kept in special air-conditioned rooms on a raised antistatic floor. Ethernet technologies developed to allow these computer systems to communicate first privately and then publicly, worldwide over the Internet. Developments in liquid crystal displays and hardware miniaturization allowed computers to become "mobile" as laptop computers. However, laptops are actually movable computers and not mobile computers. Radio signaling systems technology, complex cell networks, and further miniaturization of computer processors and memory made mobile phones possible. As mobile phones become more like computers and begin to process data and run applications, and as other small devices use mobile networks for communications, it is possible to see computers and wireless communications converging.

This new class of devices includes data-capable mobile phones, wireless personal digital assistants (PDAs), and even in-vehicle computers. What makes these devices fundamentally different from other computers is their inherent *mobility*. They provide anytime, anywhere instant access to applications. They travel with you. You don't wait for them to boot. You don't wait for them to dial an Internet service provider (ISP). Most important, you frequently use them while you are doing something else. In-vehicle navigation devices are used while driving. Mobile phones are used while walking to an appointment, waiting in a lobby, or riding an elevator. Anyone who has tried to use a laptop while driving a car knows that it is almost impossible. Traditional computers have interfaces that are designed for focused attention and stationary use.

In a stationary computing environment, it matters little where a user actually is other than to set the correct time zone and language. When real mobility is added to computing, a new world of applications and capabilities are enabled that take advantage of knowing precisely where a user is. These applications include dynamic navigation and real-time traffic, advanced emergency services and roadside assistance, instant concierge and intelligent travel services, and the ability to use location in a variety of other services to significantly improve personalization.

Mobile location services are actually a subset of a larger set of new capabilities enabled by advanced personalization technologies: *context-based services*. Applications that are context-enabled not only are able to customize themselves based on where a user is, but also on who the user is and the *role* the user might be playing at a given time. An electrical engineer seeking information on microprocessors is likely to be interested in a different level of detail than a human resources professional. However, roles are more than personalizing based on job titles and areas of expertise. Individuals themselves have multiple roles. An individual can be an employee, a university professor, a father, or all three. Both time and location give powerful indicators to the context in which someone is using an application—and therefore how it should be personalized.

THE GIS AND WIRELESS INDUSTRIES MEET

Mobile location services are at the intersection of the geographic information systems (GIS) industry and the wireless networking industry. Spatial analysis technologies developed in GIS have been repurposed for the speed and scalability required for mobile location services. Mobile operators' wireless data networks are used for application deployment, and positioning technologies leverage wireless and satellite technologies to perform complex measurements to pinpoint the location of a mobile user—a critical piece of information in many mobile location-based applications.

This is a new area for both GIS vendors and telecommunications companies. Traditionally GIS has been designed for powerful computers that perform spatial searches locally and do not require millisecond response times to route generation. Telecommunication companies and mobile operators excel at providing voice services, but are new to providing data services, content, and applications.

Location-based information and analysis enable valuable decision-making tools in a variety of applications. GIS applications have long been used for rapid response by fire departments, resource management by governmental agencies, cell tower placement by mobile phone companies, product marketing by direct marketing businesses, and many similar niche applications. How are mobile location-based services different, or are they?

Advances in communications, GIS, and Internet technologies have presented an opportunity to incorporate location information into valuable, easy-to-use solutions for the average person. When GIS was introduced to the Web in mapping and driving directions services like MapQuest, they were very popular. However, Web services are unable to leverage a key element that distinguishes mobile location services: the ability to personalize in real time by knowing where a user physically is.

So what exactly are mobile location services? In the popular context, mobile location services have come to mean solutions that leverage positional and spatial analysis tools (location information) to deliver consumer applications on a mobile device. Application opportunities can be grouped into the following categories:

- Navigation and real-time traffic
- Emergency assistance
- Concierge and travel services
- Location-based marketing and advertising
- Location-based billing

Solutions in these categories will be provided to millions of users by large players in the telecommunication, automotive, and media industries. In many cases, the solution will contain several or all of these applications. This does not mean there are not other niche opportunities. Specialized location-based applications have been developed and are in use by companies such as FedEx, UPS, and Lojack. This corporate market includes the following services:

- Dispatch and delivery route optimization
- Fleet, asset, and individual tracking
- Security and theft control

There is always room for improvement, but these markets are well served by existing solutions. These corporate applications do not have the scalability and open-standards requirements that products built for the consumer space need, requirements that traditional GIS products were not designed to meet.

The term *mobile location services* has developed in parallel with *telematics*, a term that dates to 1980, to mean the combining of computers and telecommunications. Telematics is actually the English version of the French term *telematique*, which was coined by Simon Nora and Alain Minc in the book *L'informatisation de la Societe* (La Documentation Francaise, 1978). Telematics is often used interchangeably with mobile location-based services, although more recently telematics is increasingly used to mean automotive telematics, or mobile location services for in-vehicle use. For clarity, this book uses mobile location services as defined earlier and telematics to mean automotive telematics.

To deploy high-quality location-based services means not only to correctly implement this intersection of GIS and wireless technologies, but to understand from Internet and media companies how to build robust and scalable services that people actually want to use, or better yet, feel they can't be without.

The market for mobile location services is being driven by two primary forces: auto manufacturers and mobile operators who understand the value of location-based applications and are investing heavily in pushing the products to market, and government regulation.

MARKET DRIVERS

Auto Manufacturers and Mobile Operators

In 2001, consumers were largely unaware of mobile location services, more so in North America than in Europe. Recognizing the incremental revenue opportunities and strategic importance of mobile location services, mobile operators and auto manufacturers have started investing heavily in building the infrastructure to deploy them.

Mobile location services can be divided into two categories: those designed for in-vehicle use and those designed for personal use. Although the applications may be similar at a basic level, the business models, client technology, and service providers are distinctly different. In-vehicle solutions will be developed and provided by the large auto manufacturers and their investments, often as original equipment provided with the vehicle, whereas personal solutions will be developed by mobile operators and deployed on small handheld mobile devices. Both in-vehicle applications and personal applications are likely to leverage partnerships with large media and travel services organizations for a complete and compelling solution.

In-Vehicle Mobile Location Services: Auto Manufacturers

There are a number of forces driving the interest of mobile locations services in vehicles. Consumers in North America are interested in the additional safety they provide in emergency situations. Consumers in Europe are interested in the navigation capabilities they provide. Auto manufacturers see mobile location services as powerful differentiation, an opportunity to improve their relationship with a consumer and generate incremental revenue, and as a tool to collect valuable diagnostics feedback that could reduce the cost of vehicle production. Auto manufacturers typically provide a roadside assistance package with new vehicles, and they see the capabilities enabled by mobile location services as a natural extension.

Auto manufacturers see significant value in improving their long-term relationship with an individual consumer because repeat customers are much more profitable than new ones. One of the best ways to do that is to create an interactive service in the car that provides a two-way dialogue between the customer and the manufacturer. It enhances the driving experience with navigation and concierge services, while at the same time reassuring drivers that they will quickly receive assistance in an emergency.

Because auto manufacturers see this customer relationship (and the customer's driving experience) as core to their business, they are prepared to subsidize the service while the market is developing and business models are still being determined, as shown by OnStar providing limited-time free service to more than 1 million users in North America. Auto manufacturers have invested substantially in independent companies such as General Motors' OnStar (*http://www.onstar.com*), Ford Motor Company's Wingcast (*http://www.wingcast.com*), and DaimlerChrysler's Tegaron (*http://www.tegaron.com*) to deliver their service. OnStar provides branded service for GM cars in North America and Opel (a GM brand) cars in Germany. Tegaron provides private-label service for various models of Mercedes-Benz, Audi, and Renault in Europe. Wingcast had planned to provide private-label service to owners of Ford, Jaguar, Volvo, and other makes, both inside the Ford Motor Company brand portfolio and outside it. Furthermore, most auto manufacturers soon plan to make mobile location services client hardware standard equipment in new vehicles.

In-vehicle solutions have the advantage of being able to leverage the sound and power systems of the vehicle—a significant advantage over handheld mobile location service solutions. In-vehicle solutions are not restrained by a small form factor, and can distribute their necessary components throughout the vehicle (i.e., GPS receiver in the trunk, central processing unit independent to the faceplate used to control the system, and an optional screen used to display rich color maps). They are also able to incorporate and leverage data such as vehicle velocity and heading to improve quality of the vehicle positioning—data that is hard to capture in a mobile phone. In-vehicle solutions are therefore likely to be able to provide much richer services that could be bundled with the same unit that provides the vehicle with DVD, TV, and even digital satellite (XM) radio capabilities—an additional incentive for consumers to subscribe.

Personal Mobile Location Services: Mobile Operators

The forces driving the interest in personal location services are less clear-cut than those driving in-vehicle location services. Because public transportation in Europe is excellent in comparison to most parts of North America and because Europe's dense urban areas often make private transportation an inconvenience (due to traffic and finding parking at your destination), many people don't drive or drive only infrequently. This is in stark contrast to North America, where leaving the home invariably means using a private car. This presents an interesting opportunity to provide personal location-based services in Europe that would otherwise be provided by an in-vehicle system.

Basic services would include multimodal directions, such as the ability to route a user on foot and by subways, trains, ferries, and even private car or taxi when necessary; the ability to locate nearby points of interest that can be reached on foot or via the public transportation network; and basic concierge and safety services. Although these services are of primary interest to Europeans and inhabitants of dense urban areas such as New York, it is possible to imagine a more powerful service that would provide a user with *mobility management*. This service might provide turn-key concierge services door-to-door for any destinations, including online travel reservations, navigation, personal safety, recommendations based on a personal profile, and even parking and event ticketing.

The challenges facing personal location services are several. The limited power supply and small form factor make rich services difficult to deliver. Mobile operators do not have the same vested interest auto manufacturers have in deploying location services: They are not capturing valuable diagnostics information from cars. Mobile operators already have an interactive two-way dialogue with their customers via Short Message Service (SMS), and are less likely to be in a position to fund free services to win adoption. At the end of the day, it is likely we will see location services as common in vehicles before we see them as common in wireless PDAs and mobile phones. In either case, a business model such as NTT DoCoMo's that enables and encourages third-party application development would allow the fastest development of a mass market.

Government Mandate

In addition to the marketing push generated by mobile operators and auto manufacturers, government regulations are playing a significant role in creating the market for mobile location services. Two instances are the provision of emergency services and the concept of road pricing.

E911/E112

The U.S. Federal Communications Commission (FCC) has enacted a series of orders to improve the reliability and quality of emergency services for wireless users, Enhanced 911 (E911). The requirements are to transmit all 911 calls and location information to public safety answering points without any intermediary validation procedures. All phones manufactured for sale in the United States after February 13, 2000 must provide override processing for 911 calls that allows them to be handled by any carrier regardless of whether they are the subscriber's preferred carrier or not. Additional implementation phases require sending location data with every emergency call and impose specific requirements for the accuracy required for this location information. These initiatives require mobile operators who wish to provide service in the United States to implement some of the most expensive components of a mobile location services architecture.

Europe has a similar initiative called Enhanced 112 (E112), but it is not a European Union mandate and it is fairly loosely defined. Responsibility for E112 is left to national emergency authorities. Although several countries are considering or have implemented mandates, E112 is not the driver for mobile location services that it is in the United States. Rather, European mobile location services are being driven by the value-add of location-enabled applications provided by mobile operators and the automotive sector.

Road Pricing

A second governmental regulation that is likely to impact the market for mobile location services is road pricing. Transportation in general has some negative side effects, including environmental pollution and congestion. Rather than building toll roads and imposing fuel taxes, the concept behind road pricing is to charge for actual road usage based on time, location, and type of vehicle.

The Netherlands has been carefully studying road pricing as a potential solution to their congestion problems. The Dutch government estimates that by 2006, they will have to accommodate 17 million people and 8.5 million vehicles in an area that is 34,000 square kilometers containing 118,000 kilometers of roads. A reference technical architecture called MobiMiles was developed in collaboration with the Dutch government as a potential road pricing solution. On June 6, 2001 the Dutch cabinet indicated in a letter to Parliament that it would begin implementing the MobiMiles system, with national coverage by 2006. The system is expected to receive final political approval in 2002, enter pilot tests in 2003, and begin rolling out in 2004.

The system in the Netherlands will be implemented through a public–private partnership, and will require a mobile location services device to be installed in every vehicle. This unique market force creates an attractive and significant market opportunity. The Netherlands' road pricing system is closest to becoming a significant market force to expedite deployment of mobile location services, but many other communities are also considering road pricing solutions, particularly in densely populated and economically developed regions. Solutions that use simple electronic toll collection are already in use in Singapore, Germany, and the United States.

The comprehensiveness of the Dutch approach will provide the necessary infrastructure to deploy some of the most sophisticated mobile location service applications. Additional details on the Netherlands road pricing system can be found at the Web site of the Dutch Ministry of Transportation, Public Works, and Water Management's National Traffic and Transportation Plan (*http://www.minvenw.nl/rws/projects/nvvp/*) or at *http://www.roadpricing.nl*.

NEW OPPORTUNITIES AND CHALLENGES FOR APPLICATION DEVELOPERS

The infrastructures being deployed by mobile operators and the automotive industry provide a platform that enables an entirely new category of applications. Focused on mobility-based computing, they have capabilities that today's stationary-based application development model doesn't even consider. Just as the most innovative applications for the Internet were not developed by Internet backbone providers, it is likely that the most successful "killer apps" will be developed by third-party software developers.

However, these new opportunities are not without challenges. Mobile location services require the combination of a number of independently complex technologies. Quality map data coverage, high-speed wireless data services, systems integration, and business models are just a few of the challenges that must be faced in building an application. This book seeks to provide you with an introduction to the concepts behind mobile location services and the technologies that enable them.

Chapter 2 develops a technical framework for a mobile location services solution. It provides a brief discussion of wireless networks, which is a useful background for understanding how to build applications that work in a mobile environment. We then discuss the systems architecture of the mobile operator that your mobile location services application will be deployed in.

Discussion of the mobile location server begins in Chapter 3. We discuss the cornerstone of any mobile location services infrastructure: the application server. Application servers are the glue that connects everything together and keeps it operating fluidly. We discuss why an application server is essential, as well as design techniques to make sure your mobile location services application will be able to scale.

Chapter 4 introduces spatial analysis, and the critical search and retrieval capabilities it provides your application. We briefly discuss digital maps and how they are developed. We then discuss geographic data types and the concepts behind coordinate systems and projecting locations on three-dimensional earth into a two-dimensional system that can be used in mobile location services. With an understanding of these basic map concepts, we discuss the core spatial functions you will need to location-enable your application: geocoding, reverse geocoding, routing, map imaging, points of interest searching, and real-time map attribute editing.

Central to the success of a mobile location services application is knowing where the user is. Chapter 5 introduces mobile positioning, with an overview and discussion of the various handset, network, and hybrid technologies, including cell of origin, angle of arrival, time difference of arrival, enhanced observed time difference, and the global positioning system (GPS). We also discuss dead reckoning and map matching, special techniques developed to improve the accuracy of positioning techniques for in-vehicle use.

Chapter 6 discusses authentication and security in the context of wireless applications. We analyze the security technologies implemented in the Global System for Mobile Communications (GSM) and Wireless Application Protocol (WAP), and what application risks you should be aware of.

In Chapter 7 we discuss personalization and profiling for your mobile location services application. We discuss emerging industry standards such as the Platform for Privacy Preferences (P3P) and Microsoft's .NET technology, and the privacy issues raised by the potential misuse of personal information.

Chapter 8 introduces the integration of sophisticated billing systems. Having the right business model is critical to the success of your mobile location services application. We discuss the challenges presented by roaming and billing systems that are designed for prepaid voice.

A discussion of mobile commerce (m-commerce) in Chapter 9 wraps up our focus on the mobile location server. We discuss what mobile commerce is, and technologies to facilitate your m-commerce processing, such as the Mobile Electronic Transactions Standard (MeT).

Chapter 10 extends our mobile location service infrastructure to include the client. We will discuss where to use various client platforms, including Palm OS, Microsoft Windows CE, and Symbian OS. We then look at some actual client devices from Pioneer and Blaupunkt. Finally, we will analyze a number of client protocols and languages, including Extensible Markup Language (XML), Simple Object Access Protocol (SOAP), WAP, VoiceXML (VXML), Java 2 MicroEdition (J2ME), and Qualcomm's Binary Runtime Environment for Wireless (BREW). Finally, because mobile location services are highly regional by nature, we discuss some general guidelines for making internationalization and localization of your application as easy as possible.

With a framework that now includes both mobile location services client and server, we look at some specific mobile location services applications in Chapter 11. We discuss navigation and real-time traffic, emergency assistance services, concierge and travel services, location-based advertising and marketing, and location-based billing.

Chapter 12 addresses advanced topics in digital map databases. We discuss the challenges map database quality and coverage can present to your application. Sophisticated techniques for combining map databases are analyzed, from edge matching to true database merging.

Building a Mobile Location Services Solution

WIRELESS AND MOBILE NETWORK BASICS

Mobile Location Services Deployment Environment

To better understand how developing mobile location services applications differ from traditional wireline applications, it is important to understand the basic principles of wireless networks. This section provides a brief and simplified overview of mobile network architectures. The discussion includes the basics of radio spectrum, cellular networks, and wireless data.

How Does a Wireless Network Work?

Radio Basics

Wireless networks are based on radio principles that are now more than 100 years old. Radio signals are electromagnetic radiation, a category that includes light and infrared waves as well. Radio signals are considered *transverse waves*, which means they have wavelength and frequency (see Figure 2.1).

The *wavelength* is the distance between the peaks of sequential waves and the *frequency* is the number of cycles per second (Hz). A transverse wave's speed can be calculated by multiplying the wavelength and the frequency, but all radio waves travel at the speed of light. When waves pass through solid material they are slowed down, but even in wireless communication systems the waves that pass through air and clouds are still traveling at about 300,000 kilometers per second.

Because the speed of travel is constant, increasing frequency shortens wavelength in a mathematical pattern, and usually only one is specified. The *amplitude* of the wave is its height, or distance from the axis to the peak. As waves move away from their source, and spread over a wider area, the amplitude decreases. This process of losing energy is called *attenuation*.

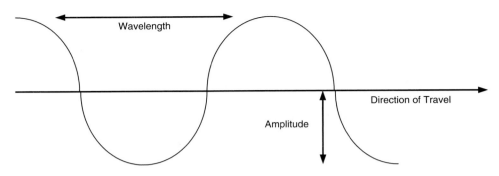

Figure 2.1   Transverse Wave of a Radio Signal.

The entire range of electromagnetic radiation is called the *spectrum*. Electromagnetic waves that are well suited to communications are known as radio, and have a lower frequency and longer wavelength than other forms of radiation (see Figure 2.2).

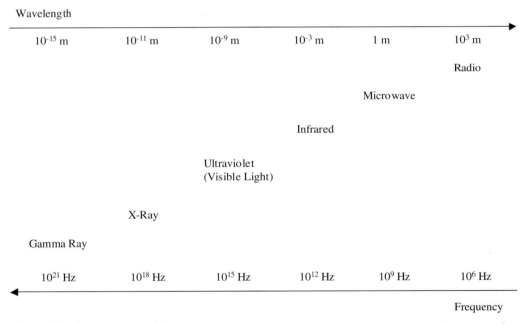

Figure 2.2   Electromagnetic Spectrum.

Generating electromagnetic waves requires accelerating a moving electrical charge by either changing its speed or direction. Radio transmitters work by vibrating electrons, the charged particles that surround all atoms. The faster the electrons vibrate, the higher the frequency of the resulting radio wave. A radio receiver uses the same process in reverse. The radio waves stir up electrons in the antenna, which create electric currents.

Wavebands

Radio signals with a high frequency have a much shorter range than radio signals with a low frequency because shorter wavelengths suffer greater attenuation. Different applications require different types of radio spectrum. Because spectrum is a scarce resource and is not unlimited, its use is subject to government licensing. Radio signals used for wireless communications are referred to as microwaves because of their very small wavelengths. The microwave spectrum ranges from .4 GHz to 100 GHz, and their high bandwidth makes them ideal for communications.

Analog and Digital Radio Systems

Traditional radio systems were designed to transmit sound. Sound is an analog signal that can be represented by a continuous wave similar to electromagnetic radiation. Simple analog broadcasting or cellular telephony systems convert sound waves to radio waves and then back to sound waves again. However, wireless networks are being increasingly used for data as much as for voice. Data is inherently digital, and a digital radio wave encodes data (the 0s and 1s used by computer systems) to radio waves and back again.

Digital radio has a number of significant advantages over analog radio. Digital systems allow a radio receiver to distinguish between static interference and the signal, so the static can be ignored. Digital signals are often also encoded with additional *checksum* data that allows the receiver to perform a mathematical calculation to make sure it received the transmission correctly. If not, it can be sent again. Digital signals can be compressed to use spectrum more efficiently and encrypted to prevent eavesdropping (see Chapter 7 for a more detailed analysis of security concerns in wireless networks and mobile location services). Finally, digital signals can take advantage of timing techniques to share communications channels in bandwidth-efficient ways.

Modulation: Encoding Radio Signals

Whether for analog voice, digital voice, or digital data, information has to be converted to radio waves before it can be transmitted. The process of altering a radio wave of a specific frequency so that useful information can be extracted from it is called *modulation*. The two primary methods of encoding a radio signal are amplitude modulation (AM) and frequency modulation (FM). AM encodes a radio signal by varying the height of the waves in accordance with an information signal. Because this method uses bandwidth inefficiently, it is rarely used in modern wireless systems.

FM keeps the amplitude constant and instead alters the frequency and wavelength of the radio signal. Because the amplitude is constant, the FM transmitter can operate at full power all the time and efficiently use the full spectrum allocated to it. Most digital wireless systems encode data using *phase modulation*, which is a special form of FM. Instead of just changing frequency and wavelength of a radio signal, phase modulation also quickly moves them to different points in their cycle, which is useful for encoding wireless data.

Cellular Networks

Mobile phone systems are comprised of a network of cells, each with a powerful radio transmitter at its center. This is both because the radio signals most effective for carrying digital voice and data are short ranged, and because the cellular design is modular and can provide redundancy and failover capabilities. The base stations are typically connected to each other via high-speed fiber, and then to the public phone system and the Internet. As a mobile device moves through the network it is passed from one base station to another, accessing services through the base station of the cell it is in.

Figure 2.3 shows a sample GSM network architecture. The mobile station (MS) is the mobile phone or handheld client device. The MS includes a subscriber identity module (SIM), used for authentication and security, and the hardware and software specific to the radio interface, called the mobile equipment (ME). The network switching subsystem (NSS) provides the basic switching, profile management and mobility management functions. The mobile switching center (MSC) provides the switching functions. The location of the MS is tracked by the home location register (HLR) and visitor location register (VLR). When an MS moves from its HLR, it is registered in the VLR of the system it is visiting. The HLR is then informed of the location of the MS. The NSS also manages subscriber authentication and security, which is handled in the authentication center (AuC). Connect-

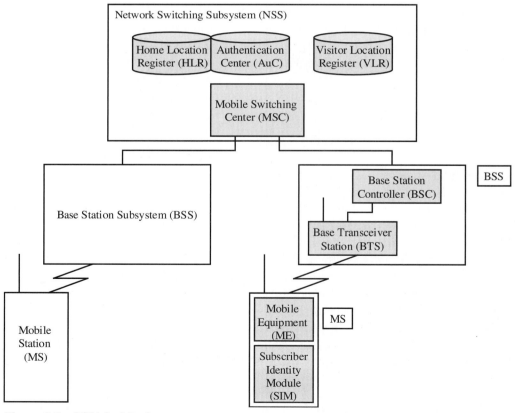

Figure 2.3 GSM Architecture.

ing the MS to the NSS is the base station subsystem (BSS). The BSS consists of base transceiver stations (BTSs) and a base station controller (BSC). The BSC performs the switching functions for the BSS and is connected to the MSC. The BSC performs hand-off management and radio channel allocation and release. The BTS contains the transmitter, receiver, and signaling equipment necessary to communicate to the MS over the radio interface.

To make most efficient use of the spectrum in a cell, mobile operators need to allow it to be shared using *multiple access* techniques. The most common method used today (including the European GSM standard) uses what is called Time Division Multiple Access (TDMA). TDMA divides a given band into a number of time slots that correspond to a communications channel. A mobile phone will transmit and send on only one slot and then remain quiet until its next turn to communicate. In GSM a time slot is 577 microseconds, so a mobile device could miss a scheduled slot and it would still not be noticeable to a human listener. The design of TDMA systems allows for them to easily upgrade to higher speed by allowing the mobile device to receive or transmit on more than one slot at a time. This is the principle behind GPRS (General Packet Radio Service). Another multiplexing technique used to make efficient use of spectrum is CDMA (Code Division Multiple Access). In CDMA, every signal is sent at the same time, but each signal is encoded differently so receivers can understand it. This is known as *spread spectrum*.

Wireless Data

Sending wireless data over a standard second-generation (2G) GSM mobile network requires full-time use of the voice channel, and allows rates of about 14.4 kbps. Third-generation (3G) mobile systems provide an always-on data connection at vastly faster rates than is possible in 2G mobile communications. An intermediate step that is a less expensive and a relatively straightforward upgrade from GSM is GPRS, sometimes also called 2.5G. GPRS is a packet-switched network, which uses bandwidth only when sending and receiving data. This allows it to be shared by numerous mobile devices at the same time, just as dial-up Internet users share one fast Internet connection from an ISP. The specification for GPRS allows it to provide up to 115 kbps.

Figure 2.4 extends the GSM architecture displayed in Figure 2.3 to include a sample GPRS architecture. The serving GPRS support node (SGSN) transmits and receives packets between the MS and the device it is communicating with over the public switched data network (PSDN). The gateway GPRS support node (GGSN) translates between the SGSN and the PSDN and supports a variety of connectionless and connection-oriented protocols, including Transmission Control Protocol/Internet Protocol (TCP/IP). The GGSN and the SGSN use the GSM location databases (HLR and VLR) to track the location of the MS to maintain connectivity as the MS moves through the mobile network.

The mobile location service infrastructure we are concerned with is accessed via the PSDN by an MS. The MS might contain proprietary client software to interact with your mobile location service, or it might use a thin client like WAP that requires no special client software. Mobile positioning technology requires additional handset or network elements and is explored in detail in Chapter 5.

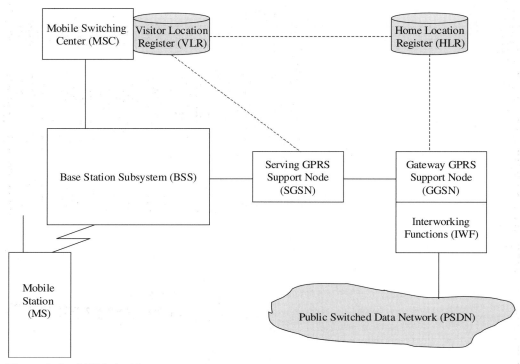

Figure 2.4 GPRS Architecture.

THE MOBILE LOCATION SERVICES INDUSTRY

Mobile location services can be divided into two categories: those designed for in-vehicle use and those designed for personal use. Although the applications might be similar at a basic level, the business models, client technology, and service providers are distinctly different. In-vehicle solutions will be developed and provided by the large auto manufacturers and their investments, often as original equipment provided with the vehicle, whereas personal solutions will be developed by mobile operators and deployed on small handheld mobile devices. Both in-vehicle applications and personal applications are likely to leverage partnerships with large media and travel services organizations for a complete and compelling solution.

Figure 2.5 illustrates the value chain for in-vehicle mobile location services. The auto manufacturers have large existing customer bases and will bundle hardware and service with new vehicles. Tier-one auto component suppliers such as Bosch (Blaupunkt) and VDO provide the hardware, and the applications and service will be provided by telematics service providers like OnStar and Tegaron. Unlike OnStar, which is a consumer brand, Tegaron plans private-label service for the auto brand. Providing mobile location services requires a mobile network and a location server to build applications with. The location server is a robust and scalable platform that provides quick operations like mobile positioning, address

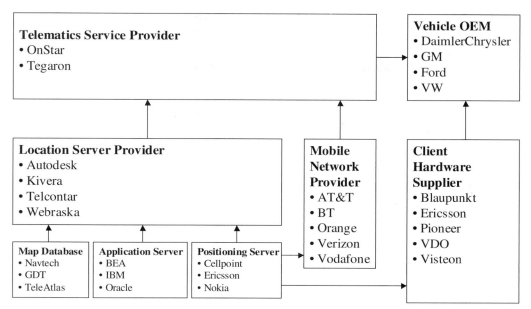

Figure 2.5   In-Vehicle Mobile Location Services Value Chain.

lookups (geocoding), driving directions (routing), map image generation, and nearest points of interest lookups. With this basic set of tools it is possible to begin building mobile location service applications.

Figure 2.6 shows how the personal mobile location services value chain is different. The existing customer base is owned by either the mobile operator or a media company such as Yahoo! or AOL. In Europe, the mobile portal is usually either an investment or a wholly owned subsidiary of the mobile operator. Unlike in the telematics value chain, services are branded and provided directly by the mobile portal. The same location server is required to provide service, and the client hardware is the mobile phone.

SOLUTION COMPONENTS: WHAT MAKES A MOBILE LOCATION SOLUTION?

Mobile location services are complex and require the integration of many disparate technologies into one seamless system. Figure 2.7 shows a sample mobile location services architecture and how the various components integrate. Central to the system is the application server, which provides the system integration framework. Services you will likely need for your mobile location services application in addition to the location server are authentication and security, user database integration, provisioning, personalization and profiling, billing, commerce, reporting, and systems management. The mobile client device communicates to the application server using the mobile network.

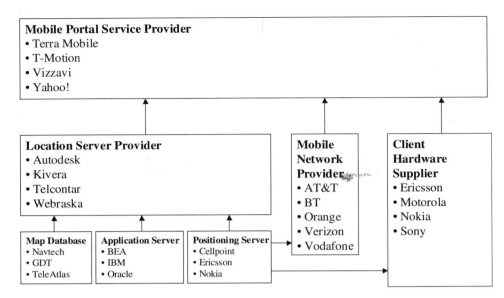

Figure 2.6   Personal Mobile Location Services Value Chain.

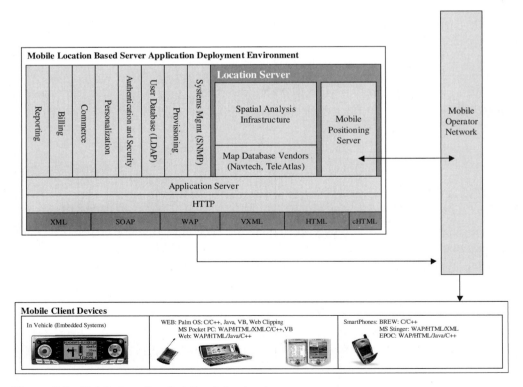

Figure 2.7   Mobile Location Solution Infrastructure.

PART 2

3

The Application Server

Mobile location service applications require the complex systems integration of many different technology components. The spatial analysis server that is designed to provide routes, maps, and points of interest is highly specialized for the scalability and performance requirements of a carrier class application. Map database quality and coverage varies significantly among map data vendors. Because most spatial analysis software vendors require data compilation in a proprietary data format, there is no assurance that the necessary coverage will be available when it is required. Positioning products are dependent on the mobile operator's network and handset technology.

It is important that a mobile location service infrastructure is designed with extensibility in mind as well as cost. This is particularly true if you don't own your own mobile network (i.e., a telematics service provider from the automotive space or an emergency services provider). An application server architecture provides a framework for extensibility. An application server allows you to develop and shelter the business logic that will differentiate you from your competitors and save you from having to rearchitect or throw out your system if a component in your architecture needs to change.

There are many reasons to approach a location service infrastructure as a series of logically discrete components integrated through business logic stored in an application server. It allows you to create infrastructure services based on industry standards for the various specialized components required. Your positioning interface might be based on the specifications recommended by the Location Interoperability Forum (LIF; see Appendix C) and your spatial analysis server interface might be based on the Geography Markup Language (GML; see Appendix B) specifications recommended by the Open GIS Consortium. The major advantage is that any one piece of your infrastructure is insulated from problems in another component. It allows you to replace components that do not deliver acceptable results without impacting the rest of the system, and it also allows you to potentially mix and match components. If your organization has a network with a positioning system that only supports assisted GPS and you acquire a company with a network that has a positioning system based on Enhanced Observed Time Difference (E-OTD), you would not necessarily have to replace the E-OTD infrastructure. Another example is your spatial analysis software. You might find one product works very well in North America, but does not handle the particularities of the European market well. Perhaps the software has poor or no map data support in a region you need coverage in. It is unnecessary and potentially expensive to be locked into a proprietary protocol.

WHAT IS AN APPLICATION SERVER?

An application server provides a server-side platform for building and deploying business logic. This business logic can be distributed across both client and server or can reside solely on the server. Most businesses use this concept today in systems that range from mainframe transaction systems to the stored procedures of client/server database management systems. All application servers have at least three discrete layers that interoperate: business logic layer, presentation layer, and data access layer. The business logic layer is the heart of the application server where all the intelligence and business rules are encapsulated in object-oriented reusable components. The data access layer allows the integration of specialized and discrete services that are made available to the components in the business logic layer. The presentation layer provides the methods and interfaces for delivering content from the application server. A simple example is the generation of a Hypertext Markup Language (HTML) Web page that is sent back to a user's Web browser. Of course the content presented could be nearly any format and any protocol, such as a Wireless Markup Language (WML) document sent over the wireless transport protocol or an XML document returned to a machine requestor via SOAP. A simplified example of a location-based services infrastructure designed in an application server environment is shown in Figure 3.1.

WHY IS AN APPLICATION SERVER IMPORTANT?

In addition to what was already discussed, an application server can provide many technical benefits. An application developer is thus able to focus on developing the business logic of his or her application rather than spending time on low-level systems features such as persistence and security.

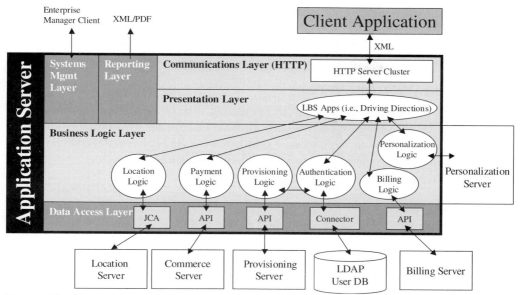

Figure 3.1   Simple LBS Infrastructure.

Intellectual Property Protection

The algorithms and processes in the business logic of a complex system are often the most valuable intellectual property. Storing these processes in the application server behind the presentation layer help prevent the system from being reverse engineered.

Reuse

Because business logic is developed independent of the presentation layer, it is possible to reuse the business logic to support new interfaces easily and efficiently. When business logic and presentation logic are mixed, it often means that significant code is repeated. As projects grow larger it becomes more and more difficult to manage a code base without separating the business logic and the presentation logic.

Manageability

Thin client applications in which the majority of the logic is on the server are much easier to manage and update than thick client applications. This is especially true in situations such as mobile location services where thousands, if not millions, of clients must be supported.

Network Communications Security

The presentation layer of an application server allows you to leverage robust Internet security protocols such as Secure Sockets Layer (SSL) and HTTPS.

Performance

Spatial analysis operations are intensive for the central processing unit (CPU), and can be more efficiently and cost-effectively processed on the server than they can be on a mobile device. The operations also perform best when located near the map database, which can be many gigabytes in size. The server cluster can take advantage of economy of scale and caching to reduce the client memory and processing utilization. This also reduces network traffic requirements and latency.

J2EE APPLICATION SERVER

There are several different approaches to integrating distributed enterprise applications. Typically they involve a high-speed network software bus operated by a transaction server such as BEA Tuxedo. A full discussion of transaction servers is beyond the scope of this book. To simplify our discussion, we focus on a mobile location service solution developed using Java 2 Enterprise Edition (J2EE).

Why Java on the Server?[1]

There are many reasons to consider using Java on the server.

It's a Better Third-Generation Language (3GL)

Java is a simpler 3GL than C++, but still provides the necessary capability to scale to solve large problems. Programming in Java can be much faster than programming in C++. With improvements in computer processing power and distributing computing, it is better to take advantage of the reduced development time than the potentially faster run-time.

It's a Better Fourth-Generation Language (4GL)

Java has been designed to be very easy to extend to reusable, high-level business abstractions. Because Java is also object oriented, it can be better suited to encapsulating business logic than a traditional fourth-generation scripting language.

Ubiquitous

Java has a massive developer community and industry support. Java's portability is an additional advantage when it is necessary to support multiple operating environments.

Robust

Java is also very robust, and can considerably reduce time to market. Unlike C and C++, Java does not provide direct access to memory locations (pointers), so memory reference errors, which are hard to diagnose and debug, are rare. Memory leaks are also rare in Java applications, due to the language's automatic garbage collection feature.

Strong Network Support

Because Java grew and developed with the Internet, it is not surprising that it has very strong network support. Java's network facilities allow high-level business object abstractions to be passed by value, allow you to change an object's underlying representations without breaking remote applications, support the ability to load new functionality with standard bytecodes, and provide distributed garbage collection.

Component-Oriented Computing

Java's component-oriented computing model allows applications to be developed with data independent of business logic and business logic independent of presentation logic.

This flexibility to reuse components makes development efforts with Java better able to support distributed development teams and long-term code changes.

It Is Fast

The performance of Java compared to native-compiled third-generation languages has improved significantly. It is important that in a distributed business application, only a small portion of the processing time is spent on business logic. The remainder is split between the database management system and the network.

What Is a J2EE Application Server?[2]

The J2EE specification states a set of minimum characteristics a J2EE application server must exhibit. These characteristics are delivered to your mobile location services infrastructure before you even begin to build. These minimum capabilities of any Java application server are described in the following sections.

Easy to Develop and Can Deploy Distributed Java Applications

A Java application server provides the structure and environment to facilitate building well-formed applications. In addition, the Java application server provides the systems infrastructure and management tools for deploying an enterprise application.

Scales to Permit Thousands of Cooperative Servers to Be Accessed by Tens of Thousands of Clients

Application servers are designed for scale. Preparing an application for deployment in an application server environment might be more complex and time consuming in the short run, but is well worth it in the long run. To deliver on scalability, an application server provides the following:

- It is fully multithreaded.
- It is parsimonious in consumption of network and other scarce resources.
- It has no architectural bottlenecks that prevent linear scaling.

Provides an Integrated Management Environment for Comprehensive View of Application Resources, Network Resources, System Resources, and Diagnostic Information

When substantial revenue and brand equity relies on consistent and high-quality service delivery, it is important to have instant access to robust monitoring and management tools. This suite of tools allows you to watch unchecked exceptions, logs, threads, sockets, network connections, and access control lists. You might want to supplement an application server's management environment with additional tools, but it is not efficient to develop your own suite of management tools. It is important that vendors for different components of your location services infrastructure (such as positioning and spatial analysis) provide systems feedback in accordance with standard monitoring tools such as Simple Network Management Protocol (SNMP) and Remote Monitoring (RMON). This data should be aggregated and monitored using a product such as HP Openview, CA Unicenter, or IBM Tivoli.

**Transaction Semantics to Protect Integrity of Data Even
as It Is Accessed by Distributed Business Components**

Because the application server environment is built to scale, a facility needs to be available to allow data to be processed in a transaction processing model with commits and rollbacks. This prevents data from being corrupted and overwritten, and systems from getting out of sync.

**Provides Secure Communications, Including SSL, Access
Control Lists, HTTP, and IIOP Tunneling to Communicate Across Firewalls**

Security is a key component of any application, and is especially important in an environment that will have thousands or millions of users. Poor security and security breaches shake confidence in your user base. A Java application server provides capabilities to operate from behind a firewall, support for various authentication systems including encrypted certificates, and encrypted communications such as SSL.

Application Server Architecture

All application servers are divided into at least three logical layers: the presentation layer, business logic layer, and data access layer (see Figure 3.2). As discussed earlier, a good application server product also has a systems management and reporting interface. A code level discussion of how these components operate is beyond the scope of this book.

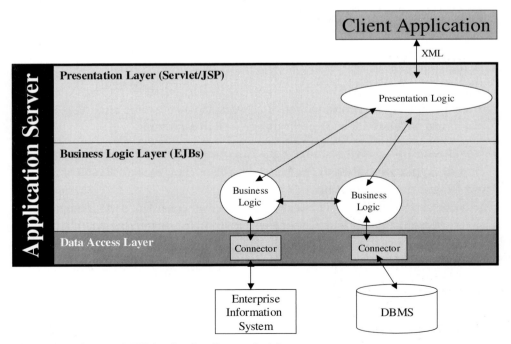

Figure 3.2   Simple J2EE Application Server Architecture.

Presentation Layer

The objective of the presentation layer is to decouple the actual user interface design and implementation from the specialized business logic of your mobile location services development process. For rapid development with minimum errors, it is most effective to have specialization.

Java Servlets and JSP   The Java servlet model is a special set of Java classes that excel in processing requests and creating responses. This model was developed with the generation of HTML Web pages transported across Hypertext Transfer Protocol (HTTP) in mind. Because HTML is related to the XML specifications, this process works equally well for applications that communicate using WML, VXML, GML, or any other proprietary XML specification. In addition, servlets are not limited to supporting the HTTP protocol— it just happens that it is the most common.

The team responsible for developing your user interface will often include people such as graphic designers, user interface specialists, and others. These people often do not have 3GL programming experience, and because servlets are 3GL Java code with graphic elements embedded in them, it can be difficult for your design team to work with. To solve this problem, Java server pages (JSPs) were developed. Java server pages are XML documents that contain tags that embed Java code in the document. When the servlet engine initializes, the JSPs are compiled into a servlet and behave exactly as a servlet from this point on. This allows your design team to operate independently on perfecting what the user will see while the presentation logic developers focus on piecing together business logic into an application and dynamically generating the user interface.

Business Logic Layer

Similar to the presentation logic layer, the objective of the business logic layer is to partition development work into manageable, reusable, and discrete components. Just as the standards used by the presentation logic team to deliver content to clients (WML for WAP phones, VXML for voice browsers, HTML for Web browsers, and specific XML formats for specific XML clients), it is necessary to have a framework and standards for the server-side business logic so the presentation logic team can use the business logic team's tools and services. Business logic should be concerned with application issues such as solving domain-specific problems and should not be concerned with system issues like managing transactions or security enforcement. Application servers provide you with a documented and proven infrastructure with a standard framework and interface for developing your business logic. This is the goal of the Enterprise JavaBeans (EJBs).

According to the EJB specification, the goals of the EJB architecture are:

- To define a component architecture for developing distributed business applications in Java
- To allow components from different vendors to be mixed to provide an enterprise-level solution
- To allow domain-specific developers to build business applications by leveraging a standard set of transaction, security, distribution, multithreading, and system-level facilities

- To address the entire software development lifecycle and provide specifications for using EJBs within interactive development environments (IDEs) during development
- To define methods for interoperability, which includes Common Object Request Broker Architecture (CORBA) and others

EJBs Enterprise JavaBeans are Java components that implement business logic. EJBs reside within EJB containers that are stored within the EJB server of the application server. A sample EJB environment is shown in Figure 3.3.

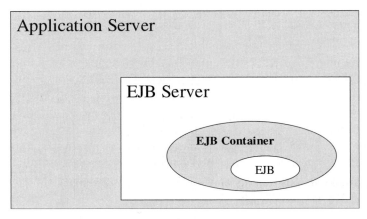

Figure 3.3   Sample EJB Environment.

Leveraging standard components of a transaction server, the EJB server provides a number of immediate benefits to the EJB developer:

- Automatic handling of resource pooling for database connections and component instances.
- EJB components have memory and create and destroy methods efficiently managed by the EJB container.
- The EJB container handles concurrency problems of multiple clients accessing the same data or functionality.
- The EJB server implements security, transactional behavior, and persistence. These are configured when the application is deployed.
- EJBs are designed to work correctly in a clustered environment.
- EJB servers support high availability and hardware failover.

Because EJBs are based on a component model, they operate within a specific environment and its interactions are governed according to a specific set of rules. There are three types of EJBs: *session beans*, which implement a client/server conversation and are typically found managing business process or workflow; *entity beans*, which represent persistent business objects such as *CustomerAccountInformation*; and *message-driven beans*, which allow clients to asynchronously invoke server-side business logic.

In a mobile location services application, session beans might include the following:

- *PositionUser*: Attempts to update the user's current position
- *Notifier*: Sends SMS message
- *GenerateRoute*: Calculates a route between two locations
- *CheckoutCounter*: Responsible for totaling all the items a user put in a user's shopping cart, processing payment, and sending a message to an inventory system to ship the items to the user's address

Entity beans might include the following:

- *Account*: Customer's account information
- *Inventory*: List of items available for purchase

Data Access Layer

An application server's data access layer provides a number of facilities to connect legacy enterprise information systems (EIS), from a relational database to a mainframe transaction processing system. This provides better management and session support than a simple socket. Although some application servers have specially tuned interfaces for proprietary systems, most have also a generic connector for the integration of the proprietary interfaces more commonly found in positioning and spatial analysis products. This is known as the J2EE Connector Architecture (JCA).

J2EE Connector Architecture (JCA) The JCA architecture enables an EIS vendor to provide a resource adapter that can be plugged in to an application server to provide the underlying infrastructure necessary to integrate with the EIS. The application server and EIS resource adapter collaborate to keep all the systems level components transparent from the application components.

JCA defines a set of system and application contracts. The application contract defines the client interface that the application uses to communicate with the EIS. The most commonly used client interface is the Common Client Interface (CCI), but a proprietary interface could also be used. The system contract is specified in the EIS vendor's resource adapter.

In many ways, a JCA resource adapter operates in a similar way to a Java Database Connectivity (JDBC) driver. In fact a JDBC driver is one example of a resource adapter. The difference is that the adapter is not limited to connecting a database using SQL. The adapter can connect to any type of EIS and use any protocol to communicate with the EIS.

Commercial Application Servers Commercially available application servers appropriate for mobile location services include the IBM Websphere Everyplace Server (*http://www.ibm.com*), BEA Weblogic Server (*http://www.bea.com*), Oracle 9iAS (*http://www.oracle.com*), Microsoft Mobile Information Server (*http://www.microsoft.com*), and the freeware Jakarta Tomcat from Apache (*http://www.apache.com*).

A Scalable J2EE Web Application Network Architecture[3]

This section presents a scalable and robust network architecture for J2EE-based applications. The architecture spreads the presentation, business logic, and database layers across different sets of physical machines. This technique provides many advantages over other J2EE application architectures, in which the application server executes both presentation and business logic.

The Architecture
A sample of this architecture is shown in Figure 3.4.

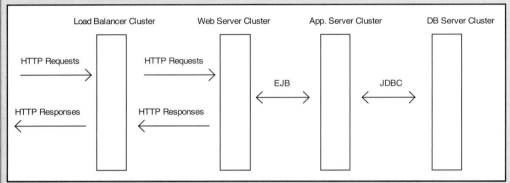

Figure 3.4   A Sample Scalable J2EE Web Application Network Architecture.

Load Balancer Cluster
The load balancer cluster contains at least one load balancer that distributes requests evenly across the Web server cluster.

Web Server Cluster
The Web server cluster contains at least one Web server machine. Each Web server should have the following software installed on it:
- HTTP server: For responding to HTTP requests and serving static content
- Servlet/JSP container: For executing servlet and JSP code to generate client output (HTML, XML, WML, etc.)
- EJB container: For communicating with the application server cluster

Application Server Cluster
The application server cluster contains one or more J2EE-compliant application server machines. Each application server is responsible for executing the business logic, which is contained in EJBs.

Database Server Cluster
The database server cluster contains one or more database server machines for storing and retrieving data.

Advantages

This architecture provides many advantages in terms of scalability and robustness.

No Single Point of Failure

Clustering ensures that if any one machine crashes, the application will still run, although performance might degrade. Clustering also provides greater scalability. For example, if a Web server is running at or near capacity, adding additional Web server boxes helps distribute the load.

Reducing Load on the Application Server(s)

With this architecture, all the presentation logic sits in the Web server cluster. This frees up application server resources to execute more business logic, requiring fewer hardware resources for the application server cluster (i.e., new machines or additional CPUs and memory for existing machines). Application servers are generally expensive and licenses tend to be based on the number of CPUs. Thus, offloading presentation-layer work onto cheaper Web server boxes can save a significant amount of money.

Higher Throughput

Because this architecture includes clustering and separates work across multiple network layers, there will be fewer bottlenecks than architectures with less clustering and no separation of the presentation and business logic layers.

Disadvantages

This architecture does have disadvantages, however. Those presented in the following section should be considered when designing systems.

Increased System Management Complexity

This increased complexity comes from two places. First, there are more machines to configure and maintain, which will stress information technology resources. Second, clustering can be complex. Although clustering Web servers is relatively simple, clustering application and database servers is difficult and requires highly skilled personnel.

Increased Software Deployment Complexity

With this architecture, different pieces of the application are spread across different machines. Servlets, JSPs, and static content reside within the Web server cluster, whereas EJBs and other business logic sit inside the application server cluster. Spreading the application across many different physical machines can make maintaining the production environment trickier. For example, when an updated version of your application is released, it can be challenging to verify that all the pieces successfully deploy to the various machines.

Increased Network Traffic and Latency

Because the presentation layer and business logic layer reside on different machines, there will be more traffic across the network than if those two layers were on a single machine. This separation might also slow individual response times slightly.

Additional Techniques

Bypass the Application Server

For nontransactional operations (e.g., reports), one can access the database layer directly from the presentation layer via JDBC. This reduces the response time for these operations, as well as the load on the application server cluster.

> *Cache Static or Nearly Static Data in the Presentation Layer*
> If there is data that rarely or never changes, caching it inside the servlet/JSP code reduces the load on the application server cluster and access times for that data.
>
> *Use EJB 2.0-Compliant Application Servers*
> EJB 2.0 improves the performance, scalability, and robustness of EJB 1.1.
>
> *James Dunn*
> *Senior Java Developer, Noosh, Inc*

Additional Server Sizing Resources

- *http://www.dell.com/us/en/slg/topics/products_size_pedge_sizing.htm*
- *http://www.sun.com/solutions/third-party/global/oracle/pdf/nca-app.pdf*
- *http://activeanswers.compaq.com/ActiveAnswers/Render/1,1027,4815-6-100-225-1,00.htm*

ENDNOTES

1. © 2002 BEA, Inc.

2. © 2002 BEA, Inc.

3. © 2001 James Dunn.

4

Spatial Analysis

For your mobile location services infrastructure, you need two location-related functions: mobile positioning and spatial analysis. These location functions are marked in Figure 4.1 as the location server. Spatial analysis requires a digital map database and a suite of tools to perform spatial operations on the data. These spatial operations include geocoding (the process of looking up a position from an address), reverse geocoding (the process of looking up an address from a position), routing (calculating a route between two positions), map rendering (translating an area of map database into a vector or raster map), and several others. Mobile positioning is the subject of Chapter 5.

SAMPLE SPATIAL ANALYSIS SERVER

A sample spatial analysis server is shown in Figure 4.2. A set of software libraries performs search and data retrieval operations on a compiled map database. The spatial operations defined earlier require specialized business logic and access the digital map database using the access libraries. An interface is provided to integrate the spatial analysis server into your location server infrastructure. If possible, selecting a standards-based interface for spatial operations, such as GML (see Appendix B for the complete specification), provides maximum extensibility for your system.

Unlike traditional GIS products, spatial analysis software for mobile location services is focused much more on speed of data retrieval and ability to support a high volume of concurrent requests. These requirements impact the way the spatial analysis software database is structured. There are three primary database structures used in spatial analysis server software: a hybrid database, a pure relational database, and a pure object-oriented database.

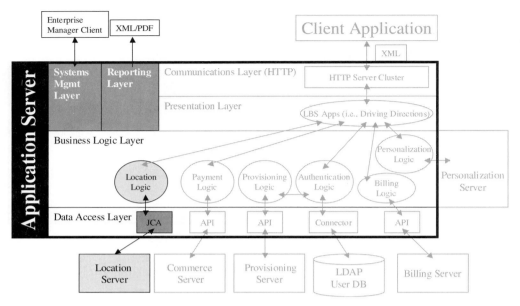

Figure 4.1   Location Server Components of a Mobile Location Service Architecture.

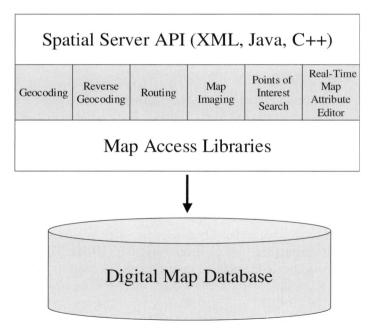

Figure 4.2   Sample Spatial Analysis Server.

Hybrid Database Model

The basic concept behind the hybrid approach is that you cannot optimize both the spatial data storage and the attribute data storage simultaneously. Thus, spatial data such as coordinates and topological data is stored using a standard file-based approach and attribute data is stored in relational database tables. The two are linked via a unique identifier.

Relational Database Model

This model stores both the spatial data and the attribute data in relational tables, linking them with a standard relational join. Storing spatial data in this manner results in much poorer system performance, because spatial data that is close together is not necessarily stored close together in the database.

Object-Oriented Database Model

An object-oriented database attempts to deliver speed without sacrificing flexibility by organizing the data around the spatial entities. These databases define data as a series of objects that have similar criteria (object classes). Relationships between objects and classes are explicitly defined. The drawbacks to this approach are that there is no standard query language for object-oriented databases and that object-oriented databases take substantially more skill and time to design than other databases.

The pure object-oriented and the hybrid database models have various advantages and disadvantages, although either model is a better choice for mobile location services than a straight relational system. The advantages of the hybrid model are that it leverages the sound bases of the relational database and file system, and it is generally easier to use than its object-oriented counterpart. Its disadvantages are that it can be slower to query, as queries require access both to the spatial data store and to the relational attribute tables, and there are potential integrity and security concerns of not storing spatial data in a database management system. The object-oriented database, on the other hand, provides faster querying because it does not have to do so many join operations, and requires less storage space than relational database systems that have many index files. However, as discussed earlier, the disadvantages are that it has no standard query language and requires significant sophistication to develop and maintain.

To understand how the spatial analysis server works and how to use it to build mobile location services applications, it is important to understand some map basics. This chapter introduces digital maps, both what they are and how they are built. We then discuss map coordinate systems and projections, followed by a brief overview of geographic data types. Finally, we discuss in detail the use of the following spatial analysis operations:

- Geocoding
- Reverse geocoding
- Routing
- Map image generation
- Point of interest searches
- Real-time map attribute editing

DIGITAL MAPS

What Is a Digital Map?

Most people are familiar with paper roadmaps. How are digital maps different and why are they so much more powerful? Paper roadmaps, and in fact all paper maps, have a major limitation in that the spatial "database" is the drawing on the paper. This imposes inherent limitations on the building and use of the map data. First, the amount of data you are able to capture on the map is severely limited by what can be clearly drawn and understood from the map. Second, the paper map is a static snapshot. It is not possible to change the map scale or easily update the map with new information. Finally, it is quite difficult to do quantitative spatial analysis with a paper map, such as calculating the fastest route from point A to point B or quickly looking up an address.

A digital map attempts to capture the underlying geographical phenomena and make it available for dynamic retrieval, spatial analysis, and representation by sophisticated software systems. There are two basic ways digital maps can represent what is present in the map and where it is. The first is based on the concept that the geographic world is composed of entities that can be positioned on the map by a geometric coordinate system and described by attributes and properties. This approach typically uses a vector data model, where entities are defined using points, lines, and polygons (see Figure 4.3 and Figure 4.4). The second is that specific attributes (e.g., elevation) vary continuously in the map as a mathematical function. Because it can be challenging to represent large geographical areas by a simple differentiable numerical function, it is common to divide the geographical space into discrete spatial units. The result is known as a *tessellation*, and can be composed of square cells if a raster model is used (see Figure 4.5).

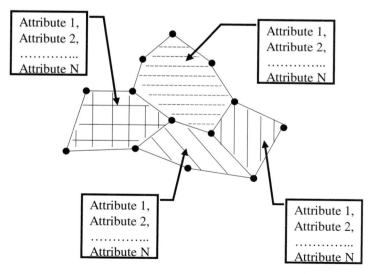

Figure 4.3   Vector Representation of Entity Data Model with Attributes for Each Polygon.

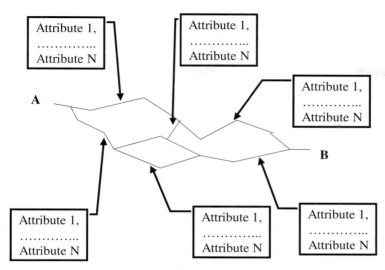

Figure 4.4   Vector Representation of Linked Network Topology in Entity Data Model.

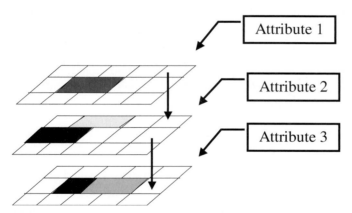

Figure 4.5   Raster Tessellation in a Continuous Variation Data Model, One Layer for Each Attribute.

The spatial analysis required in mobile location services (e.g., routing through a linked network topology; i.e., roads) is well served by the entity model. This is the most common map data model used by spatial analysis software vendors, and this is the data model focused on in the remainder of this book.

How Are Digital Maps Created and Maintained?

Building a digital map is a continual, highly labor-intensive, expensive process. The map vendors that produce highly detailed maps suitable for mobile location services (referred to as large-scale maps) typically update their map products between two and four times per year. The following steps provide a basic overview of the digital map creation and maintenance process.

Gathering Source Information

Digital maps begin with raw data collected and aggregated from local vendors. Map data companies typically employ source acquisition specialists who continually screen multiple data sources for completeness and accuracy. Their function is to continually update relationships with local vendors to collect map source materials from public and private suppliers. These relationships include those with local government agencies, which provide updated materials such as aerial-rectified photos and differential GPS field surveys. This information allows the map data vendor to extend digital maps with information like new roads, postal codes, and address ranges.

Collecting Data From the Field

After the core map database is created (or when an update is required), a group of specially trained professionals drive the roads to compare reality with the digital data, and to collect new features and attributes for the map. This information could be new turn restrictions, road geometry, and signage information. Additional information that might not be available in the core map database can also be collected. This could include one-way streets, exit signage, prohibited turns, tunnels, bridges, vehicle restrictions, and address ranges. This information is delivered to a production unit, who will make the necessary improvements to the core map database.

Developing Products

The source map data needs to be converted from its source format into various product formats optimized for specific applications. These product formats could include MultiNet GDF, Shapefile, MapInfo, MapBase, MapAccess, Spatial Data Engine (SDE), Oracle Spatial, KIWI, geocoding-specific formats, or lighter formats such as those for mobile location services.

GEOGRAPHICAL DATA TYPES

To understand how map databases store information, it is important to understand the basic geographic data types used to represent real-world geographic phenomena. In the vector data model, geographic data is stored in discrete points, lines, and polygons. Points, lines, and polygons are respectively zero-, one-, and two-dimensional static representations of real-world phenomena in terms of simple x, y coordinates.

Points

Points refer to objects that have location and attribute information, but are not large enough to be represented as areas (see Figure 4.6). Whether an object is represented as a point or a polygon depends on the level of abstraction and scale of the map. Cities that would be represented as points in a country-level map might be represented as polygons in a regional map. Other examples of points include points of interest (POIs) and parcel centroids, such as postal codes or addresses. Points have no dimensions.

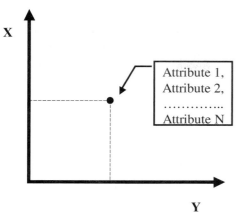

Figure 4.6   Point Primitive.

Nodes

Nodes are a special type of point that represent a junction or the endpoint of a line. Nodes are the same as points in all other respects. Nodes provide information that includes connectivity between lines and information about adjacent polygons and enclosed islands.

Lines and Arcs

Lines are one-dimensional objects that have length but no area (see Figure 4.7). Lines must begin and end with a node. In the case of a road, a line will represent the street center line. The actual width of a road will be an attribute of the line. Other attributes of a road might include street name, address range, speed, and direction of travel. At the country level, a line is an adequate representation of a road. At the regional level, a line might not be adequate and a road would be represented using a polygon area of paving. Lines might also be used to represent rivers and railways. An *arc* is a multisegment line.

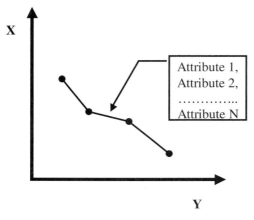

Figure 4.7   Line Primitive.

Polygons

The simplest definition of a *polygon* is a set of connected lines that form a closed mathematical figure (see Figure 4.8). Polygons can have any number of points and can be any shape or size. Other ways to represent a polygon include the set of x, y coordinates that form its boundary or the area contained by the boundary. Polygons can have holes, contain other polygons, and be directly adjacent to other polygons.

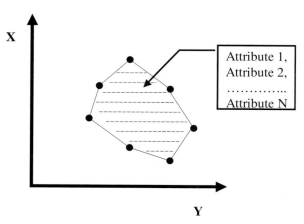

Figure 4.8   Polygon Primitive.

Linked Attributes

Each geographic primitive has attached information that describes it. In mobile location services, the most important attributes relate to the road network, such as speed, street names, turn restrictions, and connectivity.

Computer Storage of Geographic Data Structures

Now that we have developed a conceptual notion of our map using points, lines, and polygons, we must consider how this data is stored in a computer. The method used to store this data has a large impact on how large the map database is and how quickly data can be retrieved and used in a mobile location services application.

Simple spatial entities have the following characteristics: points will have an (x, y) coordinate, a simple line with only two nodes will have two (x, y) coordinates, an arc will have *n* number of (x, y) coordinates, and a polygon could have either *n* number of lines or *n* number of points. To achieve sufficient precision, it is necessary to use a 32-bit or 64-bit real data type for each value in the coordinate pair. Thus, because there are 8 bits in a byte, a point would require 8 or 16 bytes of storage, a line requires a minimum of 16 bytes of storage, and a polygon requires a minimum of 24 bytes. Because map databases can contain many millions of spatial entities and each can have linked attributes, it is easy to see how a map database can become very large.

PROJECTIONS AND COORDINATE SYSTEMS[1]

The world is not the flat, paper roadmap you might use to navigate with in your car. The earth is spherical, and there are actually several approaches to approximate the earth's shape and provide a system for identifying a position on the earth's surface. The earth's shape actually resembles an ellipsoid, as it revolves easterly on its axis, creating centrifugal force and causing a flattening at the North and South Poles and a bulge at the equator.

The ellipsoid is only a mathematical approximation of the earth, and there are several different models to provide a frame of reference for calculating coordinates on the earth's surface, each designed to closely fit the earth's surface in a particular geographical area. As shown in Figure 4.9, each datum defines the position of the ellipsoid relative to the center of the earth.

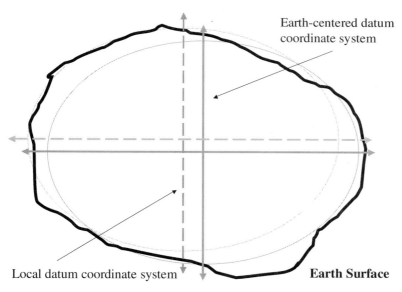

Earth-centered datum coordinate system

Local datum coordinate system   **Earth Surface**

Figure 4.9   Datums.

A point x on the earth's surface has an east–west position (longitude) and a north–south position (latitude), as seen in Figure 4.10.

Latitude is defined as the angle formed by the intersection of a line perpendicular to the earth's surface at the point and a plane passing through the equator. North of the equator, points have a positive value. Points south of the equator have a negative value. The value range for latitude is from –90 degrees to 90 degrees. Lines of latitude are also called *parallels* because lines of latitude run parallel to the equator.

Longitude is defined by the angle between the prime meridian and a plane that passes through the point and the North and South Poles. The most common prime meridian is a plane that passes through the North and South Poles and Greenwich, United Kingdom. Longitude values range from 0 degrees at the prime meridian to 180 degrees at the international dateline. West of the prime meridian is negative longitude and east of the prime meridian is positive longitude.

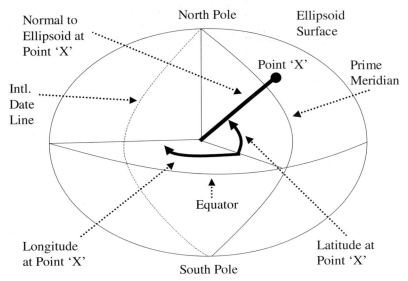

Figure 4.10 Earth as an Ellipsoid.

To be most useful in software applications (and to make flat maps), it is necessary to get the spherical three-dimensional model into two dimensions. This process is called a *projection*, and involves a mathematical transformation of a geographical location (longitude and latitude) on a sphere to a corresponding location (x and y) on a flat, two-dimensional surface. The x-axis represents position values from east to west (longitude), and the y-axis represents position values from north to south (latitude).

Different projections create different map distortions.[2] Some projections work best for large-scale, detailed maps of a very small area, whereas other projections work best for small-scale maps that need to display large, country- or continent-sized areas. There are four primary types of projection: conformal projections, equal area projections, equidistant projections, and true directions projections.

Conformal projections focus on preserving shape, which is done by preserving angles at the expense of size. An *equal area projection* preserves the area of displayed features by distorting shape, angle, and scale. *Equidistant projections* maintain the distances from the center of the map, but do not maintain scale. *True directions projections* maintain the distance between two points on a spherical surface as a straight line on a flat surface.

The best coordinate system and projection method depends on the region in which your mobile location services applications are to be deployed. It is valuable to have the flexibility to change projections based on the application and scale you are planning to use. The Universal Transverse Mercator (UTM) is an international metric coordinate with worldwide coverage. UTM is a conical projection and provides coverage between 84 degrees north and 80 degrees south (see Figure 4.11). It has the advantage of being well defined and mathematically consistent for the entire earth, and could be a good choice for applications that require continent-wide or worldwide deployment.[3]

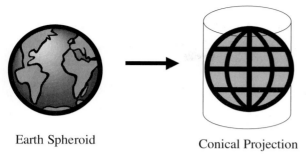

Earth Spheroid   Conical Projection

Figure 4.11 UTM Conical Projection.

GEOCODING

A fundamental service required by many mobile location service applications is the ability to look up an address on a map. This is known as *geocoding*. For example, if your location service application requires a route calculation between two locations, it is necessary to first geocode the start and end points of the route to make sure that they are valid addresses and that they can be found in the map database. For mobile location services it is necessary to have an address level geocoder (one that can geocode to a specific street address). There are many products that geocode to either a postal code or street level of accuracy, but neither provides sufficient accuracy for mobile location services.

There are many challenges in geocoding, most caused by the complexity of address schemes and the process of trying to clean up and interpret address input by a user. One challenge, especially in mobile location services, is that address schemes differ significantly by geographic region. In North America, it is common for street addressing to be sequential, with odd numbers on one side of the street and even numbers on the other. In Europe, it is common for street addressing to increase up one side of the street and decrease down the other side. This could mean that a building with a street address of 15 or 16 might face a building with a street address of 435 or 436. Instead of a single address number as is common in the United States, it isn't uncommon for buildings in Europe to actually have an address range, such as 46-50 Coombe Road.

Whereas it is common in English (United States and United Kingdom) to write an address before the street name, in many other European languages it is intuitive to put the address number after the street name and the postal code before the city, such as Lenbach-platz 3, 80333 München. If you have traveled to Japan, you might recall yet another addressing scheme, where the first building built in a particular region is numbered 1, the second 2, and so forth. Differences such as these require highly complex rule systems to analyze the address and good map data to make geocoding effective.

Technical Definition of Geocoding

A more technical definition of geocoding is the process of associating an address with geographic features. The geographic features are often represented by a line, such as a street center line database. Typically, each segment of the street center line has attributes such as high and low address range (or left and right address range), street name(s), the city, postal code, and many others. A simple example shows street addressing in two sample street center line segments (Figure 4.12).

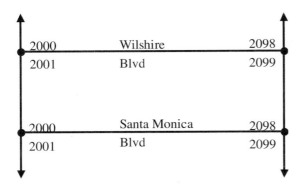

Figure 4.12 Street Address Range Example.

How Does Geocoding Work?

Geocoding generally is done using a four-step process. First, the address is input to the geocoding system. The address is then analyzed, parsed, and placed into a standard format. Third, a soundex search is done for the city and street name and an address range search is done for any matches found in the soundex search. Finally, a scoring system is used to rank the possible matches. If a match is found, the geographic coordinate (e.g., projected latitude and longitude) is returned. If multiple matches are found, they are returned ranked by the scoring system so the user can select the best match. If no matches are found, the geocoding system logs an error and returns an error message.

Address Input

Geocoding systems need a way to receive the address from the user. This might include a Java constructor, an XML document over HTTP, or a proprietary protocol. The most advanced systems allow a near free-form address input, such as one text input for address and street name and another text input for city, region, and postal code. The major advantage of this system is that the mobile location services application developer does not have to do the complex parsing and error correction that is best left to the geocoder, and the user inputs the address the same intuitive way he or she might address a letter. Less sophisticated systems require a text input for each element. The more discrete inputs required, the more unsophisticated the geocoder.

If the application developer is required to do the validation and error checking, input will be checked to make sure that it conforms to the data type that the geocoding engine is expecting, but more sophisticated error checking is too time consuming to implement, resulting in a less than satisfactory mobile location service application.

Address Standardization

Once the geocoding engine has received the address, it attempts to parse it and standardize it. If the geocoder is region specific (e.g., only designed to work with U.S. map data) the standardization is simpler. Address interpretation might appear as follows:

Address list before parsing and standardization:

1. 1000 Main Street, Suite 100
2. 555 California Avenue
3. 500 South 300 West
4. 1121 3rd Street
5. 501 Avenue G
6. 15 Jefferson Apt# 1

Address list after parsing and standardization:

Address1	Name1	Suffix	Direction	Address2	Name2
00001000	MAIN	STREET		0000100	
00000555	CALIFORNIA	AVENUE			
00000500	300		WEST		
00001121	3RD	STREET			
00000501	G	AVENUE			
00000015	JEFFERSON			0000001	

Perform Soundex and Address Range Search

Once the address has been standardized, the geocoding engine attempts to find a match in the map database. If an exact match is not available, the system might do a soundex search to find street names that are similar so the user may choose the best match.

Soundex is a technology originally developed by the U.S. government to assist in matching surnames in the analysis of U.S. census data. A soundex index is based on the way a word sounds rather than the way a word is spelled. Each entry in the index is a combination of one letter and three numbers. The letter is the first letter of the original word. The three numbers are the number encoding for the letters of the word.

Soundex Coding Guide

1. B,F,P,V
2. C, G, J, K, Q, S, X, Z
3. D, T
4. L
5. M, N
6. R

The letters A, E, I, O, U, H, W, and Y are ignored. Double letters are treated as a single letter. Side by side letters that have the same soundex value are treated as a single letter. Words with a prefix are coded both with and without the prefix. More details and rules are available from the U.S. National Archive and Records Administration at *http://www.nara.gov/ genealogy/soundex/soundex.html.*

Soundex Encoding Examples

SMITH	S-530 (S, 5 for the M, 3 for the T, 0 added)
SMYTH	S-530 (S, 5 for the M, 3 for the T, 0 added)
WASHINGTON	W-252 (W, 2 for the W, 5 for the N, 2 for the G)
JACKSON	J-250 (J, 2 for C, K ignored, S ignored, 5 for N, 0 added)

If the soundex search does not provide any matches, the user would be given an error and asked to enter another address. If the geocoder is able to find one or more street name matches, it performs an address range search to make sure that the address requested is valid for the street name. If the address range is valid, the geographic position is assigned in the appropriate format (e.g., projected latitude and longitude).

Apply Scoring Rules

Now that the geocoding engine has a set of potential results, each result is scored according to certain criteria, which might include the following:

- Whether the street name was an exact match
- Whether the street type matched (Avenue or Street)
- Whether the direction matched, if the street had a directional attribute (e.g., north or southwest)
- Whether the city, zone, or postal code matches

For example, the scoring system might run from 1 to 100, with 100 being a perfect match. Every match candidate would start at 100, and points would be subtracted for failure of various tests. Items such as street name not found in a soundex search might subtract 10 points, and a postal code that does not match might subtract 50 points. Once the matches are ranked in the scoring system, business logic can determine whether the geocoding engine will return one or multiple proposed matches.

What Makes Geocoding So Difficult?

Address Cleanup

Address cleanup is one of the greatest challenges in providing a high-quality geocoder. Typical problems include the following:

• Numeric street names	10 1st Street
• Addresses with more than one directional	123 W Main Street East
• Alphanumeric addresses	100A Mission Street
• Fractional addresses	45½ Bee Street
• Coordinate addresses (Utah)	520 East 400 South

- Addresses with dashes (Hawaii and Queens, NY)   101-123 Kaanapali Road
- Street names with numeric components   1234 10 Mile Road
- Street names that are directionals   South Street
- Street names that are suffixes (Brooklyn, NY)   Avenue G
- Spelled out address numbers   Two Second Street

Differing Address Standards

As previously mentioned, address standards vary drastically from region to region. To be effective, geocoders must be locally adapted, tested, and tuned. Language has a significant impact on how addresses will be input, and in many regions it is necessary to support many different ways to enter addresses. A geocoder in Germany must know that München and Munich are the same place and understand an address input with the address number before or after the street name, and a postal code either before or after the city name.

Soundex Mismatches

Soundex is not a perfect technology, and there are many proprietary enhancements that could increase its effectiveness in matching an address. Bad matches add processing time and could present the user with unintuitive choices. This is particularly true given that soundex was developed for analyzing surnames, and has been adapted to work with street names.

Static Map Database and Dynamic Communities

Map database releases are typically done two to four times per year, but new roads and buildings are constantly being constructed. Applications that have a central map database and thin clients have the advantage of being more up to date than systems that require map databases on CD, such as the onboard navigation systems common in 2001 and newer cars. Users become frustrated when they are directed to places that don't exist. It is unlikely users will buy and install new CDs four times per year, or that a CD-based navigation system could be released four times per year. Offboard navigation systems that use the mobile network to process spatial requests on a remote server are a better option.

Rural Delivery and Post Office Boxes

Rural delivery and post office boxes present another series of complications for geocoding. Depending on the application, the geographical position found might not be useful if the physical location is required.

Site Address and Billing Address

Ambiguity is possible when a site has both a physical address and a billing address. Certain applications might need the physical address, whereas others require the billing address. A method to distinguish the two is necessary.

Why Is Geocoding Important to Mobile Location Services?

Significant functionality in mobile location services applications depends on being able to accurately pinpoint and direct users to very specific locations. Users save time by relying on the intelligent business logic and the large knowledge bases built into mobile location services systems. However, users have very little patience with systems that direct them to the wrong place. A user might not mind (or know) if a route that has been calculated is the absolute fastest, but the user surely knows when he or she is directed to the wrong place or a place that does not exist. For location services applications to be successful, it is crucial that map data be current and that a high-quality geocoding product is used (and properly integrated if necessary). Equally important is reverse geocoding, the process of taking a geographic position (e.g., projected latitude and longitude calculated by a positioning system) and transforming it into the nearest road segment in the map database. Success in these basic location service functions is the cornerstone for success in developing higher level applications such as real-time traffic.

REVERSE GEOCODING

Reverse geocoding is the process of identifying the nearest road segment in a map database given a latitude and longitude pair. This latitude and longitude data would typically be generated by the mobile device's positioning system. Once the nearest road segment is available it is possible to process driving or walking direction requests or POI lookups.

The operation is specific to a given map database, and different map databases of the same area could yield different results. The operation usually searches for road segments within a specified range, which might or might not be configurable by the user. You must be cautious with the search range, as a range that is too small can return errors if the user is trying to use an application from a park or other open space area that has no road segments, but is located in an urban area.

Information returned by a successful reverse geocode operation might include how far the nearest road segment is from the point, where the point is in relation to the road segment (near the beginning of the address range, 15 percent; in the middle of the address range, 50 percent, or near the end of the address range, 85 percent), and the actual road segment vector data and its associated attributes. See Figure 4.13 for an example.

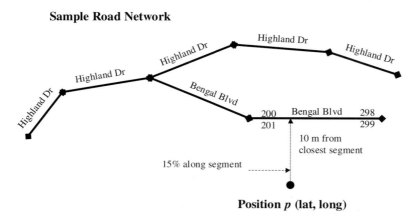

Figure 4.13 Reverse Geocoding Example.

ROUTING

A routing engine calculates the optimum path between an origin and destination, subject to certain criteria. Common criteria include "use freeways," "avoid freeways," or "fewest turns." The most common algorithm for calculating routes is based on the A* (pronounced *A-star*) algorithm developed in the artificial intelligence (AI) community.

The algorithm works by extending each known possible path to the destination by adding an intelligent guess (a heuristic estimate) and computing the total cost of the real traversed path plus the heuristic estimate.

Understanding routing requires understanding the basics of problem solving using AI techniques. The effectiveness of a combinatorial and problem-solving technology is significantly dependent on the way that the software represents the problem's states, goals, and conditions. Typically, these problems are represented using graphs and trees. The most common routing problems are referred to as shortest path, traveling salesman, multiple traveling salesman, single depot–multiple vehicle, and multiple vehicle–multiple depot routing. The simplest and most often used is the shortest path between two points, which we explore first.

Routing Problems Defined

Shortest Path Problem

Figure 4.14 shows a graphical example of the shortest path problem.

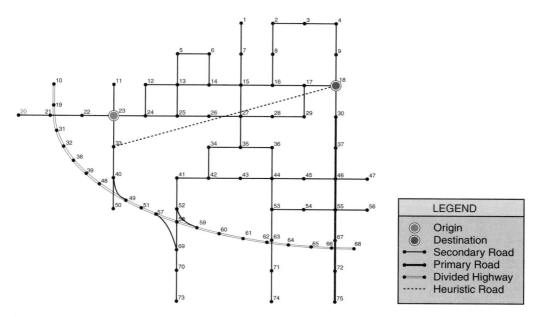

Figure 4.14 A Simple Road Network with Each Node Numbered. The Dotted Line, Indicating a Straight Path Between Current Location and Destination, Is a Common Heuristic Used to Complete the Path. © 2002 Kivera, Inc.

The A* Algorithm in Action   Spatial analysis software vendors often use a two-sided A* algorithm for routes, expanding from both origin and destination simultaneously. For simplicity, because the principles remain the same, a single-sided example is examined here.

Starting at the origin, node 23, the algorithm explores four possible paths. At the end of each link the path to the destination is completed with the addition of the heuristic path. Thus, the first four paths to try would be, in no particular order:

Known Path		Heuristic Path
23-33	+	33-18
23-22	+	22-18
23-11	+	11-18
23-24	+	24-18

These are then sorted by cost (time):

23-24	+	24-18	Fastest
23-33	+	33-18	
23-11	+	11-18	
23-22	+	22-18	Slowest

Next, the algorithm expands from the fastest path so far:

23-24	+	24-12	+	12-18
23-24	+	24-25	+	25-18

Add these paths to the total number of paths traversed so far and order again from fastest to slowest:

23-24	+	24-25	+	25-18	Fastest
23-24	+	24-12	+	12-18	
23-24	+	24-18			
23-33	+	33-18			
23-11	+	11-18			
23-22	+	22-18			Slowest

As the potential paths grow, there will be opportunities to select roads belonging to faster arterial levels. At node 40, for example, the algorithm will always "jump" to the faster arterial level. This jump to a higher arterial level often results in a jump to a larger cell size as well.

These cells cover a greater area and have fewer road segments per unit area. As seen in Figure 4.15, jumping up to cells that cover more area results in having to read in fewer cells and consequently faster routing. Figure 4.16 shows the many more cells, each with a higher road density and therefore more data to read that have to be examined if there is no opportunity to jump to higher levels.

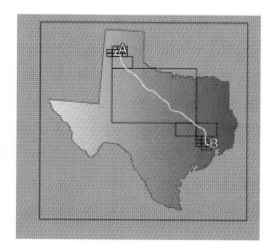

Figure 4.15 Jumping to Cells That Cover a Larger Area Results in a Routing Algorithm That Requires Fewer Cells and Operates Faster. © 2002 Kivera, Inc.

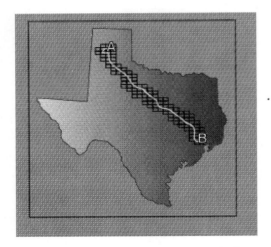

Figure 4.16 A Routing Algorithm That Is Constrained to One Level of Cells Must Read in Many More Cells. © 2002 Kivera, Inc.

It is common to use the A* algorithm for routing. However, this does not mean that all implementations of the A* algorithm are created equal. About 10 independent parameters serve as input into the algorithm, and even small changes in these parameters result in large changes in what route is taken.

Traveling Salesman Problem

The traveling salesman problem is the least time-consuming path that passes through each node of a connected network once. A simple example is shown in Figure 4.17, where a salesman must visit each of five cities, labeled A through E. There is a road between every pair of cities, with the distance given next to the road. From point A, the objective is to visit each city once and then return to A.

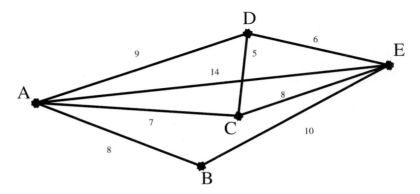

Figure 4.17 Traveling Salesman Map.

The simple method to solve this problem involves creating a search tree that explores possible paths until the search criteria have been met (see Figure 4.18).

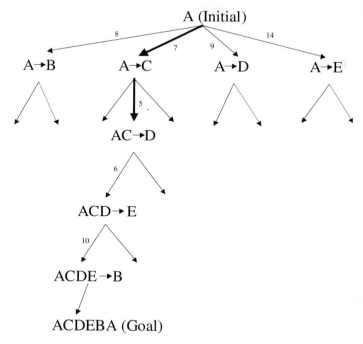

Figure 4.18 Traveling Salesman Search Tree.

Multiple Traveling Salesman Problem

This problem is the same as the single traveling salesman problem, except it involves multiple salesmen (vehicles) leaving from and returning to the same location.

Single Depot–Multiple Vehicle Node Routing

This problem involves vehicle routing, providing a solution for a set of delivery routes for vehicles stored at a central depot.

Multiple Depot–Multiple Vehicle Node Routing

This problem requires a solution similar to the single depot–multiple vehicle problem, except with multiple depots. Vehicles must leave from and return to the same depot.

What Impacts Route Calculation Performance?

A route engine typically considers six segment attributes in calculating a route:

1. Segment speed
2. Length of segment
3. Travel time

4. Turn restrictions

5. One-way indications

6. Real-time attribute editing to provide for road conditions such as traffic congestion (discussed at the end of this chapter)

Preprocessing and reducing the number of attributes the route engine has to consider both improve route performance.

MAP IMAGE GENERATION

An important capability of any spatial analysis software product is the ability to generate a map. There are two primary map types: a raster map, which is a digital image often in Graphics Interchange Format (GIF), and a vector map, which is the data required for an application to properly render the map using Vector Markup Language (VML) or some other data exchange format.

For mobile location services, speed under load is usually the primary focus. However, it is still important that if a raster map image is being returned, the image be visually attractive and in the format that the user expects. The quality of the map data significantly impacts the quality of the map image, but even good map data will not compensate for a poor quality map-rendering engine.

People Are Visual

In many cases, a user will interact with a mobile location services application using a map. Addresses that have been geocoded will be marked on the map, streets in a route will be highlighted, and traffic incidents on the travel path might be marked. People are visually oriented, so a good map can communicate a lot of information very quickly. Alternatively, a poor map or a map that does not meet a user's expectations will quickly cause frustration.

Maps Aid Decision Making

People use maps to help them make quick decisions. Ordinary paper maps require the user to do more work, because unlike a digital map, they do not mark where you are, highlight a route to where you're going, or allow you to zoom or pan. Digital maps allow you to deliver to users a more powerful set of tools to help them make decisions. Other capabilities include adding real-time traffic and highlighting specific POIs to your user. The map is personalized specifically for the person looking at it.

Maps Consumers Can Use

Consumer Expectations: Roadmaps

Most consumers' experience with maps is limited to the rather basic characteristics of roadmaps. Furthermore, because consumers are so familiar with roadmaps, they actually expect their digital maps to be very similar to them. This has several implications, as users in different regions expect their maps to look differently.

North America   Users in North America expect their maps to appear like the example shown in Figure 4.19.

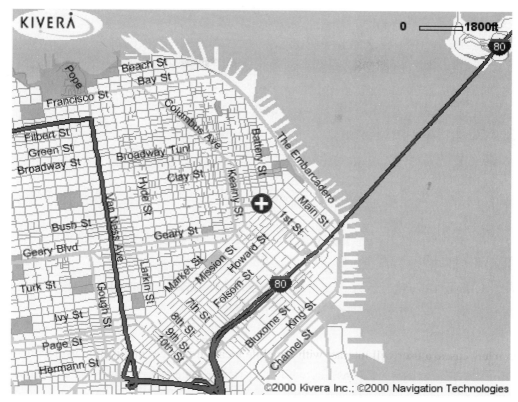

Figure 4.19 Sample North American Style Map. © 2000 Kivera, Inc. and
Navigation Technologies, Inc.

Characteristics of North American road networks are wide roads and rectangular blocks. European road networks are very different, and have particular requirements for the development of user-acceptable maps.

Europe   European consumers expect maps to be displayed as shown in the example in Figure 4.20.

You will notice in this example the different characteristics of European road networks. Roads are very narrow and close together, and blocks are often not rectangular. In addition, Europeans expect to have their roads drawn as a polygon with the name of the road inside the polygon, curving as the street curves. Notice Shaftesbury Street in Figure 4.20 as an example of the European style road representation. North American maps typically represent streets as lines with the road name on one side or the other. The color schemes used in the maps are also very different. Yellow is a very common background for North American maps, whereas in European maps, it is often gray.

To maximize user acceptance and penetration of your location-based application, it is important to be aware of the regional expectations of how maps should look and customize your map display accordingly.

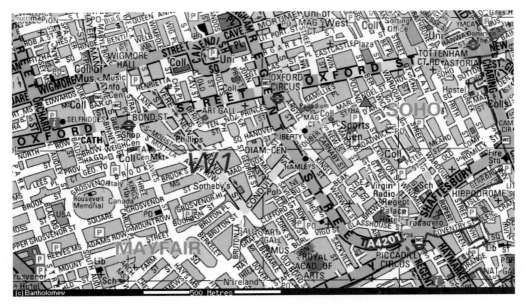

Figure 4.20 European Style Map of London. © 2002 Maporama S.A.

Geographic Literacy

The average user is not familiar with cartographic principles or using special-purpose maps designed to facilitate decision making. Accordingly, the map user interface in many of the most popular location service applications does not introduce geographic concepts such as scale, thematic shading, or even a north-pointing arrow. When a user wants to change scale, the better location service applications accomplish this through zoom-in and zoom-out functions. They don't assume the user understands the concept of scale and they don't ask the user to manually set the scale. The colors used in the map are important as well. Colors that might be best suited to provide for decision making in traditional GIS applications, such as thematic shading, might be unattractive to many users. Applications must adapt to a typical user's expectations.

Raster Maps

A raster map is most often built using a regular grid. The grid is a two-dimensional geographic surface divided into square cells that are called *pixels*. Depending on the level of detail needed for display, the size of the square cells will vary. Different geographic features are represented using mathematical functions and approximated by painting the associated squares with the appropriate feature color. POIs and text data, such as city and street names, are also overlaid by painting the squares that approximate their dimensions. A very basic raster map is shown in Figure 4.21.

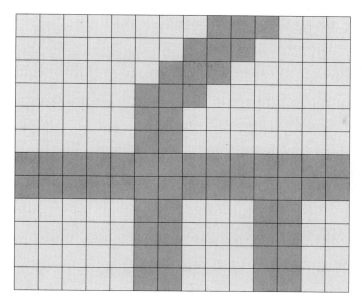

Figure 4.21 Simple Raster Roadmap.

A more realistic example of raster imaging can be seen when you look at the way an aerial photo is digitized and then represented using a raster map. A sample aerial photo of an address is shown in Figure 4.22.

Figure 4.22 Sample Aerial Photo from GlobeXplorer. © 2001 GlobeXplorer, AirPhotoUSA.

Compare the raster equivalent shown in Figure 4.23.

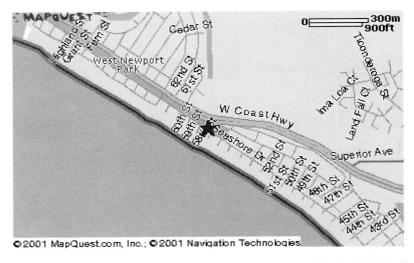

Figure 4.23 Sample Raster Map. © 2001 MapQuest and Navigation Technologies, Inc.

Advantages

Raster map images have many advantages.

Simple Data Structure   The raster map data structure is very simple. It consists of pixels on a grid. There are many industry standard and published formats that can be chosen for data storage. Consequently, there are also many commercial tools available for editing and displaying raster maps.

Easy Overlay   Map overlays are very easy with a raster data format. An overlay is made by updating the pixels that represent the approximate shape of the overlay, in the position of the overlay. Some formats, such as GIF89a, allow for storing multiple layers of data. Thus, overlays can be added and removed with no impact to the original image.

Various Kinds of Spatial Analysis   Although not suitable for all spatial analysis operations, simple operations such as areas, perimeters, and sums can actually be more efficient with a raster map image than in a vector data model because they are simple counting operations.

Uniform Size and Shape   Raster maps are easier to produce because they are a collection of pixels. Generating a map is a process of painting the pixels of the region requested for display. A vector map, on the other hand, must deal with fairly complex issues based on the size and shape of the requested map. As an example, if a road segment or link continues past the border of the requested map, the software must clip the link and update the vector data being returned. An even more complex example is when both of a segment's nodes are outside the map area being generated, but not the road itself.

Cheaper Technology   Raster maps are much cheaper technology because they only require the client device to paint pixels in a grid of varying resolutions. A vector map requires a much more sophisticated client, able to quickly process and render point, line, and polygon data. More sophisticated software and a more powerful processor is required of the client.

Disadvantages

Large Amount of Data   Depending on size and resolution, raster maps can contain a tremendous amount of data. Each pixel contains attributes, and in a high-resolution map there can be millions of pixels. In mobile location services, size can cause data transfer problems because bandwidth is very limited. It can also cause storage issues if you hope to cache frequently used maps on the client.

Less "Pretty"   In the past, the pixel grids of raster maps were often not as attractive as a vector rendition. Processor speed has improved and storage space has become much cheaper, so resolution has significantly improved. It is uncommon to be able to discern individual pixels with the naked eye in most displays.

Difficult Projection Transformation   Changing projection with raster graphics is best done by generating a new map from the server. Using vector graphics, it is much easier to do the projection on the client and save the bandwidth and server processing cycles.

Lost Information Due to Generalization   Mobile location services typically use a map database that is vector based, meaning that all data is stored as points, lines, and polygons. Because this data is in most cases a mathematical approximation of the actual geographic phenomena, converting this data to raster information increases the potential for errors.

Vector Maps

Advantages

Good Representation of Reality   Vector maps are best for representing the precise shape of discrete features such as roads.

Compact Data Structure and Speed   Vector data is small in size and can be more effectively transferred over limited-bandwidth connections than raster data. Vector data only has to store point, line, and polygon information.

Topology Can Be Described in a Network   It is possible with vector-based maps to encode topology information. With vector data, line topology encodes which lines are connected to a node and polygon topology records which polygons are on either side of a line.

Disadvantages

Complex Data Structures   A result of the very small data size of vector maps is that their data structures are very complex. They are not well suited to drawing continuous phenomena or areas that do not have distinct boundaries.

Simulation Can Be Difficult   Continuous variation simulations are easy to perform on raster maps, as applying simple mathematical formulas to modify a range of pixels can do the simulation. It is much more difficult to do ad hoc simulations with a vector map.

Difficult Spatial Analysis   Vector maps are not as good as raster maps for geographic analysis that involves spatial coincidence, surface analysis, proximity, or least cost path. However, vector maps are very good for topological map overlay, network analysis, address geocoding, and logical and spatial query. Given that the analysis that vector maps are best at is the analysis mobile location services most frequently use, it is understandable that most spatial analysis software products use a vector map database and use raster maps only for basic end user display.

POI SEARCHES

Most map databases include some basic POIs, such as airports, train stations, parks, schools, and others. The exact list depends on the map database. There are various search types that make it easy for your application to quickly find the right POI for your user.

Search Types

Name Search

A name search handles various matches that might include exact match, wildcard, or soundex. A name search might also check both primary and alternate names. Name searches are typically most valuable when the user knows the brand that he or she is searching for.

Around a Point   A name search around a point allows a user to find all the brand X locations within a radius of a point. This might be the user's location, as identified by a location positioning method, or another location that the user has input. The user might be allowed to specify the search radius, or this could be handled by the application itself.

Along a Route   Searching for POIs along a route is a much more sophisticated and difficult solution. In this scenario, a user who was very loyal to a given gasoline company (because they receive a frequent shopper discount, because they are a shareholder, etc.) might want to know about all the gas stations within a half-mile of his or her route. This would allow the user to plan a driving route and make sure he or she is able to shop with the preferred gas station.

Globally   Global name searches would typically be used only for statistical or testing purposes. It is less frequent that a user would want to do a search on every brand in Europe or North America, as an example. In most cases, the user is better off with a city/region/postal code search or a search around a point. These searches are far less taxing on the server and generate a much faster response.

Within City, Region, or Postal Code   This search type is similar to the search around a point except that the extents are a city, region, or postal code boundary rather than distance from a point.

Category Search

A category search is similar to a yellow pages search in North America. The user does not necessarily have a specific brand in mind, but wants to find the nearest dry cleaner, post office, police station, hospital, or other landmark. The search types possible for category searches are the same as those for a name search.

Phone Search

It is not often that a consumer would want to search for a POI by its phone number, so you might wonder why this search type would be useful. Phone searches are most useful in business and customer service applications. Using telecommunication services that allow computer telephony applications to receive a caller's phone number, it is possible to develop mobile location services that do a database lookup and present a user's personal and location information to a customer service agent as he or she answers the telephone. One interesting application might be to route callers by geographic region to a customer service representative with local knowledge or language skills.

POI Databases

The POI data provided by map data vendors is not rich with details. Typically the information is limited to name, location, and category. Third-party POI databases often have specific and proprietary attributes that are not handled in a generic spatial analysis software system. For many mobile location service applications, these additional attributes are very important. As an example, an off-the-shelf map database and spatial analysis engine might provide a POI database that includes banks. But if your application is targeted to people who need to find a place they can get cash, they need to know where to find an ATM. ATMs can be found in many locations that are not banks, and some banks do not have ATMs. In addition, your user might want to know what bank networks the ATM supports, bill denominations the ATM dispenses, and hours of operation. Although less common in North America, many ATMs in Europe close at night. The additional information not included in the compiled map database is best stored separately and linked with a unique identifier.

Figure 4.24 shows a description of the process of building and integrating your own POI database as handled by Kivera (*http://www.kivera.com*), a provider of mobile location service technology.[4]

1. Unique Identifiers

Only a small subset of the fields in the original POI data becomes part of Kivera's CSF database. CSF is the proprietary, compiled map data format Kivera uses. Information that does not become part of the database and yet would be useful to reference is placed into a supplemental relational database. To enable data stored in Kivera's database to refer back to the data stored in the supplemental database requires the existence of a unique set of identifiers. If one does not exist, Kivera's extraction program creates it.

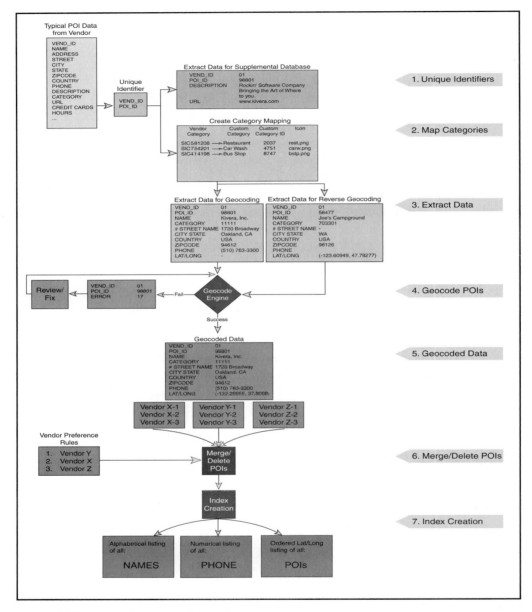

Figure 4.24 Process for Geocoding POIs.

2. Map Categories

Although the raw POIs are associated with a specific category, customers often want to redefine these for their own needs. For each vendor category that is to be used in Kivera's database, a mapping to a custom category, custom category ID, and icon is made. A field mapping (not shown) is also required to relate Kivera's required fields (name, city, state, etc.) to the fields in the vendor's data.

3. Extract Data

Geocoding is the process of determining a latitude/longitude coordinate from a given address. Most POIs do not have an associated latitude/longitude and require geocoding. However, some POIs, such as campgrounds, have an associated latitude/longitude and do not require geocoding.

4. Geocode POIs

Geocoding is the process of assigning a latitude/longitude coordinate to a place. Kivera's geocoding engine works with a variety of inputs, including the following:

- Address (number and street)   OR
- Cross street   OR
- Phone number   OR
- City and state   OR
- Zip code

When not all of the given information exactly matches a location in the database, an indication of the level of confidence of the geocode is returned. Based on an operator's application model a variety of steps are then possible:

- Accept geocode and flag with level of confidence
- Fail geocode with this dataset and, if the customer is eligible to search more expensive data, try another vendor's data
- Fail geocode

One of the functions of Kivera's Geocoding Engine is to obtain a latitude/longitude coordinate given an address. This is accomplished by finding the link that has the desired address within its range and interpolating the actual latitude/longitude coordinates. A geocode operation is said to fail if an address match cannot be found. When this occurs, a list of failed geocodes is created with an error code, defined in Kivera's Location Server API Guide, as to why the geocode was not possible. Because the unique VEND_ID, POI_ID is associated with all of these failed geocodes, it is possible to review these records and correct the incomplete, inaccurate, or ambiguous data.

5. Geocoded Data

Each successive attempt to geocode data creates a new file. Because there can be tens of millions of POIs in a database and multiple attempts can be made to geocode this data, there might exist numerous files. These files are stored in a directory representing the corresponding data vendor. Also, because there can be multiple sources of POI data, there might exist many directories, each with many files.

6. Merge/Delete POIs

To absorb these files into Kivera's database, duplicate POIs from the same POI vendor are removed. If there is duplicate POI information from different vendors, Kivera's Merge POI Engine uses a preference table to select the data to be entered into the database. For example, if both vendor X and vendor Y have similar data about a specific POI, and vendor X is more expensive, the rules can instruct Kivera's Merge/Delete Engine to automatically include only vendor Y's data in the CSF database.

Besides being able to read spatial data from different vendors, Kivera's data neutrality applies to POI vendors as well.

7. Index Creation

Finally, three indexes are created and ordered so that a binary search algorithm can be used to quickly search the data. Each record also refers back to POI data using the unique VEND_ID and POI_ID references.

With these indexes, the following types of queries are possible:

- Search by any word in a name (Oakland Hilton Hotel would show up in a search for "Oakland," "Hilton," or "Hotel").
- Search by phone number, with or without area code.
- Search by category within a specified radius.

Routing With POIs

There are applications, such as routing, with which it might be valuable to provide users with directions that involve POIs. For example, rather than provide a user with directions that say "Left turn on Mission Street (0.5 miles). Left turn on 4th Street," an application could provide directions that say "After passing Moscone Center, turn left on Mission Street. Turn left on 4th Street after passing the Sony Metreon Complex." The additional detail in directions is particularly valuable when landmarks are very obvious and the street names are difficult to find or nonexistent. To make directions like this possible requires coding all POIs with a flag indicating whether it can be used in generating directions and also a value for how visible the landmark is. Certain POIs, such as a dentist's office, would be ignored, because they are not visible landmarks. The ranking and category would allow the application developer to select the most appropriate landmark to include in the directions. It would usually make sense to not include more than one POI per maneuver.

POI Sources

Sources of POIs are varied and highly regionalized. Most major map databases ship with a very basic set of POIs included. Sources of more detailed content include companies like InfoUsa, Dun & Bradstreet, Acxiom, and Experian in North America. In Europe, a common data source is ViaMichelin, which is used by Orange in France and T-Motion in Germany and Austria.

REAL-TIME ATTRIBUTE EDITING

A map database is a static snapshot. In the real world, conditions change constantly: Roads become congested, new roads are built, roads are converted to one-way traffic, and toll roads are added. There are many changes that impact users of mobile location service applications. Traffic conditions might make a route generated from a route engine that is not traffic aware unusable. So how does spatial analysis software account for these changes?

Given that map database vendors only ship updates several times a year, a better method to make interim updates is needed. Although it would be quite challenging to add new spatial data, such as a new road, changing the values of the attributes associated with existing spatial data can accommodate most updates.

Traffic Applications

Real-time traffic is the most obvious case for updating attributes in real time. Traffic incidents have a significant impact on routing. Information collected from a traffic incident would be used to change a road's travel speed, or even block it completely.

Other Map Updates

There are many other applications that can take advantage of the ability to make real-time attribute changes. Attribute editing allows an application developer to further improve the data quality and make interim updates between map vendor release dates.

ENDNOTES

1. Additional map resources can be found online at the following sites:
   http://www.nima.mil
   http://www.noaa.gov
   http://www.usgs.gov
   http://terraserver.homeadvisor.msn.com/
   http://everest.hunter.cuny.edu/mp/

2. An excellent graphical explanation is available at: *http://www.nationalgeographic.com/features/2000/exploration/projections/index.html.*

3. See Appendix C for a more detailed discussion of UTM.

4. The balance of this section is © 2002 Kivera, Inc.

5

Mobile Positioning

In addition to having a high-quality map database and a massively scalable suite of tools to perform spatial analyses such as routing and POI lookups, for a mobile locations services solution to be effective it must be possible to look up and pinpoint a user's location. A discussion of mobile positioning completes our focus on the location-specific components of your mobile location services infrastructure, as shown in Figure 5.1.

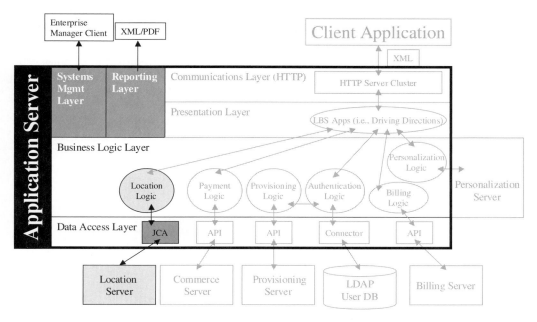

Figure 5.1   Location Server Components of a Mobile Location Service Architecture.

There are a variety of positioning solutions available, which can be broken into three groups: handset-based positioning methods, network-based positioning methods, and hybrid positioning methods (a combination of handset- and network-based positioning). The various solutions have a general trade-off between speed of location determination and accuracy. Depending on the application you are developing, accuracy might be more important than speed or vice versa. For example, if you have an application for deployment on mobile phones that allows a user to look up the nearest coffee shop to his or her location, it might be sufficient to use the latitude and longitude of the cell that the user is calling from as the user's location. It might also be preferable to return a list of potential matches as soon as possible rather than wait for a GPS reading. In some cases, both speed and accuracy might be important. A mobile location services application designed for an in-vehicle system in Europe requires very accurate positioning responses because roads are so narrow and close together. It also requires very fast responses because of the speed of the moving vehicle.

Figure 5.2 builds on the GSM architecture diagrams introduced in Chapter 2 and illustrates a sample positioning server architecture. Your applications will interface with the gateway mobile location center (GMLC) to request the location of a mobile device in the network. Selecting a standards-based interface such as the Mobile Location Protocol from the LIF is best (see Appendix C for the complete specification). The GMLC is able to request registration information from the HLR so that it communicates with the correct serving mobile location center (SMLC). The SMLC coordinates and schedules the resources required to position the mobile device and calculates the final location estimate. The SMLC also controls several location measurement units (LMUs) that obtain radio interface measurements from the MS. The LMU devices are either connected to the BTS via the normal GSM air interface or connected to the BSC via the interface used to connect BTSs. Once an MS has been positioned successfully, the GMLC returns the location estimate to the mobile location services external client.

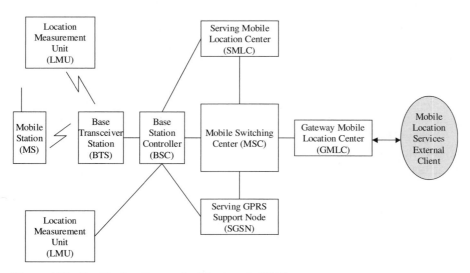

Figure 5.2 Positioning Server Architecture in GSM.

This chapter discusses the various positioning methods and their advantages and disadvantages. The discussion is fairly elementary—further technical reading can be found in the References section. We also discuss briefly the differing requirements for accuracy and speed between mobile location service applications. Finally, we discuss the positioning server standards proposed by the LIF and how to integrate a positioning server into your location service infrastructure.

CELL OF ORIGIN

Cell of origin (COO) or Cell-ID is a purely network-based location positioning solution. The solution uses the latitude and longitude coordinates of the base station serving the mobile device as the location of the user. As such, COO has the highest response time and was the most widely deployed positioning solution in 2001. It can, however, be very inaccurate.

Accuracy is dependent on the size of the network cells. The simplest cell networks have transmitters that are omnidirectional, transmitting equally in all directions and producing a circle. Because circles don't tessellate well, mobile network architects try to approximate them to hexagons, as shown in Figure 5.3.

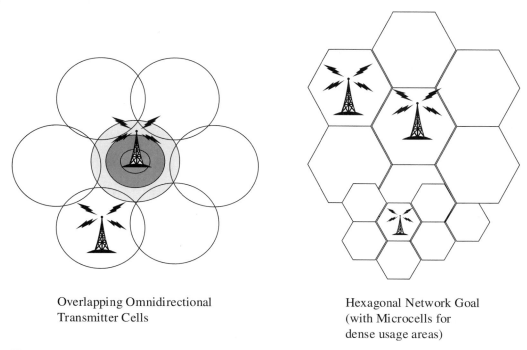

Overlapping Omnidirectional
Transmitter Cells

Hexagonal Network Goal
(with Microcells for
dense usage areas)

Figure 5.3 Base Station Configuration in a Cellular Network.

Cell size in a typical large urban network is from 100 to 1,000 meters, which is the approximate accuracy of COO. A typical response time is from 2 to 5 seconds, with 2.5 seconds being a normal response.

COO is often also referred to as cell global identity (CGI) and complemented with timing advance (TA) information. The TA is the time difference between the start of a radio frame and a data burst, which is used to approximate where the user is within the cell.

Commercial vendors that provide products that support COO positioning methods include Cellpoint (*http://www.cellpt.com*), Ericsson (*http://www.ericsson.com*), and Nokia (*http://www.nokia.com*). Products that support COO also typically support various other more accurate positioning methods that should be used when possible.

GPS

The GPS is a passive, satellite-based navigation system maintained and operated by the U.S. Department of Defense. Its primary purpose is to provide global positioning and navigation for land-, sea-, and air-based tactical forces—but it has also been made available for commercial use. Until May 2000, a selective availability mask distorted the satellite clock signals to reduce the accuracy available for commercial applications. Now that this has been removed, the accuracy of GPS is between 5 and 40 meters, provided the GPS receiver has a clear view of the sky.

How GPS Works

Absolute positioning is the fastest and most common use of GPS for real-time navigation (see Figure 5.4). GPS receivers are range measurement devices that measure the distance between the receiver antenna and various satellites, and determine the position of the receiver by the intersection of the range vectors. This is similar to the method used in time of arrival. A simplified two-dimensional model shows the location at the intersection of three range circles. The circles represent spheres in space and the center is the satellite (see Figure 5.5).

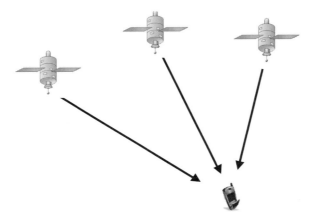

Mobile Device with GPS Receiver

Figure 5.4   GPS Absolute Positioning.

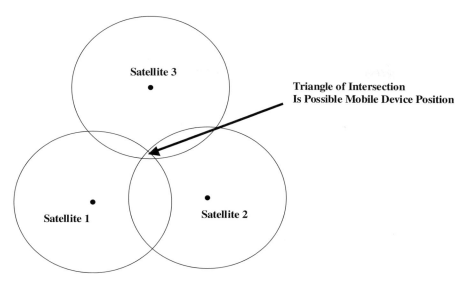

Figure 5.5 GPS Positioning Based on Time of Arrival Principles.

Faster and more accurate positioning methods have been developed that involve post-processing, network assistance, and relative or differential positioning. These methods are known as differential GPS (D-GPS) and assisted GPS (A-GPS). D-GPS uses relative positions to correct position estimates and can be accurate to within 1 meter. A-GPS uses a GPS reference network to accelerate and improve accuracy of positioning.

A-GPS

The idea behind A-GPS is that a wide area differential GPS network is set up with receivers that operate continuously and have a clear view of the sky. This network is connected to the GSM network, and when a mobile device requests a position fix, assistance data from the reference network is transmitted to enhance the performance of the GPS receiver (see Figure 5.6).

The information from the GPS reference network can enhance several aspects of the positioning performance. The A-GPS process allows the GPS sensor to initialize and locate satellites much faster, it increases the accuracy of the positioning, and it requires less power than a standard GPS system. Additional information can also be given to the GPS sensor from the network to further improve performance, including differential GPS corrections and base station location. A time to first fix in standard GPS can take as long as 10 minutes, as a GPS receiver that does not know where it is has to search the entire frequency space (–4 kHz to 4 kHz) and the entire code phase space (1 to 1,023 chips) to locate visible satellites.

A-GPS requires both a GPS receiver in the handset and a reference GPS network that can provide information to assist the GPS receiver in the positioning process. Commercially available solutions are available from Ericsson (*http://www.ericsson.com*), Sirf (*http://www.sirf.com*), and SnapTrack, a Qualcomm company (*http://www.snaptrack.com*).

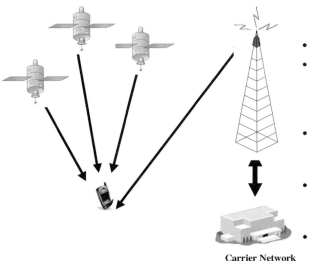

- Location request initiated
- Approximate handset location (location of nearest Cell ID) is given to Location Server
- Location Server tells handset which satellites should be relevant
- Handset takes GPS reading and returns information to Location Server
- Location Server calculates position

Carrier Network

Figure 5.6   A-GPS.

Augmenting GPS With Dead Reckoning

Dead reckoning, also know as inertial navigation, is a technique often used in vehicle navigation to improve the accuracy and reliability of positioning when using GPS. Dead reckoning uses speed and direction sensors to calculate position based on the last known position. It works well when GPS signals are blocked by tunnels and when GPS signals are either blocked or reflected by tall buildings in dense urban areas.

ENHANCED-OBSERVED TIME DIFFERENCE

Enhanced-observed time difference (E-OTD) is similar to time difference of arrival (TDOA), but is a handset-based positioning solution rather than a network-based solution. E-OTD takes the data received from the surrounding base stations to measure the difference in time it takes for the data to reach the terminal (see Figure 5.7). The time difference is used to calculate where the mobile device is in relation to the base stations. For this to work, the location of the base stations must be known and the data sent from the different base stations must be synchronized.

Base stations are typically synchronized using fixed GPS receivers. Accuracy of E-OTD is expected to be as good as 50 meters using GSM and even greater with 3G networks. E-OTD requires additional memory and processing power in the handset, and can be used both when the terminal is idle and when the device is handling a call. E-OTD does have one major advantage over simple GPS in that it works indoors and in overcast weather conditions.

Commercially available solutions are available from Ericsson (*http://www.ericsson.com*) and Cambridge Positioning (*http://www.cursor-system.com*).

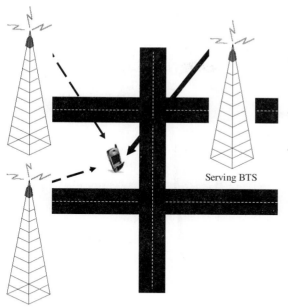

- Handset listens to bursts sent from neighboring base transceiver stations (BTSs)
- Handset records arrival time for bursts
- Position triangulated from:
  - BTS coordinates
  - Arrival time of BTS bursts
  - Timing differences between each BTS

Figure 5.7   E-OTD.

ANGLE OF ARRIVAL

Angle of arrival (AOA) is a network-based method of determining position that does not require a mobile device upgrade to operate. In AOA, a mobile device's signal is received by multiple base stations (see Figure 5.8). The base stations have additional equipment that determines the compass direction from which the user's signal is arriving. The information from each base station is sent to the mobile switch, where it is analyzed and used to generate an approximate latitude and longitude for the mobile device.

An advantage of AOA is that it supports legacy handsets. Disadvantages include the fact that every base station needs to have an equipment upgrade. Users might also be concerned about privacy issues because they are not able to disable positioning from the handset.

A commercially available solution is available from Trueposition (*http://www.trueposition.com*), supporting both AOA and TDOA.

TIME DIFFERENCE OF ARRIVAL

The positioning method known as uplink time of arrival (TOA) is based on the time of arrival of a known signal sent from the mobile device and received by three or more base stations (see Figure 5.9). The signal is the access burst created by having a mobile device perform an asynchronous handover.

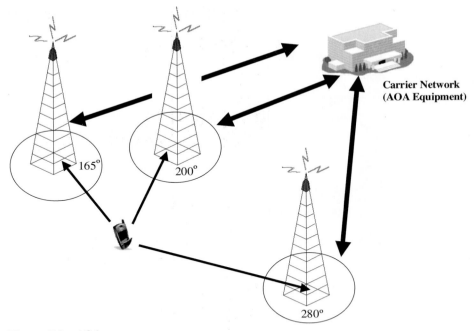

Figure 5.8 AOA.

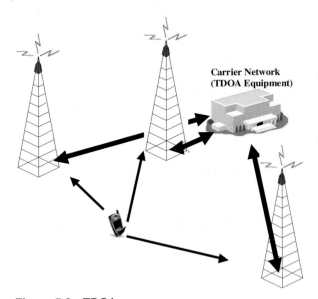

- Base transceiver stations (BTSs) receive signal from handset
- TDOA equipment
  - Measures the difference in the time the signals arrive at the BTS
  - Translates this data into latitude and longitude

Figure 5.9 TDOA.

The TDOA values are calculated by pairwise subtracting the TOA values at the SMLC. The position of the mobile device is then calculated by hyperbolic trilateration (see the earlier section "How GPS Works"), provided that the geographic coordinates of the measurement units are known and the timing offset between the measurement units used in the measurement are known.

Additional technical details are presented by the 3rd Generation Partnership Project (*http://www.3gpp.org*) in GSM 03.72: "Digital Cellular Telecommunications System (Phase 2+); Location Services (LCS); (Functional Description) – Stage 2":

> Access bursts are used for detecting the TOA at the listening measurement units. At a positioning request, the units which should measure the TOA of the Mobile Station (MS) signal are selected and configured to listen at the correct frequency. The MS is then forced to perform an asynchronous handover. Under such circumstances, the MS is transmitting up to 70 access bursts (320 ms) with specified power on a traffic channel (which may be frequency hopping).
>
> The TOA measurements are performed at each measurement unit by integrating the received bursts to enhance the sensitivity, and therefore increasing the detection probability and measurement accuracy, and by applying a multipath rejection technique to accurately measure the arrival time of the Line of Sight component of the signal. The presence of diversity, e.g., antenna diversity and frequency hopping will improve the multipath rejection capability and therefore the measurement accuracy.
>
> When an application requires the position of a mobile, it has to send a request to SMLC for the identification of the mobile and the accuracy level parameter. Depending on this accuracy level, SMLC decides how many measurement units to include in the positioning request. The measured TOA values together with the accuracy parameter of the TOA value are collected and transmitted to the SMLC. The SMLC utilizes the TOA measurements in combination with information about the coordinates of the measurement units and the RTD (Relative Time Difference) values (a and b above) to produce a position estimate. The SMLC delivers the position estimate together with an uncertainty estimate to the application.

The uplink TOA method requires additional hardware (LMUs) to accurately measure the arrival time of the bursts. Different implementation options exist for this positioning method. For instance, it is possible to either integrate the measuring units in the BTSs or implement them as stand-alone units. If the measurement units are implemented as stand-alone units, the communication between the measurement units and the network is preferably carried out over the air interface. The stand-alone units can have separate antennas or share antennas with an existing BTS.

Similar to AOA, TDOA is a purely network-based solution and will support legacy handsets. The drawback, of course, is the installation of equipment in almost every base station, a potentially much more expensive proposition for the mobile operator. Another challenge with TDOA is that it requires that the mobile device be in range of at least three base stations. This is often not the case in rural or even some suburban areas.

A commercially available solution from Trueposition (*http://www.trueposition.com*) supports both AOA and TDOA.

LOCATION PATTERN MATCHING

Location pattern matching is a technology patented by U.S. Wireless Corporation (*http://www.uswcorp.com*) that can determine the location of CDMA devices based on patterns of radio frequency reflections. The technology works with existing handsets without requiring an upgrade. When a positioning fix is required, the mobile device radio signal's distinct frequency patterns and multipath characteristics are compared against a database of previously identified frequency "signatures" and their corresponding locations. This positioning method complies with the U.S. FCC's E911 initiative requiring network-based positioning methods to achieve 100-meter accuracy for 67 percent of calls and 300-meter accuracy for 95 percent of calls. The solution is cost-effective and does not require line of site triangulation, which can be difficult to achieve in dense urban areas where buildings block the line of sight.

MAP MATCHING

Map matching is a technique to improve and correct dead reckoning (see the earlier section "Augmenting GPS With Dead Reckoning"). It uses distinctive features of a mobile device's movement and a road network, for example, to find a corresponding point. This comparison takes place when conspicuous movements such as turning take place. A second method of map matching calculates the distances between the estimated position and the edges of the polygon the device is traveling on. This method is primarily used to relocate position before the estimated position moves too far from the road network.

In mobile location services, estimated position might be generated by GPS or A-GPS, and then sent from the mobile device to the processing center via SMS. This generates what is sometimes called a string of pearls, used for map matching (see Figure 5.10). Recent developments include algorithms to allow map matching for personal navigation by extending the techniques used for map matching in vehicle navigation.

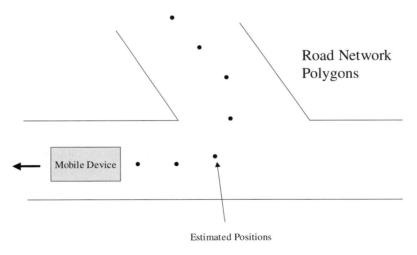

Figure 5.10 Map Matching.

POSITIONING ACCURACY AND SPEED REQUIREMENTS FOR MOBILE LOCATION SERVICES APPLICATIONS

Mobile location service applications vary widely in their requirements for speed and accuracy. Table 5.1 was developed by Mobile Streams (*http://www.mobilestreams.com*) to compare the requirements of common location service applications.

Table 5.1　Comparing the Requirements of Common Location Service Applications

Application	Entry-Level Accuracy Requirements	Mass Acceptance Accuracy Requirements	Custom Device Required?	Objective	Location Frequency
Location-sensitive billing	Cell/Sector	250m	No	Competitive pricing	Originated calls, received calls, midcall
Roadside assistance	500m	125m	No	Send help	Originated calls
Mobile yellow pages	Cell/Sector	250m	No	What's near me?	Originated calls
Traffic information	Cell/Sector	Cell/Sector	No	What's traffic like?	Originated calls or every 5 minutes
Location-based messages	Cell/Sector	125m	Short message or data capable	Advertise, alert, inform	Originated calls or every 5 minutes
Fleet tracking	Cell/Sector	30–125m	No	Resource management	Every 5 minutes or on demand
Track packages	Cell/Sector	Cell/Sector	Yes	Locate and direct	On demand
Driving directions	125m	30m	Yes	Guidance	Every 5 seconds

Source: © 2001 Mobile Streams Ltd.

6

Authentication and Security

Many people think of computer security in terms of secrecy or confidentiality. Equally important to computer security are data integrity, authentication, and systems availability. In mobile location services, it is not only important to keep customer information and commerce transactions confidential, but to keep account records from being corrupted or modified and to make sure that the system is resistant to *denial of service* attacks. Denial of service attacks can prevent service delivery, and are the most potentially devastating attack a mobile location service application can face. Security breaches damage consumer confidence and brand equity.

When considering security for your mobile location services architecture, it is important to think about the following components that interact to provide a complete security infrastructure:

- Process
- Physical
- Platform
- Network

It is important that these components be integrated to improve the quality of the security system. The process category includes corporate security policies and procedures for creating, using, storing, and disposing of data, including the networks and systems on which the data resides. Examples of physical security include key cards, door locks and keys, identification badges, security cameras, cages, and security guards. The platform category focuses on the application-level access controls of the client and server software. Finally, the network category includes routers, switches, firewalls, and remote access devices, which are used to monitor and protect data traversing the network or using an application. A complete security infrastructure is beyond the scope of this book. Although wireless technologies have various limitations with regard to security, for the most part, standard security procedures still apply.[1] This chapter focuses on how wireless is different, and the specific wireless authentication and security issues to be aware of in developing a mobile location service application. Figure 6.1 illustrates the application server security component of the platform layer of your mobile location services architecture.

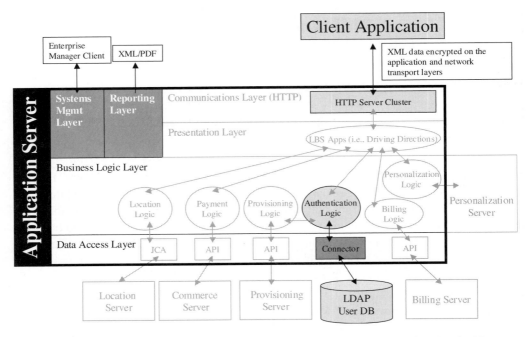

Figure 6.1   Authentication and Security Components of a Mobile Location Service Architecture.

SPECIAL CONCERNS IN WIRELESS NETWORKS

The two primary considerations application developers must consider when developing mobile applications are that radio signals travel through open space and can be intercepted by people who are constantly moving, and that wireless solutions are dependent on public, shared infrastructure where you have less control and awareness of security employed.

Unfortunately, not all radio systems are equal when it comes to security; some are significantly more secure than others. For example, the spread spectrum employed by CDMA is potentially much more secure than TDMA's simple time multiplexing. Spread spectrum techniques were first developed during World War II because the U.S. military was concerned that an ordinary radio signal's carrier waves were very easy to detect. The goal was to make radio signals appear like random noise so it was difficult to tell a transmission was taking place. The carrier waves of these radio signals are not random in reality and are actually agreed on by the transmitter and receiver in advance. The spread spectrum technique used by CDMA is Direct Sequence Spread Spectrum (DSSS).

In addition to the underlying radio signaling of a mobile network, commercial operators employ several methods to ensure the confidentiality, integrity, and availability of communications, including encryption and digital signatures. To authenticate users on the network, GSM uses a digital signature that is stored on the SIM card located in the phone. To masquerade as another user requires gaining access to the SIM card and copying it. GSM uses an algorithm called A5 to encrypt all communications from the mobile device to the base station. Wireless data protocols such as WAP that ride on top of a mobile network such as GSM should also encrypt at the protocol or application level to better secure the communications.

GSM and Security

To understand the security mechanisms that exist in mobile networks, it is worthwhile to look at how GSM handles security. GSM has several security mechanisms that include user anonymity, user authentication, and data and signaling encryption.

When a user first turns on a mobile device, his or her real identity is used for authentication and then a temporary ID is issued. All further communications between the handset and the mobile network use the temporary ID.

A mobile device must authenticate to the network using an encrypted challenge and response mechanism. A random challenge is sent to the mobile device, which it then encrypts using the GSM authentication algorithm and the key issued to the mobile device, and then returns it. The authentication algorithm is implemented in the SIM card installed in the user's mobile device. The mobile operator then verifies that the response to the challenge was correctly encrypted with the key issued to the mobile device.

Once the operator has authenticated the mobile device, a key is generated from the mobile device's response that is used to encrypt all further communications between the mobile operator and the mobile device. This encryption algorithm is known as A5.

Other security precautions include an International Mobile Equipment Identifier (IMEI) to prevent the use of stolen or nonapproved mobile devices. Operators are able to monitor a central equipment identity register for enforcement. GPRS security is equivalent to the security used in GSM. WAP provides additional security mechanisms that we analyze next.

Wireless Transport Layer Security (WTLS): Security for WAP

The Wireless Transport Layer Security (WTLS) provides a security layer for WAP wireless data in addition to the security the mobile operator might employ in its radio signaling. Similar to other security mechanisms, WTLS focuses on:

- Privacy: Encrypting data to prevent unauthorized users from seeing content transferred using WAP.
- Data Integrity: Using digital signatures to make sure content is not tampered with.
- Authentication: Verification that both sender and receiver are who they say they are.
- Non-repudiation: A provision to prevent users or providers from denying they having performed a transaction.

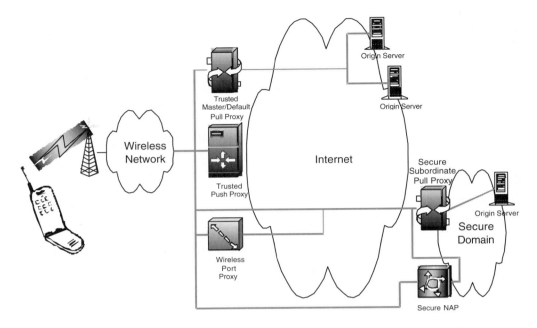

Figure 6.2   Transport Layer: End-to-End Security Overview. © 2002 Open Mobile Alliance.

The basic WTLS end-to-end security process is shown in Figure 6.2.

Figure 6.3 provides an example of the sequence involved in WTLS end-to-end security. As detailed in Figure 6.3, a typical WTLS process is as follows:

- The user selects a service provider (i.e., a bank) on a client device.
- The client user agent sends a Wireless Session Protocol (WSP) method that requests the selected URL (using its default pull proxy).
- The default pull proxy then forwards the method request to the origin server.
- The origin server sends back an HTTP status 300 response to the default pull proxy. The reply includes an XML navigation document in the body section.
- The master pull proxy functionality of the default pull proxy gets the error reply message and analyzes the navigation document to ensure that it contains valid parameters that comply with the policies defined in this document and the master pull proxy owner policies. The master pull proxy also checks the cache control headers and directives for the document.
- Once validated, the navigation document is forwarded to the handset user agent (using an HTTP 300 error).

- After being accepted by the user agent, the navigation document is then cached according to the cache control headers and directives specified for each document, and the appropriate configuration data is made available to the proxy selection mechanism.
- When the user next requests a URL, the user agent uses the proxy selection mechanism to determine which subordinate pull proxy it should use to complete the request.
- If no secure session exists between the user agent and the selected subordinate pull proxy, the user agent will establish a WTLS session with the selected proxy.
- The user agent then informs the user that the session is secure. It also shows the information available in the certificate.
- The originally requested method is then sent on to the selected subordinate pull proxy.
- The subordinate pull proxy forwards the request to the origin server.
- The origin server then replies to the user agent via the subordinate pull proxy.

When the document finally expires, the navigation document and its associated configuration data (if there is any) are invalidated.

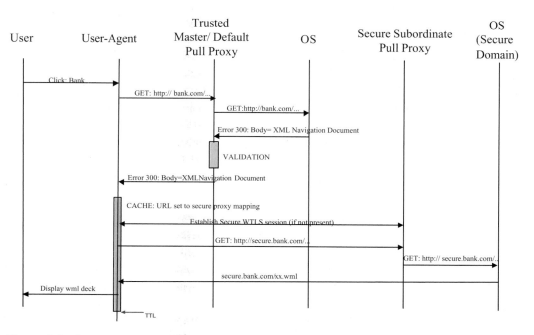

Figure 6.3   Sequence Diagram Overview. © 2002 Open Mobile Alliance.

ENDNOTES

1. Good security resources include the following:
- General security: King, M., Dalton, C., and Osmanoglu, T.E., *Security Architecture*, RSA Press, 2001.
- Application server security: *http://dev2dev.bea.com/resourcelibrary/*
- M-commerce security: Chapter 9

Personalization
and Profiling

It is easier than ever to provide packaged products and services, and consequently consumers are presented with an overwhelming array of brands, offers, and opportunities. The objective of profiling and personalization is to provide the best possible set of services to a user. The better the personalization, the more likely the user is to be highly satisfied. Profiling is the process of collecting, aggregating, and analyzing information so that personalization logic can be implemented. Personalization can range from very simple things like interface customization based on frequently used applications to much more sophisticated systems, like intelligent agents that deliver information a user is interested in.

The goal of this chapter is to analyze what exactly is meant by personalization and profiling, and how to build a technical architecture for personalization. Figure 7.1 illustrates where the personalization engine fits into our mobile location services architecture. Our discussion includes the important privacy issues that are raised by profiling, and some of the technologies such as P3P and Microsoft .NET that provide possible solutions.

WHAT IS PERSONALIZATION AND WHY IS IT IMPORTANT?

Personalization is not new. Think of a personal shopper or tailor. Why do you keep going to the same one? Because they know you, they know your measurements, they know what fashion styles you like and don't like, and they are able to deliver what you want quickly and effectively. You are rewarded with better than average service and the vendor is rewarded with repeat business—a mutual benefit for both parties. The same concept applies to many other services you might require, including realtors, barbers, auto mechanics, and others.

The development of the Internet changed software development forever. Web sites that began as brochures and media publications began to evolve into applications. Software applications that were traditionally shipped once a year on CD now are delivered over the

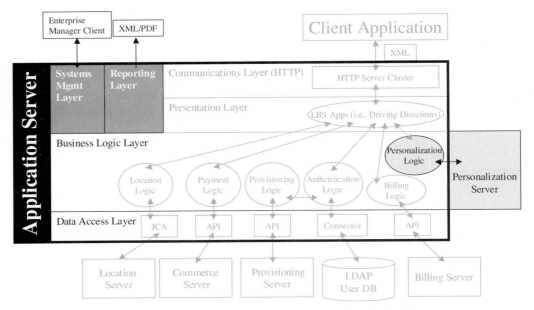

Figure 7.1   Personalization Component of a Mobile Location Service Architecture.

Internet and mixed with relevant content that is updated continuously. These developments spurred the term *transactive content* to describe this new medium. In this new, always-on, always-connected medium, it is possible to enhance product offerings using personalization techniques borrowed from traditional direct marketing: the collection and analysis of data about who users are, what their preferences are, and what they have actually purchased in the past. Opportunities that were never available previously are now possible because of personalization and profiling. For example, suppose that of the people who bought Stephen King's last book, 20 percent purchased it from Amazon.com. Amazon.com has a direct relationship with each of those customers. It might make sense for them to approach Stephen King and say, "Write your next book and we will publish it." In principle, the reader gets a cheaper book, and Amazon.com and Stephen King keep a much larger percentage of each sale.

In a mobile location services application, personalization becomes even more interesting. You might recall our earlier discussion about location services being one component of a new class of opportunities that are based on a user's *context*. This context includes not only where a user might be, but also when he or she is using an application and what his or her role in using that application is. Role can be derived from time, location, and other factors to help understand if someone is using an application as an employee, mother, or one of many other potential roles. The answer to a given question when a user is operating in the role of software engineer is likely to be very different from the answer given when a user is operating as general manager or father.

The term *user profiling* often raises concern over privacy abuses. Personalization is not possible without profiling. A major benefit of good technology is that it allows the individual using it to be more efficient and more effective. Trust is a critical factor, and the importance of trust cannot be underestimated. If users are comfortable that their information is not going to be stolen or abused, if they trust the service provider, there are real benefits to participating in a profiling and personalization relationship. Specific privacy issues and industry self-regulation initiatives are discussed later in this chapter in more detail.

A PERSONALIZATION AND PROFILING SYSTEM

If you plan to develop a profiling system that allows you to provide personalization services, where do you begin? Figure 7.2 shows an example profiling process and gives you an idea of the type of data that could be collected and used.

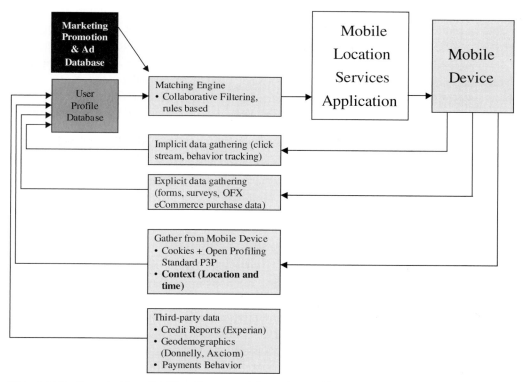

Figure 7.2   Sample User Profiling Process. Adapted from *Net Worth* by John Hagel and Marc Singer, © 1999.

A more detailed and specific study of data that can be collected can be found in John Hagel and Marc Singer's 1999 book, *Net Worth*. User mobility allows you to capture and profile location data, which could range from tracking a user to categorizing them based on their patterns of travel to storing favorite places (so a user does not have to reenter locations constantly), favorite routes (so a user can quickly check traffic conditions), and even favorite maps.

Provided that a user agrees in advance, it is possible to subsidize these sophisticated and personalized product enhancements. Marketers often know which type of person is likely to buy a product or service, and they find that investment returns on direct marketing are much higher than investment returns on broadcast advertising. The more specific the targeting is, the more a marketer is willing to pay. Different marketing services that could make use of location-based profiles include targeted direct marketing, market research, targeted advertising, agent services, and message filtering.

Generation and Control of Location Information

As discussed in Chapter 5, there are many different methods employed to locate mobile devices in a mobile network. There is an increasing shift away from pure network-based positioning methods to handset-based positioning (e.g., E-OTD) or hybrid positioning (e.g., A-GPS) systems because they are fast, accurate, and involve the least cost to the mobile operators. To protect user privacy, most handset-based positioning systems allow the user to disable or mute it. The Qualcomm gpsOne product allows a user to set the positioning unit to operate either by default, on a per-use basis, or only in emergency situations.

There is often a debate about who owns or controls the customer's location data. This is particularly true in telematics, as in-vehicle navigation involves both the automotive company who provides the applications and physical infrastructure and the mobile operator who carries the signals. The answer is that the customers own their data, and they might or might not be willing to share it based on the value they receive for doing so. If customers feel uncomfortable with the way their data is used, as, for example, having to worry about getting automatic fines if they drive their rental car 5 miles per hour over the speed limit, they will simply not use a product. Trust is critical in the relationship between the location-based service provider and the customer, and it is up to the service provider to develop a product that both parties are comfortable with and benefit from.

Integration With Customer Relationship Management

As we've discussed, profiling and personalization in the context of mobile location services can be much more than traditional customer relationship management (CRM). Leading CRM providers such as Siebel (*http://www.siebel.com*) are installed in many communications and automotive organizations today, and provide open interfaces to integrate with the application server used in your location services infrastructure. Solutions are increasingly integrated to meet the needs of large carrier-class solutions, as shown by the increasing synergies between Portal Software's (*http://www.portal.com*) billing platform and Siebel's CRM environment. More specialized products and services designed specifically for the type of profiling and personalization discussed in this chapter are available from niche providers such as Personify (*http://www.personify.com*) and DoubleClick (*http://www.double-click.com*). Additional resources on CRM can be found at *http://www.crmcommunity.com/*.

P3P

One of the challenges in personalization and profiling is providing a simple method for users to tell service providers which information they're willing to share and under which conditions they're willing to share it. Similar to having a multitude of user names and passwords, imagine having hundreds of profiles. How many times have you typed your name and address into an e-commerce Web site? Clearly, it would be better if at the time you were ready to purchase an item, a vendor could make one request: "Please provide your address." Your mobile device could allow you to scroll a list of addresses and submit one with a single click. You might be willing to type your address many times at a Web site, but anyone who has ever keyed a full address into a mobile device, especially a mobile phone, will agree it is not something they would like to do often. To make this a reality requires a standard platform for storing and sharing profiles. P3P is an attempt to provide this standard for the Web. Its principles can be applied to mobile devices and location service applications.

What Is P3P?

P3P is an emerging industry standard promoted by the World Wide Web Consortium (W3C) for automating and giving users more control over the use of personal information at the Web sites they visit. The basics of P3P involve a standardized set of multiple-choice questions covering the major aspects of a typical Web site's privacy and profiling policies. Web sites that are P3P-enabled make this information available so that a user's browser can read it and compare it with the user's own set of privacy preferences before delivering a Web page.

P3P covers nine aspects of online privacy, five of which are related to the data the site is tracking:

- Who is collecting the data?
- Specifically what information is being collected?
- What purpose is the information being collected for?
- What information is being shared with others?
- Who are the recipients of the information being shared?

The other four topics involve the Web site's internal privacy policies:

- Is it possible for users to make changes in how their data is used?
- How are disputes resolved?
- How long is data retained for?
- Where can the site's privacy policies be found in "human-readable" form?

How It Works

Figure 7.3 gives a basic overview of what is required to make a site compliant with P3P. A simple P3P transaction is shown in Figure 7.4.

The developers and participants sponsoring and involved in creating P3P include the following international organizations:

- America Online
- AT&T
- Center for Democracy & Technology
- Citigroup
- Crystaliz

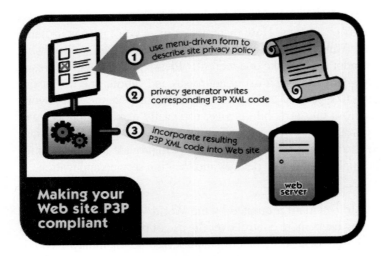

Figure 7.3   Making Your Site P3P Compliant. © 2001 WC3® (MIT, INRIA Keio).
All rights reserved.

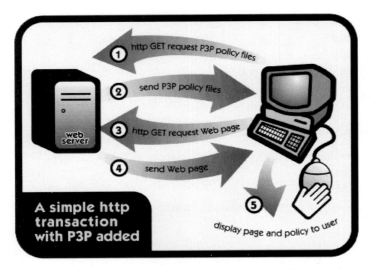

Figure 7.4   Simple HTTP Transaction With P3P Added. © 2001 WC3® (MIT, INRIA Keio).
All rights reserved.

- Direct Marketing Association
- Electronic Network Consortium
- Geotrust
- Gesellschaft für Mathematik und Datenverarbeitung (GMD)
- Hewlett-Packard
- IBM
- IDcide
- International Security, Trust, and Privacy Alliance
- Internet Alliance
- Jotter Technologies
- Microsoft
- NCR
- NEC
- Netscape
- Nokia
- Ontario Office of the Information and Privacy Commissioner
- Phone.com
- Privacy Commission of Schleswig-Holstein, Germany
- TRUSTe

The specifications for P3P can be found in Appendix D, as can an explanation of how to implement P3P in your mobile location services application. Additional reading on P3P is available on the Web at *http://www.w3.org/P3P/*.

Microsoft .NET and Passport

One of the most ambitious projects to address the problem of user profile and authentication management is Microsoft's .NET initiative. Microsoft recognizes the challenges inherent in providing single sign-on access across the multitude of disparate and unrelated Web sites, and has offered Microsoft Passport to provide a single sign-on service. Furthermore, Microsoft has recognized the power of transactive content in making applications more valuable, and that applications are increasingly being delivered via the Internet as a service. This new development minimizes the value of a Microsoft desktop and server operating system platform and therefore many software developers who would otherwise be building Windows applications are building Web applications.

The idea of an initiative like Microsoft's .NET is to view the entire Internet as a large network operating system. All network operating systems need a basic set of infrastructure services that systems within the network require to be valuable and efficient. Microsoft has proposed a set of such services, which it calls Hailstorm. These services provide an enabling infrastructure and make it much easier to develop and deploy applications on the network. Hailstorm Web services are delivered using the XML-based SOAP, and can embed content in HTML, WAP, or Windows Forms Applications. Microsoft Passport provides a standard platform for authentication and profile preferences, similar to P3P.

The basic set of Web services include MyAddress, MyProfile, MyContacts, MyNotifications, MyInbox, MyCalendar, MyDocuments, MyApplicationSettings, MyFavoriteWebSites, MyWallet, MyDevices, and of greatest interest to us, MyLocation.

MyLocation is designed to store basic information such as a user's favorite routes, favorite maps, and favorite locations. We have already discussed how this information can be valuably leveraged, but when you consider that many applications require additional location-oriented operations, such as GetUserPosition(), GeocodeAddress(), CalculateRoute(), or LookupPOIs(), there is an opportunity to provide these operations as a more sophisticated pay-as-you-go package to application developers. Figure 7.5 shows how you might implement a pay-per-use or subscription-based system to provide a more sophisticated service to a mobile location-based application.

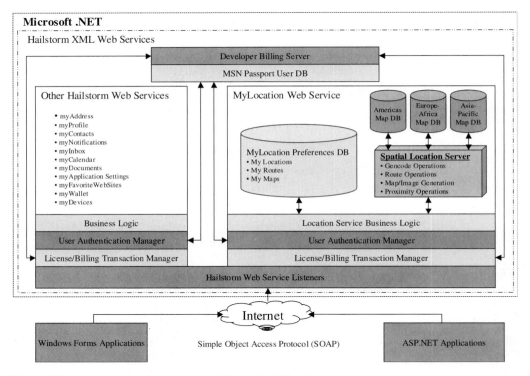

Figure 7.5   Possible Implementation of Extended Web Services.

Additional technical details on Microsoft .NET and Microsoft Passport can be found at *http://www.passport.com/sdkdocuments/sdk21/default.htm* and *http://www.microsoft.com/net/ whitepapers.asp.*

PRIVACY ISSUES

We have discussed the many advantages location-based personalization can provide to consumers. However, location-based services also provide an unprecedented opportunity for information to be abused. Without clear privacy rules, a user's every movement can be recorded and later used by commercial and governmental organizations. Consumers are already overwhelmed by unsolicited marketing directed at them via mail, phone, and e-mail. More important, location information can reveal details that might be protected by regional and national privacy laws. Access to a user's location profile could reveal physical destinations such as government offices and medical clinics. In an extreme case, location data provides real-time information about an individual that could place them in physical danger if improperly disclosed.

The U.S. Wireless Communications and Public Safety Act of 1999 provides consumers with a certain level of privacy protection. Section 222 of the act requires express prior authorization before a user's location information may can be accessed or disclosed. The express prior authorization required must be based on an "opt-in" standard. It is important that mobile location service applications are designed with these guidelines in mind so that consumers' privacy expectations are met. Additional commentary has been submitted by the Center for Democracy and Technology to the U.S. Federal Communications Commission (FCC) and can be read at *http://www.cdt.org/privacy/issues/location/010406fcc.shtml.*

Industry Self-Regulation Efforts

In an effort to prevent unnecessarily restrictive legislation and to educate marketers and service providers, self-regulation initiatives have been launched by the Cellular Telecommunications & Internet Association (CTIA) and the Mobile Marketing Association (MMA).

CTIA

The CTIA is an international organization that represents all elements of wireless communications, including cellular, personal communications services, enhanced specialized mobile radio, and mobile satellite services. The organization represents manufacturers, service providers, and many other groups. One of their primary functions is to represent its members in a constant dialogue with the FCC and in Congress.

The CTIA has developed and published a set of guidelines for appropriate use of location information. The guidelines suggest the following components:[1]

Notice: Customers must be informed about location information collection and use practices before any disclosure or use takes place.

Consent: Express authorization must be obtained prior to collection.

Security and Information Integrity: Location information must be protected against unauthorized access and the provider must ensure that third parties to whom the information is provided adhere to the provider's location information practices.

Technology Neutral: The privacy guidelines should be consistent, whether the service depends on handset-, vehicle-, or network-based location determination techniques.

MMA

The MMA is an international industry trade association devoted to handheld device manufacturers, carriers, and operators; software providers, agencies, retailers, and advertisers; and service providers of mobile wireless marketing and advertising. The activities of the MMA include evaluating and recommending standards and practices, fielding research to document the effectiveness of the wireless medium, and educating the wireless (mobile) advertising industry about the effective, responsible use of wireless advertising.

The MMA has developed a set of guidelines related to privacy and spam. The goals of these guidelines are to put MMA members on record as being in favor of consumers controlling their personal information, to instill trust in consumers that their privacy information is being appropriately handled, and to enable robust and diverse content and service offerings to consumers. The MMA guidelines recommend the following:[2]

- Wireless subscribers are notified of what information is being collected.
- Subscribers should be given notice and choice about how their personal information is going to be used, and it should not be used for purposes other than explicitly agreed to via an opt-in.
- Every effort should be made to ensure that personal information is accurate and secure, and subscribers should be given the opportunity to correct or delete it.
- Wireless push advertising and content should not be sent to subscribers without explicit consent.

ENDNOTES

1. Additional information on the guidelines and CTIA can be found at *http://www.wow-com.com/*.

2. Additional information on the MMA guidelines and MMA can be found at *http://www.waaglobal.org/*.

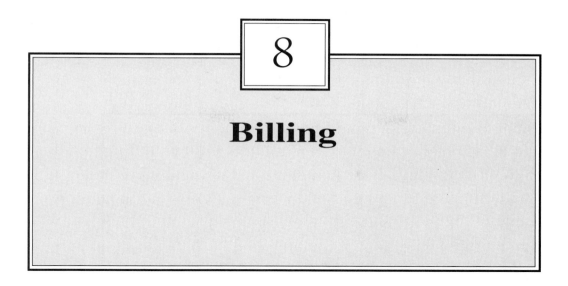

8

Billing

This chapter focuses on the billing system required for mobile location services, as shown in Figure 8.1. Mobile operators and auto manufacturers are both interested in the potential incremental revenues of mobile location service applications. To make them successful, it is important not only to have the right technology, but also to have the right business model. The pricing models available to most mobile operators today are very simple. Wireless data, whether GPRS or WAP, is typically charged for based on either number of minutes or bytes transferred, and there is no visibility into the content itself. Business models used by the telematics service providers in the automotive space are also very simplistic, and often based on a flat monthly fee.

Like the many unsuccessful advertising-based business models tried by dot-coms in 1999, business models that do not cover the cost of providing service and generate profits will not survive. It is clear that the business models for wireless data and location services, whether provided by mobile operators or telematics service providers, must be much more flexible and creative, and are likely to change dramatically. Many European mobile operators have attempted to provide applications and services on their mobile network using a "walled garden" approach, which limits wireless data access to a small number of applications developed by the mobile operator's content division. In many cases, access to competing services is blocked. These initiatives have not been very successful, and it is difficult to argue that mobile operators have a core competence in this area. If Microsoft had designed Windows 3.1 with a closed interface and tried to build all the software applications that were bundled at the product's launch themselves, Windows 3.1 would not have been the success it was.

For wireless data and location-based applications to be truly successful and build a mass market, open standards and an open application programming interface is required. NTT DoCoMo has experienced considerable success with its open wireless data business model that charges subscribers per byte transferred, with a percentage being paid back to the content provider.

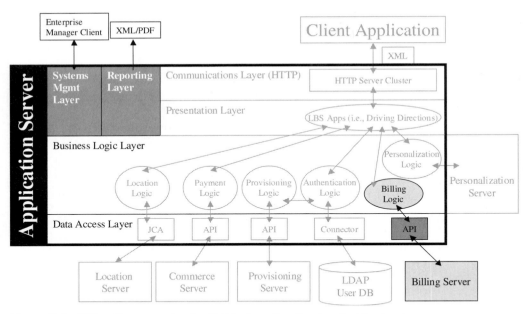

Figure 8.1 Billing Components of a Mobile Location Service Architecture.

Location-based services are highly complex and they require a very expensive infrastructure. For mobile operators and telematics service providers to recoup their infrastructure investment requires an even more sophisticated and more flexible business model that can involve third-party application developers. It is important to be able to provide pricing models that bill subscribers and compensate application developers, and involve more than just the number of bytes transferred (e.g., charging application developers on a per-usage basis for specialized services like subscriber positioning, and providing them with prepackaged business models and billing services).

TECHNOLOGY AND BUSINESS MODELS

To provide open interfaces and flexible billing models, service providers are shifting away from their traditional highly customized and homegrown systems. These systems are designed for postpay voice, are typically not object oriented, and have often been built by systems integrators that are no longer available to improve them. They are a speed to market impediment to launching new products and services.

In addition to being costly, the challenges these systems present to service providers include being inflexible and unable to scale. To make wireless data and location services attractive, billing systems must be able to charge for more than call duration and time of day. Billing systems must be able to handle pre-pay customers, and billing based on number of bytes transferred, URL accessed, application accessed, and even quality of service delivered. Not being object-oriented and based on open standards makes integration and scaling very difficult. As an example, imagine if a service provider in Germany wanted to charge different rates for service among the approximately 30 regions in Germany. If every zone has a different rate, there are 30^2 combinations, a crippling number to many legacy billing systems. Providers such as Telenor, who have adopted sophisticated billing products like those available from Portal Software (*http://www.portal.com*) are already providing mobile location service applications, such as restaurant and taxi locator services. Subscribers are charged for the service and a positioning surcharge is passed on to the application provider.

Billing systems like Portal's are designed to plug into service provider infrastructures, with flexible integration with application server products like IBM's Websphere Everyplace Server and others. They allow authentication to be handled internally in the billing system or to be passed off to another specialized authentication system. They also allow close integration with CRM and personalization servers such as those provided by Siebel.

ROAMING

The need to allow customers to roam on other provider's networks has long been a challenge for mobile operators, and the requirements of mobile location services only make it more difficult. It is important to understand how roaming for voice services works, because roaming for mobile location services will be similar but more complex.

To allow roaming for voice services in a postpay system, a mobile operator needs to have network elements of its roaming partners updated in the regions in which the subscriber wishes to travel. The network elements detail the subscriber's mobile number and authorization for roaming services and provide the means for authenticating the subscriber. When a subscriber makes or takes a call while roaming, the call detail record (CDR) is rated and the revenue due the roaming partner is calculated and billed to the home mobile operator. The CDRs are also sent to the home mobile operator so they can reconcile. In many cases, a billing clearinghouse is also involved, which collects the roaming CDRs for the mobile operators and provides the reconciliation for billing and settlement purposes.

Providing roaming for voice services in a prepay system is significantly more complicated. Unlike in North America, prepay systems are very common in Europe and could account in part for the very high rates of mobile service penetration. Prepay systems must provide the same authentication, authorization, and reconciliation service as a postpay roaming system, but must in addition make sure a subscriber has not run out of money. It is virtually impossible to provide roaming services in a handset-based prepay system. Fortunately, most prepay systems are now network based, and are able to share information from their intelligent network (IN) node to provide account balance details and call control (start/stop).

Addressing the roaming issue in wireless data is important, especially given consumers' expectations of seamless cross-border coverage (e.g., the United States and Canada in North America and pan-European in western Europe). Location services further complicate the roaming issue, particularly if mobile operators implement different location positioning methods. Handset-based positioning is a particular problem for roaming. Unfortunately, many mobile operators are choosing positioning systems like A-GPS and E-OTD that require network/handset cooperation for best results. The challenge in billing for mobile location services while roaming provides another case for the value of open standards.

WHO WILL PAY?

Because the market for mobile location services is in the very early stages (in 2001, most consumers were completely unaware of mobile location services), it is often debated who will pay for them. General Motors' OnStar has built a large subscriber base in the United States by giving away service with new vehicles, but as we know from the Internet, this is not a sustainable business model. It is expected that to build the market, services must initially be underwritten by the large automotive manufacturers and telecommunication providers.

Given the heavy reliance on cars for personal transportation in North America, mobile location services will likely be dominated by automotives. In Europe and many parts of Asia, by contrast, with so many people living in large cities and having access to excellent transportation, mobile operators have a very interesting business opportunity and will deliver personal telematics products to compete with the in-vehicle telematics products. It is crucial that the mobile location service applications deliver compelling, "must-have" benefits so that consumers can be upsold to a profitable, transaction-based business model. The automotive providers have a significant advantage over the mobile operators for several reasons. The form factor of the in-vehicle device provides substantially more flexibility than a mobile phone. In-vehicle systems are able to embed the GPS receiver in the vehicle, utilize the vehicle sound system, and tap into the vehicle's power supply. Most important, auto manufacturers are able to capture diagnostics information that can be leveraged to produce cheaper and more reliable cars—and more accurately diagnose and refer problems with a car to the driver and a repair facility. This information can substantially reduce an auto manufacturer's costs. Mobile operators have no comparable incentive.

In addition to the large auto manufacturers and mobile operators, mobile location service provision is likely to also come in the automotive space from today's emergency service and roadside assistance providers, such as the American Automobile Association, Cross Country, and ATX Technologies in the United States and Altea Europe in Europe. Given the commitment of a large number of auto manufacturers to make telematics and navigation standard equipment in most new vehicles beginning as soon as model-year 2004, and the enormous investments they have made in launching telematics service providers such as OnStar and Tegaron, it is clear that they see mobile location services as core to their business and intend to dominate it. They furthermore have the deep pockets to buy their customers by subsidizing service until real products and a real market develop.

Case Study: ATX and Portal Software

An example of how telematics service providers are implementing billing technology today can be seen by ATX's announcement to integrate Portal Software's billing software.[1]

CUPERTINO, Calif. (July 2, 2001) — Portal Software, Inc. (Nasdaq: PRSF), a leading provider of business infrastructure software for next-generation communications services, announced today that ATX Technologies, Inc., the leading independent telematics service provider to the automotive market, is deploying Portal's Infranet® customer management and billing platform to support its line of location-based mobile services. Infranet, the foundation of the Infranet Wireless product-line for convergent mobile service providers, will support ATX's telematics services such as automatic collision notification, location-based emergency response, and GPS-based navigation. Infranet also has the capability to manage future ATX services such as digital entertainment, mobile commerce, and telephony.

Telematics is a rapidly growing in-vehicle class of services that integrates wireless communications, location identification technology, and off-board database and computing functions to provide drivers with a variety of safety, security, navigation, entertainment, and convenience services. Infranet will enable ATX to quickly and easily launch and support branded services, develop new service and price plans, manage its customer base, and collect payments. With Infranet, ATX could eventually support such innovative services as streaming music or loyalty programs for mobile commerce transactions.

"ATX's services are rapidly evolving from a value-added luxury to a mass-market safety and convenience feature in automobiles," said Norm Feldman, vice president of IT for ATX. "To support our subscriber growth and service innovation, we need a strategic customer management and billing platform that gives us the scalability to manage a large and diverse subscriber base and the flexibility to support innovative new services and sophisticated price plans. After thoroughly researching a number of solutions, we chose Portal Software for its leadership and depth of experience in supporting next-generation mobile services."

Portal Software's Infranet Wireless product line provides mission-critical business infrastructure for mobile operators offering voice and mobile Internet services. Infranet Wireless features modular, off-the-shelf enhancements to the Infranet platform for supporting switched voice services such as GSM, mobile Internet services such as WAP and GPRS, interconnect billing, and business intelligence modeling.

"ATX's business model is predicated on their ability to continually pioneer inventive new mobile services. Because the Infranet platform features an open and flexible architecture, it can easily support the advanced location-based telematics services offered today by ATX, as well as whatever next-generation services they experiment with in the future," said Vijay Iyer, vice president of market development for Portal Software. "We're excited to support their next-generation mobile services, and look forward to a working with ATX as their services evolve and customer base grows."

ENDNOTES

1. Courtesy of Portal Software. © 2002 Portal, Inc.

9

Mobile Commerce

Billing systems handle a service provider's accounting, bill presentment, and payment for basic telecommunication services, but commerce systems focus on transaction processing, payment, and fulfillment for products that must then be delivered to a customer, whether electronically or physically. Figure 9.1 illustrates the commerce components of our mobile location service infrastructure. This chapter discusses mobile e-commerce, from specific mobile e-commerce opportunities enabled by location-based information to technologies under development to facilitate mobile e-commerce.

WHAT IS M-COMMERCE?

The process of doing electronic commerce via a mobile device is known as *m-commerce*. Many m-commerce applications are simple, direct translations of Web e-commerce applications to the mobile environment. However, the type of shopping a user is likely to do on a mobile phone is very different from the shopping he or she might do on the Web. It is unlikely that a mobile user is going to browse large catalogs and shop in a mall-style environment. Because of the small-form factor, wireless push advertising is for the most part overly intrusive, and users are unlikely to accept the amount of advertising they receive today on the Web. Successful m-commerce will require applications that are specifically focused on the requirements of mobility and leverage additional information, such as the user's location, to make highly efficient and compelling user services.

In general, m-commerce applications should focus on the fact that a user is likely to be mobile and short on time. Applications that focus on event-specific services are likely to be well utilized. For example, an application might focus on Mother's Day, with integrated and highly focused service offerings from a flower provider, a gift company, a restaurant finder, an e-card provider, and perhaps even movie and concert listings. The service

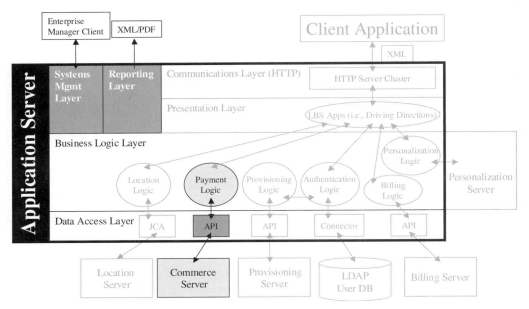

Figure 9.1   Commerce Server Components of a Mobile Location Service Architecture.

provided to users is a one-stop shop to make necessary arrangements if they have forgotten an event. Event-specific services need to utilize only the most basic location information, such as a region or metropolitan area.

A more sophisticated use of location information in m-commerce would to enable mobility management and real-time ticket-based applications. The concept behind mobility management is to provide an end-to-end environment to help users make multimodal travel arrangements and provide assistance at every step in their journey. This requires the ability to route on foot, by car, and via trains, ferries, planes, and buses. It also leverages the fact that the application knows where the user is and can always provide relevant information. For example, if a user is at the hotel he or she reserved through a mobility management application, the user's mobile device can direct him or her to a customized set of services that the hotel provides. If the user is at the train station on a particular segment of a journey, he or she can be presented with services available at the train station. A service like this would build a strong relationship with the subscriber, and the subscriber would reward the service provider by using them to book all travel arrangements and providing many opportunities to fulfill other service opportunities.

Another opportunity for m-commerce that leverages location information is real-time ticket-based applications. Think of the challenge of finding parking in many downtown areas—in some areas there are more people driving around looking for parking spaces than spaces to park in. Whether a user is a daily commuter or a one-time visitor, an application that calculates a route and allows a user to reserve (and pay for) a parking space near the destination would be a great value. The user's route would be modified to help provide directions to the parking spot, and the user would be routed on foot from the parking garage to the final destination. Other examples of ticket-based m-commerce application would be reserving and purchasing movie, concert, train, or plane tickets.

MOBILE ELECTRONIC TRANSACTIONS STANDARD

To help facilitate the development of m-commerce, a standard industry framework called the Mobile Electronic Transactions Standard (MeT) is under development by a working group that includes Nokia, Ericsson, and Motorola. MeT is an initiative designed to provide a framework for secure mobile transactions and to provide a consistent user experience that is independent of mobile device, service, and network. The idea behind this initiative is to ensure that interoperable mobile transaction solutions are developed around the world, with the goal being that consumers have seamless access to goods and services wherever they may be. An example of how this system is expected to work is shown in Figure 9.2.

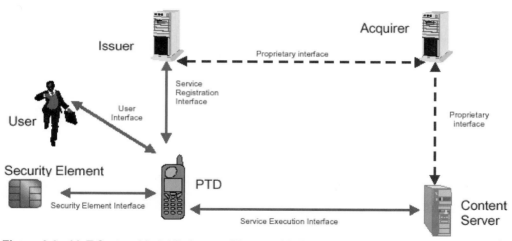

Figure 9.2 MeT System Model Reference Diagram. MeT defined interfaces are shown as solid lines; proprietary interfaces are shown as dashed lines. Security Element interface is proprietary to PTD manufacturer for non-removable (built-in) Security Element. © 2002 Mobile electronic Transactions Limited.

The MeT provides infrastructure guidelines for the mobile device to be used as a personal trusted device (PTD). The mobile device and the mobile operator's network already provide the technical capabilities necessary to authenticate a user and provide secure key storage, cryptographic processing, and transaction processing. The core functions of MeT include the following:

- Initialization: Providing the PTD with public/private key pairs for signing and authentication.
- Registration: The PTD will be provided with both service certificates and root certificates.

- Establishing a secure session: WTLS is used to provide a secure session for remote and local environments.
- Authentication: The client is authenticated using a client certificate and PIN number.
- User authorization: The PTD uses a signature key to create a digital signature of a string of text provided to the user. The user accepts the transaction by entering a signature PIN that signs the text.

PART 3

10

Client Platforms and Protocols

The location services client is the interface through which users interact with your system. It is the point at which all technologies are brought together—and usability and performance of your client determine a user's opinion of not only your service, but in many cases your company and its brand as well. Mainstream consumers will judge products on issues such as the aesthetics of the form factor and ease of use in addition to application performance and delivery.

The innovation generated by Internet and World Wide Web technology has impacted the way client/server software is developed in many positive ways. Supporting massive installed bases has never been easy, but distributed environments, object-oriented computing, connectionless application protocols, and text-based markup languages have made it easier than ever before. Because business logic is stored on the server, it can be continuously improved—removing many of the headaches of software upgrades and multiversion support. Application functionality is downloaded on the fly as needed.

A key technology that has made many of these advances possible is XML, which allows you to develop a single interface to your location server that uses different style sheets to customize output for the format the client is expecting. HTML, HDML, cHTML, WML, and VXML are all based on the XML specification. If your location server's interface is XML based, supporting a new client (or protocol) requires nothing more than creating a new style sheet. It does not require engineering a new interface.

This chapter discusses common client platforms and network communication methods. Given the highly localized nature of mobile location services and keeping with our focus on extensibility, the chapter concludes with a discussion of developing a mobile location services application ready for internationalization and localization.

PLATFORMS

There are a variety of platforms available for location services clients, from standards-based platforms like Palm OS and Windows CE to proprietary platforms like those typically found in low-end mobile phones and some in-vehicle navigation systems. Standards-based platforms make development easier because they provide more tools for development and more engineers with experience on the platform. Proprietary platforms can often be optimized to achieve better performance, but are very costly to develop, support, and upgrade over time.

No matter what type of platform you choose for your mobile location services application, the small-form factor requires adhering to certain user interface guidelines to ensure usability. In addition to standard user interface guidelines, it is important to keep screens short and minimize the use of colors and graphics. A short screen is faster to refresh and reduces the need to scroll (see Table 10.1 for platform screen guidelines).

Table 10.1   Client Platform Screen Details

Mobile Device	Screen Size	Screen Colors
Mobile phone	12 characters × 4 lines (minimum size)	Monochrome
Palm/Handspring	160 pixels × 160 pixels	4 shades of gray/256 colors
Microsoft P/PC	320 pixels × 320 pixels	16 shades of gray/256 colors
Microsoft H/PC	640 pixels × 240 pixels (minimum size)	256 colors

It is best to use menus (pick lists) as often as possible and to reduce the need for users to enter text. This might mean using more sophisticated personalization, including prepopulating responses with answers from previous visits. For mobile phones in particular, it might even mean that in an area in which you might ask for a user name and password in a desktop application at every visit, you authenticate far less frequently.

This section gives an overview of Palm, Windows CE, and the Symbian OS operating systems. We then discuss some sample location services clients.

Palm OS

With the majority of the PDA market and more than 10,000 third-party software products, handheld devices based on the Palm OS are some of the most popular. Palm OS is particularly popular among corporate enterprise users, providing excellent integration and synchronization with products like Microsoft Outlook. A significant advantage of the Palm OS is the graffiti handwriting recognition used for data input. It allows the Palm OS device to keep a small-form factor (by not requiring a keyboard) and eliminates the frustrating key-based text input method found in mobile phones.

Wireless network support is provided in North America for Palm OS devices via the Palm.net network, a wireless modem, and cellular digital packet data (CDPD) service or a mobile phone connection. Other areas of the world, including Europe, have wireless network support through a mobile phone and mobile Internet connection kit. The lack of integrated, inexpensive, and always-on wireless network support outside North America makes Palm OS really only feasible for mobile location service applications designed for use within the United States and Canada.

There are a number of different choices of programming languages for developing on the Palm OS platform, including C/C++, Java, and Palm's proprietary *web clipping* format. C/C++ applications must be compiled to the native Palm Resource File Format (PRC). This is accomplished using third-party development tools. The resulting application is installed via HotSync. Java applications require the installation of a Java virtual machine to run. Web clipping applications are designed with pages built using a subset of HTML 3.2. They are then compressed into a format called PQA and added to a Palm OS device that has network capabilities. Web clipping applications launch an internal Web browser that displays both static pages from the PQA file and dynamic pages generated by an external Web server.

Examples of mobile location service applications designed for the Palm OS platform include an ATM Locator Service from Visa (*http://www.visa.com*) and an Emergency Services application from RoadMedic (*http://www.roadmedic.com*; see also Part 2 of this book). Additional technical details on Palm OS can be found at *http://www.palm.com*.

Microsoft Windows CE

Microsoft Windows CE is an open, scalable, 32-bit operating system designed for intelligent network devices. Windows CE provides application developers with the familiar Microsoft Win32 APIs, ActiveX controls, Component Object Model (COM) interfaces, and the Microsoft Foundation Classes (MFC) library. In addition, Windows CE provides built-in support for multimedia, security, and communications (e.g., TCP/IP).

Windows CE is a powerful platform for mobile devices that have sufficient processing and memory available to run the operating system. Unlike some other mobile device operating systems, Windows CE supports real-time processing and preemptive multitasking and multithreading. For equipment manufacturers interested in developing mobile location service applications, Microsoft offers the ability to create custom shells, the ability to encrypt all network communications, and a special suite of power management capabilities.

Because Windows CE requires a much more powerful processor to operate, the battery life of Windows CE–based applications is much shorter than for Palm or EPOC devices. Furthermore, the more powerful hardware required usually means that the Windows CE form factor is larger than other PDAs or mobile phone communicator products. Although perhaps not ideal for mobile phone applications, the Windows CE operating system can work well in the automotive environment. Devices in the car have far fewer restrictions on size and power, and they benefit from the additional processing power and memory.

In fact, Microsoft has developed a specialized version of Windows CE called Windows CE for Automotive (WCEfA) to build capabilities specific to the needs of the automotive industry.

Added to the standard capabilities of Windows CE, WCEfA provides a number of automotive-specific features. In addition to a suite of APIs to handle the integration of positioning and navigation software applications, WCEfA provides the following features:

- Real-time, 32-bit, memory-protected OS kernel, with support for a variety of CPUs
- Fast boot capability and deterministic interrupt response times under ten microseconds
- Windows (Win32®) Application Programming Interface subset, including file and memory management, thread and process management, networking stacks, etc.
- Hands-free (speech) communication interface
- High performance graphics support through DirectX® and GDI-Sub
- Multimedia support through DirectShow® with support for MP3, Windows Media™ Audio, and DVD
- Customizable and skinnable graphic driver interfaces
- Driver Distraction Control
- Critical Process Monitor
- Advanced power management
- Customizable developer tools
- Multilanguage support
- Internet Explorer Web browser, available in both high-function Generic Internet Explorer (GenIE) and small footprint (Microsoft Mobile Explorer™) formats
- XML, the native language of Microsoft's Car .NET telematics initiative

WCEfA provides automotive suppliers with a computing platform to power an automotive computing device. The automotive computing device is comprised of an integrated computing architecture that allows it to reside on top of the automobile mechanical system. The automotive system architecture enables a software application to run in an automobile by providing several layers of technology between the application and the automobile mechanical system. The automotive computing device is installed in the dashboard of an automobile.

The user interface is a faceplate mounted in the dashboard of the automobile that allows the user to control all the functions of the automotive computing device. Microsoft provides a sample hardware bezel that can be used as the faceplate of the automotive computing device or the automotive supplier can build a custom hardware bezel.

WCEfA 3.5 provides a tremendous amount of flexibility to customize the operating system and choose the components you need to fit any number of potential architectures. WCEfA can require as little as 800Kb of space for a simple device. Memory requirements increase depending on exactly which operating system features are selected for the final product. Processors supported in WCEfA 3.5 include the Intel ARM (SA1100), Hitachi SHx (SH4), Intel x86 (i486), and MIPS (R4100 and R4300).

Additional information on developing mobile location services applications using the WCEfA APIs is available at *http://www.microsoft.com/automotive*.

Symbian OS

Symbian is a software development and licensing company headquartered in the United Kingdom and owned by Ericsson, Nokia, Matsushita (Panasonic), Motorola, and Psion. Symbian develops and supplies Symbian OS, an open, standards-based operating system for data-enabled mobile phones to the major handset manufacturers.

The Symbian OS was designed with a number of goals in mind. Mobile phones are small and mobile and require high availability. They cannot afford to go through boot sequences, and with limited power and processing resources, must maximize efficiency. For example, Symbian OS implements multitasking with event-driven messaging instead of multithreading, which is much more complex and expensive. Mobile phones must be highly reliable. They should almost never lock up, and service packs and patches should be rare. Because of wireless network issues, such as spotty coverage and many different protocols, a mobile phone operating system must gracefully handle the loss of network connectivity and user notification. Finally, because of the many different hardware platforms and to enable third-party software development, Symbian OS is open and standards based. Application development on Symbian OS can be done in C++, Java, WAP, and Web, and the operating system supports standards that include Unicode for internationalization, POSIX, TCP/IP, IMAP, POP3, SMTP, SMS, Bluetooth, OBEX, I-mode, J2ME, JavaPhone, and SyncML. Figure 10.1 shows the Symbian OS components and system dependencies.

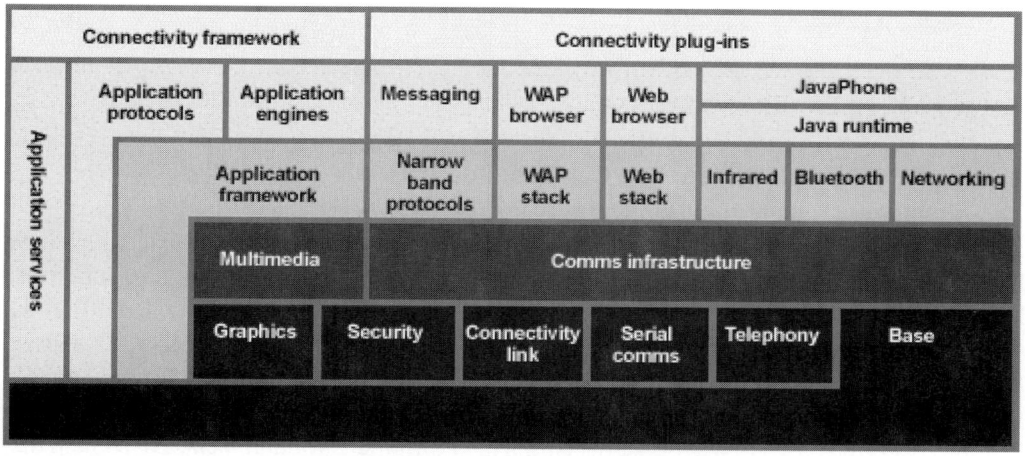

Figure 10.1 Symbian OS 6.x System Dependency. © 2002 Symbian Ltd.

A sample Symbian OS mobile location services application available for the Nokia Communicator phones is the TomTom CityMaps product from the Netherlands-based Palm-Top Software. Figure 10.2 shows an example map with TomTom CityMap and Figure 10.3 shows an example route.

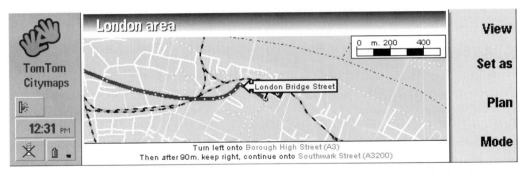

Figure 10.2 TomTom CityMap Mapping Example. © 2001 Palmtop BV, The Netherlands.

More technical details on the TomTom CityMap product are available from *http://www.tomtom.com*. Additional technical details and specifications on the Symbian OS are available from *http://www.symbian.com*.

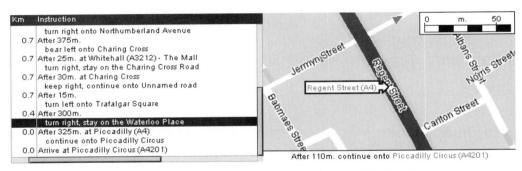

Figure 10.3 TomTom CityMap Routing Example. © 2001 Palmtop BV, The Netherlands.

Sample LBS Client Devices: The Car Dashboard

The capability of client hardware now available in vehicles is exciting. Particularly in Europe, original equipment and after-market suppliers that specialize in car audio have launched combined DVD/navigation systems. Many of these systems in production are known as *onboard* systems because they store all map data on CD or DVD and do all the spatial processing in the vehicle. This is inconvenient for users when data does not stay up to date and when they must change CDs when they cross a border. The LBS client devices we discuss are for systems available today in Germany. These LBS client devices are broken into two types: LCD-based systems with a next turn indicator and VGA-based map display systems. It is not uncommon for both system types to display traffic information that is broadcast by radio using RDS–TMC technology (see Chapter 11 for more details on RDS–TMC).

The TravelPilot DX-N system produced by Blaupunkt,[1] shown in Figure 10.4 and Figure 10.5, provides automatic and dynamic route guidance leveraging for onboard map CDs and RDS–TMC. Destination input support includes both addresses and points of interest. A user is guided to his or her destination by use of maps, arrow indicator, or voice output. The system has a memory that stores the last 12 destinations and has an additional programmable memory for up to 50 destinations.

Figure 10.4 Blaupunkt TravelPilot DX-N (Germany). © 2002 Bosch Group.

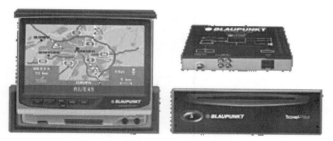

Figure 10.5 Blaupunkt TravelPilot DX-N Multimedia. © 2002 Bosch Group.

Similar to the TravelPilot DX-N, the TravelPilot DX-R70 provides a driver with a dynamic navigation and routing system that leverages traffic information available over RDS–TMC to detect traffic incident delays and either route through the incident or evade it altogether. The TravelPilot DX-R70 has a much simpler and more compact LCD interface. Routing is done with the arrow turn indicator and voice output—there is no map display (see Figure 10.6).

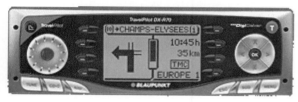

Figure 10.6 Blaupunkt Radio Navigation System TravelPilot DX-R70 Titan. © 2002 Bosch Group.

Additional information and technical specifications on Blaupunkt location service systems can be found at *http://www.blaupunkt.de.*

The Pioneer AVIC-8DVD is a voice-based navigation system that includes map data for all of Europe on a single DVD (see Figure 10.7). The system has an integrated TMC processor, has voice recognition capabilities, and integrates traffic information for dynamic routing. If new traffic information is received while the user is driving, the route is automatically recalculated and an alternative is proposed to the driver. The system has a destination memory, and destination input can be handled by inputting an address, cross streets, postal code, or POI. Figure 10.8 and Figure 10.9 show more sophisticated Pioneer systems that use a VGA interface for map display.

Figure 10.7 Pioneer Voice DVD Navigation System AVIC-8DVD. © 2002 Pioneer Europe NV.

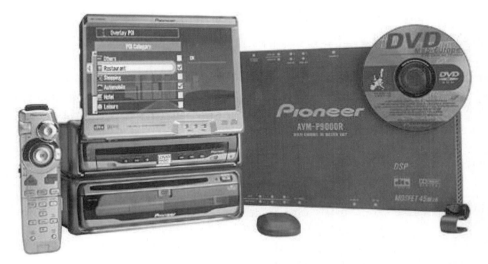

Figure 10.8 Pioneer DVD Navigation/AV System AVIC90DAV. © 2002 Pioneer Europe NV.

Figure 10.9   Pioneer DVD Navigation System AVIC70D. © 2002 Pioneer Europe NV.

Additional details and specifications on Pioneer location services products can be found at *http://www.pioneer.de.*

The Blaupunkt and Pioneer systems just shown are closed systems that operate without mobile network connectivity, but similar and much more powerful interfaces will soon be available in the car that leverage the open interfaces we discussed earlier, such as Microsoft's WCEfA. The current in-vehicle navigation systems in cars produced by companies like Pioneer and Blaupunkt are likely to merge with the emergency services and applications provided by telematics service providers such as OnStar (*http://www.onstar.com* and *http://www.onstar.de*) and Tegaron (*http://www.tegaron.com*).

CLIENT PROTOCOLS AND LANGUAGES

XML

XML is a set of rules for designing text formats that allow you to structure data. Examples of structured data include presentations, spreadsheets, address books, financial transactions, maps, and driving directions. Using XML makes it easy for a software application to create, organize, and share information with other software applications. Furthermore, XML is extensible, it is platform independent, and it supports internationalization.

How Does XML Work?

Similar to HTML, XML uses tags (words delimited by < and >) and attributes (such as name="value"). Unlike HTML, which specifies what tags and attributes mean and how they will be represented in a Web browser, XML uses tags to delimit data and leaves the interpretation of the data to the application. Also unlike HTML, XML is case sensitive and has a very rigid set of syntax rules.

Specifying a set of rules for structuring a certain type of data is done in a document type definition (DTD). There are specific DTDs for structuring geography data (GML), voice data (VXML), and countless other types of data. The family of XML technologies also includes cascading style sheets (CSS), Extensible Stylesheet Language (XSL), and XSL Transformations (XSLT). CSS are used to attach style elements (fonts, spacing, etc.) to structured data to separate content from presentation. CSS works the same with XML as it does with HTML. XSL is a language for expressing style sheets, and XSLT is a language for transforming XML documents into other XML documents. The value of using style sheets in this manner is that spatial and positioning data retrieval can operate without specific knowledge of how the data will be employed by the location server client.

An engineering team tasked with building a location services infrastructure could be split into core spatial operations, core positioning operations, and application delivery operations. Application delivery operations could generate routing queries to the spatial engine and positioning queries to the positioning engine without having to know anything about their underlying algorithms. The application delivery operations could then package the results in a standard XML format, customized for the intended target client using CSS and XSL/XSLT. If data needs to be sent to the client in GML, the GML style sheet is employed and data is transformed into GML. If data needs to be sent to the client in WML, the WML style sheet is employed and data is transformed into WML. This modular approach allows spatial operations to focus on fast, reliable spatial data retrieval without worrying about delivering data to the client application. The same is true for the positioning operations team. In this environment, the client development team knows that information will be received according to the agreed XML specification, and can proceed with development in parallel using dummy data.

SOAP

Based on XML, SOAP is a lightweight protocol for exchanging information in a distributed, decentralized environment. SOAP is used for application-to-application communication, and consists of three parts: an envelope that defines a framework for describing what's in a message and how to process it, encoding rules for expressing instances of application-defined datatypes, and a method for representing remote procedure calls and responses.

SOAP enables more flexible mobile location service applications by making servers more modular and not relying on a proprietary interface. SOAP is typically used in conjunction with HTTP. Example 10.1 shows a SOAP request and response loosely based on the Standard Location Immediate Report DTD defined in LIF's Mobile Location Protocol (see Appendix C for details).

Example 10.1 SOAP Message Embedded in HTTP Request.

```
POST /GetVehicleLocation HTTP/1.1
Host: www.mobilelbs.com
Content-Type: text/xml; charset="utf-8"
Content-Length: nnnn
SOAPAction: "Some-URI"

<SOAP-ENV:Envelope
 xmlns:SOAP-ENV="http://schemas.xmlsoap.org/soap/envelope/"
 SOAP-ENV:encodingStyle="http://schemas.xmlsoap.org/soap/encoding/">
 <SOAP-ENV:Body>
   <m:GetLastPosition xmlns:m="Some-URI">
     <VehicleID>1010101010</VehicleID>
   </m:GetLastPosition>
 </SOAP-ENV:Body>
</SOAP-ENV:Envelope>
```

Example 10.2 shows a sample response HTTP message with the SOAP message as the payload.

Example 10.2 SOAP Message Embedded in HTTP Response.

```
HTTP/1.1 200 OK
Content-Type: text/xml; charset="utf-8"
Content-Length: nnnn
<SOAP-ENV:Envelope
 xmlns:SOAP-ENV="http://schemas.xmlsoap.org/soap/envelope/"
 SOAP-
ENV:encodingStyle="http://schemas.xmlsoap.org/soap/encoding/"/>
 <SOAP-ENV:Body>
 <m:GetLastPositionResponse xmlns:m="Some-URI">
 <pos>
  <msid>461018765711</msid>
  <pd>
   <time utc_off="+0300">20000623110205</time>
   <shape>
    <circle>
     <point>
      <ll_point>
       <lat>37.789200</lat>
       <long>-122.415700</long>
      </ll_point>
     </point>
     <rad>15</rad>
    </circle>
   </shape>
  </pd>
 </pos>
 </m:GetLastPositionResponse>
 </SOAP-ENV:Body>
</SOAP-ENV:Envelope>
```

WAP and WML

WAP has been developed by an international group of wireless infrastructure providers that include Nokia, Ericsson, Motorola, Openwave, Microsoft, and Palm. WAP is designed for the markup and presentation of wireless content on small-form factor devices, such as mobile phones and handheld devices. The WAP standard defines a very sophisticated environment for wireless applications (see Figure 10.10). At its base is the underlying bearer network of the mobile operator. Built on top of the bearer network layer is the transport layer, which is based on TCP/IP's UDP protocol and is designed to address some wireless-data-specific issues such as latency. Above the transport layer are the Wireless Transport Layer Security (WTLS) and the Wireless Transaction Protocol (WTP) layers that enable secure transactions between hosts. Above the WTP layer is the Wireless Session Protocol (WSP), which provides both a connection-oriented and connectionless service that uses WDP. Relying on WSP is the WML, which uses WSP to facilitate communications between the server and the mobile device.

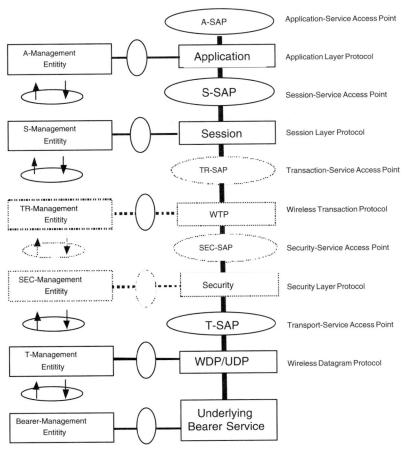

Figure 10.10 Wireless Application Protocol Reference Model. © 2002 Open Mobile Alliance.

WAP is a very efficient standard, and has been designed specifically for wireless use. Accordingly, it is much more efficient than Web-based technologies. WML content is compiled before communications with a client take place, and WML content is generally more compact than HTML content. WTP and WSP are also more efficient than TCP/IP or HTTP.

Development using WAP uses the paradigm of a deck of cards rather than the paradigm of individual pages used in the World Wide Web. A card is a grouping of user interface elements that is presented as a single screen on a mobile device, although scrolling might be required, depending on the contents of the card. A deck is a collection of cards, or a WML file. The benefit of structuring mobile applications this way is that it reduces network traffic and compensates for inherent problems in wireless networks, such as service availability.

WAP was heavily hyped in 1999 and 2000—with a general disappointment when users failed to flock to services. This failure is part marketing and business model failure and part technology failure. Many Web applications were translated directly to WAP applications without considerable thought as to whether they made sense. Mobile applications must allow users to get information and handle problems quickly. Interfaces are fundamentally different. They must be simple enough to be used in one hand while the user is walking or engaged in another activity. Applications designed with mobility in mind, applications that are heavily personalized and enhanced with location-based information, and applications that solve real user problems are the applications that will be highly successful. Of course, a legitimate business model is also needed. Possible models include transaction-oriented or ticket-based applications that pay the provider a commission for facilitating them.

Compounding the problems with WAP was the consensus among mobile operators to control the contents of the WAP start page, often selling placements on this page to the highest bidder. Most mobile operators do not allow you to create your own WAP start page, and even worse, they often block WAP sites that you might actually want to use (e.g., Yahoo!) for fear of competition. This is a problem, particularly in Europe. Few of the principles that made the Internet a wildly successful new paradigm were followed with the launch of WAP, so it is no surprise that WAP has struggled.

Voice and VoiceXML

Driver distraction problems and the challenges of inputting data with a mobile phone interface like WAP have generated a significant amount of interest in voice-based user interfaces. These interfaces are comprised of voice-recognition technology and prerecorded messages. The best and most natural-sounding voice user interfaces use prerecorded speech segments whenever possible to create a natural-sounding dialogue.

It is easy to underestimate the substantial amount of work required to deliver a high-quality voice interface. In the mobile location services environment we have been discussing, a voice user interface would require additional software integration in both the client and the server components of a location services infrastructure. The client is a voice browser that the user connects to via a mobile phone connection. This could be by dialing a specific phone number, or pressing a button on a mobile phone system or in-vehicle navigation system. The voice browser processes VXML, the markup language for structuring voice data. Input is sent to a speech recognition engine, which uses a system such as Nuance's, modified by pronunciation and grammar dictionaries and acoustic models that have been tuned for the target audience. Output is sent to an audio processing engine, which programmatically sequences and streams audio files back to the user.

VXML is designed to allow developers to create audio dialogues that utilize speech recognition, digitized audio, telephony, speech recording, and synthesized speech. The major goal of VXML is to provide the advantages of Web-based development and content delivery to voice applications. Example 10.3 shows part of a VXML document that processes a request for driving directions.

Example 10.3 Sample of a VXML Document.

```
<form id="get_from_and_to_cities">
<grammar src="http://sample.mobilelbs.com/grammars/from-to.grxml"
   type="application/grammar+xml"/>
 <block>
   Welcome to the Driving Directions By Phone.
 </block>
 <initial name="bypass_init">
  <prompt>
   Where do you want to drive from and to?
  </prompt>
  <nomatch count="1">
    Please say something like "from London to Oxford".
  </nomatch>
  <nomatch count="2">
    I'm sorry, I still don't understand.
    I'll ask you for information one piece at a time.
    <assign name="bypass_init" expr="true"/>
    <reprompt/>
  </nomatch>
 </initial>
 <field>

...etcetera

</form>
```

Voice-enabling mobile location service applications requires a number of highly specialized processes, mainly involving prerecording city, street, and POI names in the map data you plan to use. This is a time-consuming task that must be repeated with every map database update to reflect the latest changes. It is also necessary in many cases to tune and trim the route explication (or the text of the driving directions) to eliminate unnecessary detail that confuses users in a voice system. Voice interfaces are expensive to develop and maintain, but when they are designed correctly, they can be highly effective. Detailed technical information on VXML can be found at the VoiceXML Forum (*http://www.voicexml.org*).

Java and J2ME

What Is J2ME?

J2ME is an end-to-end platform developed by Sun Microsystems to allow the development of network-enabled applications and products for the consumer and embedded systems markets. J2ME technology consists of a Java virtual machine and a suite of interfaces that are designed to provide customized run-time environments for embedded and consumer electronics.

J2ME has *configurations* specifically designed for different types of devices. The Connected Limited Device Configuration (CLDC) is designed for small, resource-constrained devices such as mobile phones and mainstream digital organizers. CLDC is built around Sun's K Java virtual machine (KVM), and is suitable for devices that have 16- or 32-bit reduced instruction set computer (RISC)/complex instruction set computer (CISC) microprocessors and as little as 160 KB of available memory. For higher end devices such as smart communicators, PDAs, and set-top boxes, Sun provides the Connected Device Configuration (CDC). CDC utilizes Sun's CVM Java virtual machine and supports devices with at least a 32-bit microprocessor and 2 MB of available memory.

Because device interfaces can be highly proprietary, a J2ME configuration relies on a device-specific *profile* to provide supplementary APIs such as user interface, database, and device-specific networking (see Figure 10.11). The APIs of the profile are layered above the low-level system APIs of CLDC or CDC. CLDC uses a profile specification called Mobile Information Device Profile (MIDP) and CDC uses a profile specification called Foundation. Applications built to run on a device supporting MIDP are called MIDlets, and are similar to applets in some ways. MIDPs are device specific, but MIDlets written to run on one MIDP will run on other MIDPs, providing application portability across a variety of mobile devices. CDC's Foundation profile is actually a subset of Java 2 Standard Edition (J2SE), with GUI-specific dependencies of the Abstract Windowing Toolkit (AWT) removed and various other modifications to tune the APIs for mobile devices.

When to Use J2ME

Developing a mobile location service application using J2ME provides you with tremendous flexibility and ability to deploy on multiple client devices. However, with this flexibility comes substantially more work than that required for an application that leverages an existing environment such as WAP. WAP and J2ME are both valuable and have different uses. WAP allows very thin client applications that are optimized for wireless delivery. J2ME allows thicker client applications with more sophisticated logic and the ability to run in "offline" mode.

It makes sense to use a technology like J2ME when necessary to overcome user interface and technology limitations of existing wireless client technology. Network-centric applications that do not require the advanced capabilities of J2ME, such as its security, are better developed using WAP's WML and Wireless Markup Language Script (WMLS).

Additional technical details and specifications are available from Sun at *http://www.java.sun.com/j2me/*.

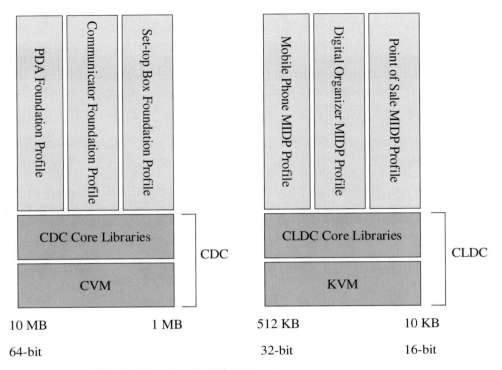

Figure 10.11 J2ME Configurations and Profiles.

BREW

What Is BREW?

BREW is Qualcomm's Binary Runtime Environment for Wireless, an application execution environment for mobile devices. BREW is a thin, fast environment that runs C/C++ applications.

BREW uses an event-driven architecture similar to Microsoft Windows or UNIX X-Windows, and efficiently uses RAM and persistent memory by loading and unloading objects on an as-needed basis. All text is managed as Unicode, and BREW provides a set of core classes that allow applets built in the BREW environment (note that these applets are not the same as Java applets). These core classes include time/timers, notifications, networks/sockets, HTTP, download/billing, display, graphics, and sound.

When to Use BREW

Device-dependent applications can gain an advantage by leveraging the high performance of BREW. It is important to keep in mind, though, that BREW is very Qualcomm-specific and is biased toward network chips. For high-level business applications, J2ME and Java might be a better choice for multidevice support. It is, of course, possible to run a Java virtual machine under BREW and split application development between high-level business logic running in Java and low-level device drivers and system logic running under

BREW. Additional information and technical specifications are available from Qualcomm at *http://www.qualcomm.com/brew.*

INTERNATIONALIZATION AND LOCALIZATION

The increasingly global world we live in means that localization needs to be more than an afterthought, both from a technical standpoint and from a business standpoint. A shift to free-market economies and free-trade zones, privatization of key industries, and the global Internet infrastructure have changed the way business is done forever. The Web has provided a low-cost infrastructure for sales, marketing, distribution, and support of products worldwide, and the increasing penetration of mobile phones will extend these capabilities with the efficiency benefits of mobile e-business: the ability to remain connected and do business at any time, at any place, no matter who you are and what language you speak. To be successful globally, products must be customized for a local market. This customization includes product packaging and presentation (shape and size, colors, and graphics), as well as consideration of many other cultural, language, and technical issues. Local language adaptation is essential, particularly for consumer applications such as mobile location services, given that not more than 25 percent of the world population has basic English skills.

The Mozilla Project (*http://www.mozilla.org*), spawned from Netscape during the AOL acquisition, provides good definitions of internationalization and localization. It defines internationalization as:

> Designing and developing a software product to function in multiple locales. This process involves identifying the locales that must be supported, designing features which support those locales, and writing code that functions equally well in any of the supported locales.

Internationalization is designing software architecture correctly so that it is easy to localize a product. This involves separating all localizable elements from the source code (i.e., in separate text files), creating a user interface that is flexible and provides sufficient space for character sets that require more space, and ensuring that the necessary character sets and regional standards are supported.

The Mozilla Project defines localization as follows:

> Modifying or adapting a software product to fit the requirements of a particular locale. This process includes (but may not be limited to) translating the user interface, documentation and packaging, changing dialog box geometries, customizing features (if necessary), and testing the translated product to ensure that it still works (at least as well as the original).

Elements in the user interface that are localizable include all text, icons, buttons, and alerts. Many programming languages, including C/C++ and Java, allow localization parameters to be configured so that the correct regional standards are utilized. Regional standardization support to consider includes differences in units of measure, using the 24-hour or 12-hour clock, using different thousands separators for number formats, calendar formats, and currency and phone number formats. The two steps that go furthest in enabling a software product to be easily localized are to adopt Unicode as the character set for all operations and to correctly implement the locale functions of the programming language you are using.

A substantial industry has developed that specializes in localizing software products. Additional details on this industry and internationalization and localization in general are available from the Localization Industry Standard Association at *http://www.lisa.org/*.

Localization: The Voice System Challenge in Europe

Developing a voice-based system using one language is challenging. Pronunciation and grammar dictionaries must be developed and constantly refined. For a true pan-European solution to be effective, a mobile location service provider would need to develop a different system in each major market, one for the United Kingdom, one for Germany, one for France, one for Italy, and so on. Even if multiple language systems were not developed, a voice-based system would need to be designed to recognize the accents of nonnative speakers. If the system were designed to handle voice commands in English, would it recognize the distinctly different accents of the English of a native German speaker and the English of a native French speaker? Despite the driver distraction problems that might be introduced by audio and video navigation systems in the front seat, voice-based systems will continue to be expensive and challenging to deploy in Europe.

Localization: The Map Data Challenge

As discussed earlier, European consumers expect a very different look and feel for their maps compared to what North Americans expect. Localization also requires that map databases have street, city, and POI information in the languages of your target customers. This is less complicated in North America than it is in Europe, where maps must be provided in German, English, French, Italian, Spanish, and perhaps other languages. To further compound this challenge is the actual map vendor support for a seamless solution. As of 2001, creating a navigable database of either North America or Europe required stitching together map data from different vendors. In Europe, Finland is a good example of a country with little map data support, which is interesting given the prominence of Nokia and Sonera in the mobile location service space. Even in regions where you don't need to stitch different map vendors' data together, there might be quality differences between regions (e.g., Germany and France). A subscriber using a solution in Germany with certain quality expectations might drive over the border to France and find inferior product performance as a result of poor map data. Details on how to address this problem are explored in Chapter 12.

ENDNOTES

1. © 2002 Bosch Group.

PART 4

Mobile Location Service Technology

11

Mobile Location Service Applications

NAVIGATION AND REAL-TIME TRAFFIC

Navigation is core to mobile location services. Navigation tools allow you to locate yourself and your destination, generate directions, and find POIs along the way or at your final destination. These tools are available today, both on the World Wide Web and in onboard navigation systems, turn-key in-vehicle systems such as Magellan's Neverlost, and others. See Figure 11.1 for a sample onboard navigation system available in Europe from Blaupunkt.

So what is the value of an *offboard* system that has a thin client and does its spatial processing remotely? Because navigation and spatial analysis are core to the many more advanced location services that require remote processing, such as concierge, emergency

Figure 11.1 Blaupunkt Radio Navigation System TravelPilot DX-R70 Titan. © 2002 Bosch Group.

assistance, and travel services, it does not make sense for spatial processing to happen in the vehicle for navigation and remotely for other services.

The results of spatial search must be consistent. Having spatial analysis split might allow the user to generate different driving directions for the same origin and destinations as those generated by a call center operator, an embarrassing situation if the user is calling because he or she is lost and needs help.

Thin clients are inherently less expensive to support than thick clients. Thin client applications can be constantly improved from the server side without upgrading software on the client. Because map data can be as large as 10 GB or more and changes frequently, this is important. Furthermore, mobile devices that are not in-vehicle units do not have the power supply and storage space ideal for visual navigation system.

Most important, offboard systems are able to leverage real-time information about traffic conditions to dynamically route a user around problem areas. In parts of Europe, traffic information is broadcast as data over FM radio. This information is typically from free public sources, and the service is not consistent in availability and quality. A better solution is to leverage information from a traffic information provider that has been normalized and aggregated from many sources.

Traffic

Of the many possible mobile location service applications, traffic is one that has generated a tremendous amount of interest. People are willing to pay for services that will save them time—and traffic information is one of them. To effectively use traffic information requires integration into your navigation system.

Awareness of traffic conditions allows the development of dynamic rather than static navigation applications. Before a route between two points is returned to a user, a check is made to see if there is traffic that will impact their proposed route. If so, a number of calculations need to be performed to decide whether it is best to route the user through the traffic or if an alternate route might be faster. A dynamic navigation system also knows where a user is on the route, so if the system receives notification that an incident has occurred on a route segment that has not yet been traversed, it is able to notify the user and suggest alternates. The quality of the dynamic navigation system is dependent on the data available in the traffic report and the logic it is programmed with. Traffic information at its most basic is a simple incident report. More complex traffic information is also able to incorporate historical and congestion information.

The key to good traffic applications is good traffic data. This means not just where a traffic incident has taken place, but how it has impacted the surrounding road network. Traffic data providers collect information through road sensors, inductive loops, and floating car data (FCD). Road sensors and inductive loops are fixed-speed measurement devices in the road network. FCD, on the other hand, is captured by vehicles equipped with speed sensors driving in traffic. These collection systems report information via mobile networks on traffic incidents and traffic flow (average speed). The best data providers, such as DDG in Germany, aggregate various disparate sources and normalize the data before providing a traffic feed. This process can be seen in Figure 11.2.

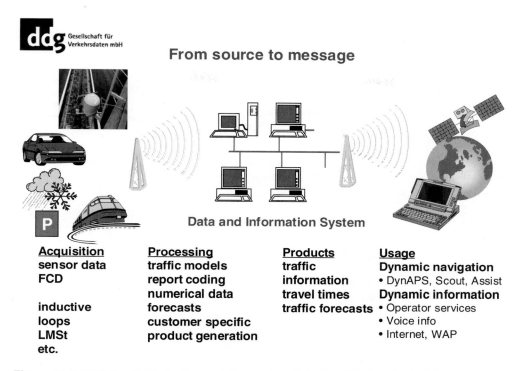

Figure 11.2 DDG Traffic Collection and Processing. © 2001 DDG Gesellschaft für Verkehrsdaten mbH.

Data processing after collection includes building traffic models, coding reports from various sources to one standard format, and producing data forecasts based on current conditions. Traffic reporting is one of the least standardized areas of mobile location services, with many local providers using proprietary formats. As shown in Figure 11.3, DDG processes many different traffic data sources, some proprietary and owned by DDG and other public and broadcast by radio or even sent by fax.

Incident-Based Traffic

Traffic reports that are incident based provide information on where an accident occurred, but give no information on what impact it has had on the surrounding road network, if any. This type of traffic information is typical of U.S.-based incident reports, because the vast road network makes coverage with induction loops and road sensors that capture flow information a very expensive proposition. It is challenging to use incident-based traffic information to build a dynamic navigation system, because an incident could have little impact on driving conditions. Without congestion information such as real-time access to average road speed, deciding whether to route someone around an incident is primarily guesswork.

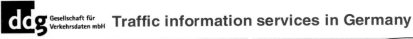

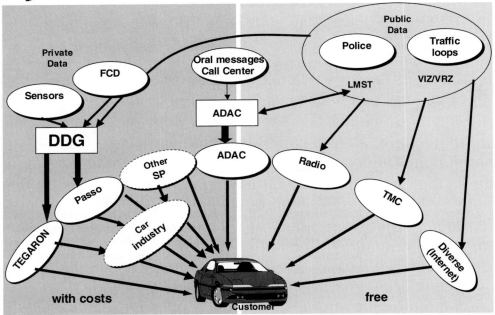

Figure 11.3 Traffic Data Normalization Process. © 2001 DDG Gesellschaft für Verkehrsdaten mbH.

Historical and Congestion-Based Traffic

To make dynamic navigation solutions really work requires the ability to know the real-time impact on the road network of a traffic incident. This is usually measured by average speed and is calculated using induction loops or road sensors. In Europe, traffic information aggregators such as Trafficmaster and DDG have developed large networks with good coverage for the major motorways. It is possible to add FCD to congestion-based traffic information to develop historical traffic models. FCD allows traffic providers to drive areas where they don't have sensors, and using GPS and speed sensors, capture road speed information at specific intervals to build a historical model for predicting future conditions. Unlike real-time traffic information, which is only valuable for current condition queries, historical traffic information can be taken into account for very long trips and when a future trip is planned. A user planning a trip that takes several days would have anticipated traffic conditions considered in his or her route. If planning a future trip, alternate travel times could be suggested to a user if anticipated traffic conditions might make a journey longer at the time requested.

Product Integration and Use in LBS

Linking Traffic to Maps   To be able to use real-time traffic information in mobile location service applications requires embedding traffic location IDs in maps as attributes of road links or providing translation tables to convert traffic locations to a map vendor's internal road link or location IDs. In Europe, traffic incidents are usually keyed to Radio Data Services–Traffic Message Channel (RDS–TMC) codes that have been established by the European Union to allow traffic information to be accurately delivered by FM radio. An exchange protocol named ALERT-C has been recommended as a standard method for exchanging traffic information throughout Europe. The development of comparable traffic standards in North America is in the very preliminary stages, although map data vendors such as Navigation Technologies have included a coding similar to European-style TMC codes in their North American maps.

EMERGENCY ASSISTANCE

In several studies, personal safety has been shown to be the mobile location service application users in North America are most interested in. Emergency services will be coordinated either directly by public safety answering points (PSAP) or indirectly by roadside assistance providers in conjunction with a PSAP if necessary. Mobile operators with operations in the United States are required by the FCC to provide location data with every emergency call made on their network.

Emergency calls made from a mobile phone would typically be life threatening as opposed to calls from an in-vehicle unit, which could range from a less serious incident, such as a flat tire, to a serious accident. Accordingly, emergency calls from a mobile phone go directly to a government-operated PSAP. Most auto manufacturers have a roadside assistance partner, with whom they bundle roadside assistance service with new vehicles. There is also after-market roadside assistance available directly from companies such as American Automobile Association and Europe Assist. The roadside assistance companies operate call centers that specialize in handling various emergency calls and dispatching the necessary services, whether it is a towing service or an ambulance.

If you've been in an accident, had a flat tire, or run out of gas in a remote area without a cell phone, you know that it can be difficult to call the roadside assistance company's toll-free number to get help. If the accident is serious enough, you might not be conscious and are dependent on another motorist seeing the accident and calling it in. Seconds can be crucial in a life-or-death situation. Wouldn't a better solution be to have the roadside assistance provider automatically notified in an accident and provided with necessary details, such as vehicle position and system status? Or at least provide the driver with a built-in interface to get help without having to find a pay phone or their mobile phone? This is the concept behind systems like OnStar's Onboard.

The OnStar Onboard system includes a three-button interface, a GPS receiver, a cellular chipset, and audio input/output capabilities. The system has a status light that is green when the system is working and red when there is a problem. While a call is in progress the status light flashes green. The button with the white dot either answers or ends calls. The blue button labeled "On" connects the user to an OnStar advisor for concierge services in the OnStar call center. The button with the red cross on it sends a priority call to an OnStar advisor, who is able to connect the user to the appropriate emergency service center. Since the infrastructure required to service emergency calls is in place, it makes sense for OnStar to offer a variety of additional services such as navigation and concierge services with the same interface.

To deliver services like these requires an expensive infrastructure. Emergency services calls will always require an operator in a call center, but concierge and navigation services can be provided with automated systems that provide a cost savings. These automated systems might be voice based or have graphical user interfaces designed for small mobile devices. In GPRS and 3G mobile networks, the emergency service provider has the advantage of moving information on the data channel rather than the voice channel. Because these networks are packet switched rather than circuit switched, they are always on. This saves valuable time in call set-up and also allows more efficient "pay-per-use" billing methods rather than paying while online, whether you are using the system or not.

Many industry players have an interest in improving the way emergency services are provided, from government agencies to managed health-care providers. All are interested in providing better service and reducing their costs. One such provider working to improve roadside medical assistance with location-based services is Roadside Telematics Corporation. Roadside Telematics is working with the Communications for Coordinated Assistance and Response to Emergencies (ComCARE) alliance to improve the way emergency services are handled (see Figure 11.4). Larry Williams, their CEO, details in the "RoadMedic Emergency Services Application" sidebar how simple mobile location services might be developed today using FM subcarrier networks instead of the cellular networks.

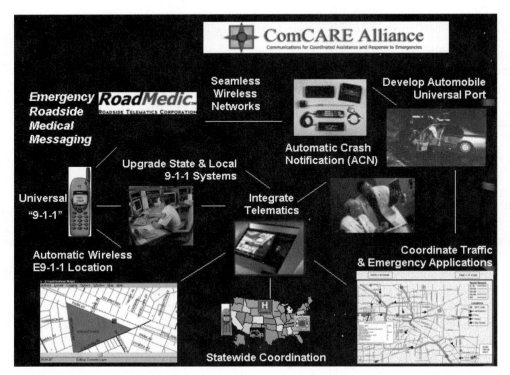

Figure 11.4 Emergency Services ComCARE Alliance. © 2002 Roadside Telematics Corporation.

RoadMedic Emergency Services Application

Distributing Safety & Security Data by FM Sub-Carrier[1]

By Lawrence E. Williams, President & CEO, Roadside Telematics Corporation

Existing Automatic Crash Notification (ACN) technologies are primarily voice-centric notification services based on a cellular network. Emerging second-generation ACN technologies are quickly expanding safety and security capabilities to include sophisticated crash data messaging.

These second-generation ACN systems will provide highly sophisticated real-time crash data such as, (1) the speed of vehicle travel, (2) the direction and point of impact, and (3) vehicle rollover information. Crash victim personal medical information such as blood type, drug allergies, and current medications will be integrated with the crash data to form the automotive equivalent of the airline industry's "black box." Responding emergency personnel will be able to instantly access critical lifesaving information regarding crash occurrence and personal emergency medical information.

As second-generation ACN systems evolve to include automated real-time wireless transmission, questions arise regarding the reliability of data transmission and delivery via a cellular network.

For the most part, issues focus on geographic coverage and delivery costs. In general, cellular network coverage is currently driven by demand flowing from population centers resulting in geographic coverage gaps in rural areas of sparse population. While only 24 percent of crashes occur on rural roads, nearly 59 percent of crash deaths occur on rural roads. Lack of transmission and delivery reliability causes a delay in delivering emergency medical services, which is considered to be one of the factors contributing to the disproportionately high fatality rate for rural crash victims, according to the National Highway Traffic Safety Administration.

Furthermore, when a cellular network achieves broader and more complete coverage, it will be at a very high cost that may prohibit broadcasting and transmitting small amounts of wireless data, such as ACN applications.

Driven by these issues, telematics service providers are now exploring alternatives for the efficient and effective deployment of emerging safety and security data applications. A highly promising broadcast (downlink) alternative, with potential for widespread global acceptance, is the FM subcarrier technology developed by CUE Corporation.

CUE currently utilizes a very extensive FM subcarrier network for the broadcast of real-time traffic information, weather forecasts, emergency weather warnings, and CRM messaging. Utilizing the FM subcarrier frequencies of approximately 600 radio stations in North America, CUE provides substantial coverage of the United States and Canada—up to six times the geographic coverage of any other wireless network in America.

Later this year, CUE will add to its telematics portfolio a unique medical information messaging service called RoadMedic™, providing automated access to vital emergency roadside medical information at the push of a button. This service can be best described as an automated version of medical alert jewelry or emergency medical information cards carried in wallets and purses. RoadMedic will be a logical evolutionary step for CUE to supplement its current real-time traffic distribution service.

This action will mainstream CUE into the safety and security market and will provide a fully thin-client approach, with all telematics subscriber personal medical information entered into the RoadMedic database that is designed to protect subscriber privacy and confidentiality through compliance with the Health Information Portability and Accountability Act (HIPAA), and French and Swiss medical information privacy standards. Given 1 out of 3 victims of serious car crashes are either unconscious or otherwise unable to give fast, accurate personal medical information, RoadMedic will enable emergency personnel to instantly access vital information, thereby reducing the delivery time for crash victim care while improving the quality of care provided.

In the future, RoadMedic will be able to automatically integrate the crash victim personal medical profile with the sophisticated ACN crash data, creating the black box for wireless uplink transmission to E911. This process will offer three key pieces of life-saving information: a crash alert, an exact crash location, and emergency medical information. Emergency personnel will know exactly where to go and what to expect on arrival at the crash scene. Future uplink alternatives to cellular include evolving FM subcarrier two-way technology and emerging satellite communications supporting 9.6 Kbps on the uplink, which is more than adequate for emergency medical information and ACN data.

The historical problem with FM subcarrier data broadcast has been a relatively low data broadcast rate of approximately 8 Kbps. CUE has vigorously addressed this weakness over the past five years through the development of a proprietary technology known as SuperDARC. SuperDARC facilitates data transmission rates of 64 Kbps, more than sufficient for real-time traffic information and personal emergency medical information. SuperDARC makes CUE technology an attractive option for broadcasting data to telematics-equipped emergency response vehicles for several reasons:

1. Ease of inclusion: Most standard radios today are already capable of handling CUE's wholly voice-based service, which is particularly relevant given the recent "driver distraction" legislative activity aimed at in-vehicle telematics systems. Although a second receiver would need to be built into most existing radios in order to make them CUE-compatible for downloading emergency medical information data to emergency response vehicle navigation units and other telematics devices, there are minimal costs associated with this hardware upgrade, and the radio itself is ubiquitous as far as inclusion in the vehicle is concerned.

2. Low data transmission costs: Broadcasting data over FM subcarrier frequencies is inexpensive, particularly when compared with the cost of transmitting data over the cellular network.

3. Global coverage: It is significant that the SuperDARC data protocol is becoming a global standard, particularly when it comes to distributing the service in other countries.

CUE is familiar and comfortable with automotive industry applications. CUE is perhaps the leader in broadcasting real-time traffic information to in-vehicle navigation systems. The company is currently working with BMW, Volvo, Honda, Pioneer, Alpine, Kenwood, Visteon, and Audiovox.

The integration of safety information with telematics will become an important issue. As in-vehicle navigation systems and other telematics devices become widespread, they create the opportunity for revolutionizing emergency medical response and the delivery process. The FM subcarrier will be a logical choice for data distribution of emergency medical information. In fact, it is very possible the FM subcarrier will emerge as a dominant player for the emerging ITS Public Safety System by providing real-time traffic and medical information broadcast services for the ComCARE Alliance, a national coalition whose goal is to promote a comprehensive "end-to-end telematics communications system" to enhance public safety and improve emergency response.

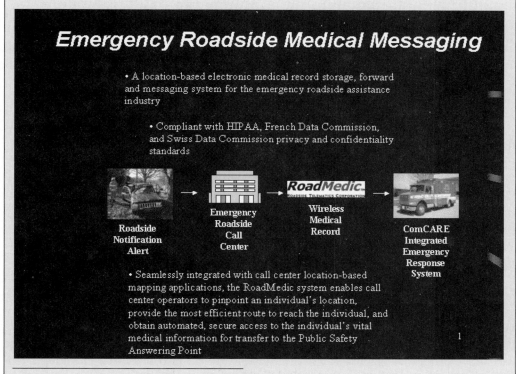

Lawrence E. Williams is the President and CEO of Roadside Telematics Corporation, a leading telematics integration company providing technology for aggregating and integrating telematics subscriber medical information with the emerging E911 (U.S.), E112 (Europe), and E119 (Japan) telematics uniform emergency response systems. The company supplies emergency medical record infrastructure and a medical messaging application, the RoadMedic™ system, to strategic channel partners for delivery to end users through in-vehicle telematics devices, in-dash navigation systems, car radios, handheld receivers, and smart card platforms. For more information, visit www.roadsidetelematics.com.

CONCIERGE AND TRAVEL SERVICES

The travel industry is likely to be a heavy user of mobile location-based technology to improve the services they provide. Basic services include enhanced 411 concierge service (i.e., "Where's the nearest restaurant?") and local travel advisory 511 service. More advanced services range from specialized travel service providers (Travelocity, last-minute.com, and Priceline) to complete *mobility management* solutions.

MetroOne (*http://www.metroone.com*) is one of the leading providers of enhanced 411 services to telecommunications carriers. In addition to the basic phone number lookup service, MetroOne provides a number of other valuable location-oriented services in the United States:

- Movie information: Theater locations, movie listings, show times, reviews, ratings, and phone numbers
- Radio station information: Allows you to find a nationally syndicated show or look up the station format, call letters, frequency, program times, and coverage area for local stations
- TV station information: Allows you to look up program listings and times for local stations
- Concierge: Full-service live operator concierge service similar to what you'd expect to receive in a first-class hotel

In July of 2000, the U.S. FCC designated 511 as the single traffic information phone number to be made available to local and state jurisdictions. There are no federal requirements or mandates requiring service similar to the positioning requirements of 911 emergency calls, but many regions already have implemented solutions, including Utah, which developed service for the 2002 Winter Olympics. Based on TellMe Network Inc.'s (*http://www.tellme.com*) voice application platform, the system integrates with the state's Web infrastructure and existing CommuterLink system (*http://www.utahcommuterlink.com*), which provides traffic, weather, and accident information via radio, television, the news media, and the Internet. The system is very easy to use and highly effective.

In many ways, the regional 511 service in the United States is similar to the idea behind the RDS–TMC service that is more common in Europe. RDS–TMC is a data broadcast via FM radio of traffic and travel advisories. The broadcasts are associated with specific TMC map locations that navigation systems can use to associate with streets on a map, which, unlike the 511 voice service, has the added advantage of allowing RDS–TMC to be used for dynamic navigation. RDS–TMC broadcasts are also broadcast in multiple languages so the user can preprogram the language he or she would like to receive.

Advanced use of location-based information is used in the service offerings of companies like Travelocity, lastminute.com, and Priceline. Travelocity offers concierge service relevant to a traveler's itinerary via its Web site and on a variety of mobile devices. Last-minute.com, a provider of last-minute "deals" on travel, geocodes POIs like hotels to allow users to do a map-based search for hotels near their destination. Priceline, which provides a "name your own price" service, uses GIS principles to create the various regions of a metropolitan area that users of its hotel application must choose between to do a hotel price match search in.

Airlines are already using SMS to notify travelers of delays and gate changes, and once location-based information is incorporated into these basic mobile applications, an entirely new class of service is possible. An interesting example is the provision of an end-to-end mobility management solution. This solution would combine many of the products already discussed and enhance them by personalizing them based on both who the user is and where the user is. A mobility management system would allow turn-key travel planning that includes all modes of transportation, including air, train, subway, ferry, and private car or taxi. Because the system would know where the user is on his or her journey, it would be able to make intelligent concierge-style recommendations either automatically or on demand. The system could be aware of traffic and travel advisories and last-minute discount opportunities, making itinerary suggestions accordingly. Using proximity-based services could allow a user with hotel instant checkin and checkout privileges to take care of this ahead of time, eliminating the need to queue in the hotel lobby. Authentication could be handled with a PIN number. During any part of a user's journey, he or she could access information on events, POIs, or even services relevant to their location. A mobility management portal might customize the interface based on where a user is and what it believes the user is doing. There is no reason why it could not provide hotel amenity information when the user is in a hotel, or train schedules when a user is on a train. The key to making the service successful is developing trust with the user and having the right balance of information delivered by alerts (which can be intrusive) and by a trusted and custom on-demand resource.

LOCATION-BASED ADVERTISING AND MARKETING

Using location-based information in advertising and marketing has been a big business for some time. Direct marketers routinely use geography-based targeting techniques in direct marketing, and marketing database companies and mailing houses use geocoding products extensively to check and improve addresses. Using mobile location services in marketing ranges from very subtle to highly overt.

An event-based service, such as a turn-key Mother's Day solution toolkit, could use location technology very subtly to improve the overall service delivered. For example, if you're using your WAP browser and notice that the Mother's Day toolkit has appeared as the second item in your mobile portal, you might remember that you haven't made any plans or bought any gifts. Clicking the link might present you with local flower shops to order flowers from, nearby restaurants you can make reservations at, or a local event guide. Being able to customize whatever application or service subscribers are using to include logical offers based on their location is location-based marketing.

The other extreme is the highly overt. The idea that when you walk by a store and a special offer advertisement might ring you on a mobile device might seem intrusive to many people. ZagMe launched an opt-in shopper alert service in the United Kingdom using a location-based system with alerting technology from Adeptra (*http://www.adeptra.com*) in late 2000. After receiving a message, users can either purchase an item immediately or have it held. This service might have limited appeal apart from teenagers, although Reebok used the technology to run a particularly interesting promotion. Reebok sent an alert offering a free pair of shoes to the first person who could get to the store and show a ZagMe message. More than 50 people had sprinted to the store within four minutes.

LOCATION-BASED BILLING

The primary target customer of location-based billing is the mobile operator, although the concept could be applied to many businesses. Location-based billing allows a mobile operator to charge different rates to mobile subscribers based on where they are. The advantage to a mobile operator is this allows them to target wireline subscribers by offering plans that are competitive with a wireline offering at the subscriber's home or office. A mobile operator might carry local calls at a flat rate when a user is at home, but at the standard rate when they leave.

Signalsoft (*http://www.signalsoftcorp.com*) has partnered with Portal Software (*http://www.portal.com*) to provide a location-sensitive billing solution. The solution integrates with multiple vendor networks and allows subscribers to self-provision their rate plans with up to 10 different rate zones.

ENDNOTES

1. © 2002 Roadside Telematics Corporation.

PART 5

Advanced Topics

12

Digital Map
Databases

"BEST OF BREED" MAPS

Map databases that contain sufficient detail to support mobile location service applications, such as routing, are extremely complex and time consuming to build and maintain. As of 2001, no single map data vendor provided a suitable worldwide map database for location-based services. In fact, even within continents such as North America and Europe, quality of data varies considerably from one map vendor to another. In Europe, as an example, one map data vendor might have excellent data for Germany, but very poor data for the United Kingdom. In North America, one map data vendor might have excellent data in urban areas, but very poor data in rural areas. Thus, it is often necessary to combine various map data sources to achieve complete coverage and adequate quality.

When Is a Best of Breed Map Required?

Best of breed maps can be necessary to achieve high-quality coverage in a market. In Europe, it is important that mobile operator or automotive solutions be pan-European. If a German customer drives to Spain or Italy, he or she expects to have the same quality of service he or she receives at home, and in his or her native language. Thus it is important that the location service is able to route seamlessly across borders and that the quality of the solution in Spain is the same as in Germany.

It might also be necessary to consider a best of breed map database to address specific business model concerns. Map data vendor pricing varies widely, and is often dependent on how the map data is used. It is common to charge based on number of transactions. It might be beneficial in very large location service systems to use less expensive data for transactions that do not require more expensive data sets. As an example, some map databases might have enormous detail that is very valuable for routing, but not necessary for other spatial operations.

Other scenarios to consider could include having primary and backup map databases to provide failover capabilities if a spatial operation (e.g., geocoding or routing) fails, or being able to route on two databases.

Map Database Conflation[1]

Introduction

One of the greater underlying challenges in delivering mobile location-based service (LBS) applications is the integration and management of high-quality data sets. From a routing and navigation perspective, where accuracy of the database and information is paramount, this is particularly true. The application solution requires high-quality information to identify the user's position, determine the appropriate path of travel, and accurately display the moving position along a map. Because pinpoint positional accuracy is essential for producing quality routes, some map vendors drive tens of thousands of road segments to verify the accuracy of links and attribute data. Any anomalies discovered during testing are corrected in the digital database to reposition road segments to better match their real-world geographic locations. The collection and verification process of these vendors ensures that the high level of data accuracy necessary for deploying sound LBS applications is achieved.

Traditionally these capabilities have been achieved using a single supplier of data. However as the LBS application space grows and increasingly demands greater geographical coverage, developers are finding that a single data source is no longer adequate for meeting these stringent requirements. The varying strengths in each source database drive the need to integrate multiple sources. Once the domain of technical experimentation, map database conflation has found its way into a variety of commercial and consumer applications. Online map providers MapQuest and Vicinity have been integrating high-quality NavTech coverage and broad GDT coverage to deliver door-to-door routing solutions over the Internet. Both Microsoft and Kivera, using radically different techniques, have extended this capability to deliver high-quality display solutions for consumer products.

Data Conflation and Compilation

The conflation process integrates multisource, disparate geographic data sets and their corresponding attribute data. Because data sets frequently vary in their formats, naming conventions, and positional accuracy, conflation is a complicated procedure. Data integrators, to achieve a high-quality data set, must be concerned with the relative accuracy and currency of the data sets to be used.

The specific requirements for base map content are often tied to the application. Each application has its own threshold for quality and accuracy. In-vehicle navigation with map matching capabilities requires a high degree of quality for map rendering, whereas an Internet product that delivers raster map images at high zoom levels will be less concerned with the database's final geometric representation.

Developing the necessary tools, processes, and algorithms to conflate map databases is a long and involved process. There are two general strategies for conflation, each with its own methods for production, qualities, and limitations. The first, database overlays, is the simpler of the two approaches. When data sets are overlaid, the original data set is essentially left intact. The logic for integration is found in the application layer. On the other hand, database merging calls for the union of two data sets to create a new single map database. In this scenario, the real effort is done during the map compilation phase. This sidebar reviews the different approaches, evaluating their strengths and weaknesses.

Overlay Strategy

The general principle of overlaying databases is to utilize two or more completely self-contained data sets. The application is responsible for knowing which database contains which information. This is relatively straightforward for geocoding and display applications. Based on some qualitative criteria such as accuracy of positional and attribute data, completeness of the data set and cost, the application sets a preference order for data access.

For instance, GDT provides a North America database that is known for its breadth of coverage and accurate address attribution. For this reason, it is often the first database searched when geocoding a user-entered address. The very same application, when displaying a map, might opt to use NavTech for display, due to its high degree of positional accuracy in detailed coverage areas.

Obviously there are significant benefits to this approach. It is relatively easy to maintain; changes in one data set have little bearing on the other data set in use. An application developer can independently license and track database usage at the application level. However, routing across disparate data sets presents a great challenge. A routing algorithm must explore across the data set(s) for the appropriate path. If the origin and destination are on different data sets, or if the data set used is disjointed, the application requires some mechanism to "connect" them together.

This mechanism is called a *standard location reference*. The standard reference identifies positions in the first data set that are the same in the second data set. The measure of likeness is based an evaluation of the geometry of the feature and its attribution. There are two different general approaches for defining a standard location reference. With an increasing cost and complexity they are capable of delivering successively higher quality routes.

Run-Time Determination of Cross-Over Points

The first approach, run-time determination, is an appropriate approach when the data coverage in one database is a subset of the coverage in the other. This was the case in North America, before NavTech began distributing its Full Coverage product. This approach calls for run-time determination of a cross-over point between the two data sets during route calculation. For it to work, the application must identify a primary database in which all route calculations are attempted first. Only in the event that a destination does not fall on the primary database will the secondary database be used. In this case, the application will determine the closest "like" point in the primary database to the destination in the secondary database. This point is then matched against a road segment in the secondary database with similar attribution. Finally, the application calculates the two subroutes joining them at the reference point.

Because this approach can be used regardless of the underlying data set, it is cost effective and easy to maintain. However, the quality of the match and route can be rather poor. This is particularly true in situations where there is no primary database coverage in proximity to the secondary database point. This scenario might force the calculation of a roundabout path in the primary database, even though a more direct path exists in the secondary database. Inconsistent naming conventions for street names across the databases can also confuse the algorithms, resulting in very poor matches. For instance, one database might refer to a street as "Mason Drive" and the other might identify it as "Mason Street." Without manual verification, it is nearly impossible for an algorithm to verify the likeness of these two street names. Finally, run-time determination presents significant problems at display time. Because the final matched points are represented at different geographical positions, the connection between them is often off the road network entirely. When displaying this portion of the route, it is common to see a zig-zag image representing the connection. Generally this is not a problem at higher zoom levels; however, if the solution calls for the display of maneuver maps for each route segment, the results are inadequate for high-end services.

Precompiled Cross-Over Reference Points

Database integrators can resolve many of the problems associated with run-time determination by precompiling a list of "qualified" cross-over points. These lists identify specific places shared between the two data sets that produce high-quality routes when joined together.

For data sets with overlapping coverage territory, this method is an improvement over the run-time calculation of cross-over points. It does so by identifying specific freeway interchanges that can serve as database jump points. This solution is effective because it relies on the primary database for high-level arterial connectivity, but makes greater use of the secondary database's network, especially in areas off the highway. Furthermore, naming conventions for freeways are more likely to be routinely verified by data suppliers and in sync with each other.

Precompiled cross-over points are also effective in routing across two data sets that do not overlap, but are contiguous. This is quite common, especially along national borders where a best of breed data integrator requires data from two different suppliers. In this scenario, the cross-over points are matched at known border crossings. Border crossings, even more so than freeway interchanges, are fixed network features that will have a high degree of accuracy.

These higher quality routes come with a cost. The solutions are specific to a data set and the list of possible cross-over points must be regenerated with each new data release. Furthermore, the introduction of fixed cross-over points creates new challenges for route optimization, especially for routes with multiple stopovers. Regardless, this approach is still a significant improvement over run-time determination. It allows for exhaustive testing and verification before releasing the data to ensure route quality and display standards are met.

Database Merging

Database merging is a significant strategy shift. Unlike overlaid databases, where the data sets are kept distinct, the goal of merging is to create a new data set that consists of the best available coverage from each data supplier. Often, merging requires a shift in the geometry of one data source such that it lines up with the other. Depending on the desired quality of the conflated database, the process of merging can be extremely complex and time consuming. The requirement to remove duplicate data and resolve conflicting information from the competing suppliers further complicates matters.

Despite the high production costs and technical hurdles associated with conflation, data integrators find that merging provides significant benefits above and beyond overlaid data. First of all, the application challenges inherent in overlaid database solutions disappear. The final data set is one single graph, delivered in a single access format. Display, geocoding, and routing applications can be optimized for this data set alone, requiring no intelligence about the quality of one data set in relation to another. Assuming a quality merge process was utilized, map rendering is seamless across the edges. Routes can be calculated across the entire network, and the shape of these route paths will line up exactly with the road network's geometry.

Of course, although all this blurring of source data is advantageous to the application developer, it creates a number of complications for data suppliers. Data suppliers, looking to enforce license agreements, have a difficult time tracking real usage of a merged data set. For delivered fixed media products, like desktop software and in-vehicle navigation, this is not an insurmountable problem. Each data supplier negotiates royalties commensurate with the value their data brings to the final solution. Obtaining the volume of units on any shipped product is no great challenge; therefore, calculating the total license cost is relatively straightforward. However, in server-based applications, where per-transaction pricing applies, license management becomes a significant problem. If the underlying source information about the merged database is sufficiently blurred, it is nearly impossible to know which data supplier was really responsible for the transaction and to whom the royalty should be sent.

Varying quality requirements have brought about two different conflation strategies for database merging. The first is grid-based merging, the process of edge-matching contiguous tiles of coverage and sewing them together. The other approach, selective area merging, uses one database as a reference grid and augments it with coverage from a secondary supplier. The former approach is far easier to achieve and maintain but produces a database of inferior quality.

Grid-Based Merging

The approach behind grid-based merging is similar to assembling a jigsaw puzzle. Each sub-section is a complete unmodified representation of the map from a single supplier. An adjacent tile from a secondary source is edge-matched to the first, inducing connectivity between contiguous features where necessary. The process typically forces a shift in geometry to ensure connectivity. This can cause distortion in the map's appearance that might be visible at detailed zoom levels. However, the distortion will be confined to the boundaries along the seam. Moreover, because the stitch boundaries are fairly well defined, this approach has a slightly reduced overhead for ongoing maintenance.

There are significant problems with this approach. Because the process calls for edge-matching adjacent tiles, it is difficult to remove duplicate data, especially along stitch borders that vary significantly in their representation. This is especially problematic for geocoding applications that might accurately resolve "100 Main Street" twice, once in each underlying data source. Routing algorithms can behave quite erratically on a data set merged this way. A single highway's attribution might change periodically as it moves in and out of a particular source supplier's coverage area. As a result, the conflated database is a blend of attributes, with variable improvements based on location.

Selective Area Merging

Selective area merging provides significant improvements over a grid-based merge. Conflation software automatically recognizes important features in the primary data set and correlates them with features in the secondary data set. Links are precisely merged, repositioned, and associated with additional attribute data. Each link in a digital map database can have literally hundreds of attributes associated with it, and hence, requires highly intelligent algorithms to test the validity of the road network's connectivity.

Start and end points connect links, and when two data sets are merged, the number of links that meet at any given node is an important measure of similarity. The program intuitively performs quality checks to determine whether all links agree with each other. Links containing the better data, from either a positional or attribute perspective, are preserved and redundant links are deleted. These deletions might cause a disconnection of links, but a set of aggressive algorithms that intuitively shifts the links back into place preserves connectivity. The end result is a single database without duplicated data, preserving the best geometry and attribution from each source supplier.

Brian Shenson
Kivera, Inc.

Example Merge: Kivera Combined NavTech/GDT Solution[2]

The goal of this project is to provide the most complete road data available for use with in-vehicle navigation systems. To accomplish this, Kivera has applied a novel technology to merge the data sets available from two different vendors: Navigation Technologies (NavTech) and Geographic Data Technologies (GDT).

The purpose of the Kivera merge process is to take information from the GDT data set and add it to the NavTech data set to fill empty spaces. This involves reading map data, clipping it into smaller, more workable files, and then establishing control nodes to align the two maps together.

The Problem

The data set from NavTech is notable for its extremely accurate, reliable information. The coordinates provided by NavTech are highly consistent with those provided by the GPS. The representations of roads in this data set are extremely accurate and very closely reflect the actual roads. Unfortunately, this data set is not complete. Most of the United States has only partial coverage, with data covering only the main highways and extremely sparse coverage of local roads.

Currently, various vehicle navigation systems rely on the NavTech data set, but the incompleteness of the data set creates a lot of customer complaints. That leads to a significant loss of sales of in-vehicle navigation systems because the data set does not cover areas where customers live and drive.

The map data set from GDT is much more complete. There is effectively 100 percent coverage of areas in the United States. However, the data has significant geographical position errors, as well as geometry problems. Using the GDT data alone in an in-vehicle navigation system would be a dubious choice because the GPS system would report very different coordinates for ground positions than the GDT map indicates. GDT data often shows extremely inaccurate representations of ramps and complicated road configurations. Because of the discrepancies between the GDT data set and reality, users of in-vehicle navigation systems using this data would get hopelessly lost quite regularly. GDT data therefore has not been used for highway navigation.

Kivera's Solution

Kivera has come up with a way to merge the data sets of the two vendors to produce a data set that takes advantage of GDT's more complete coverage and NavTech's higher accuracy. In effect, Kivera "sews" the data sets together like a patchwork quilt, as shown in Figures 12.1 through 12.3.

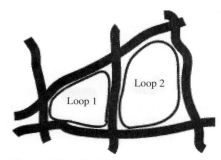

Figure 12.1 NavTech Area Showing Major Roads and Two Loops.

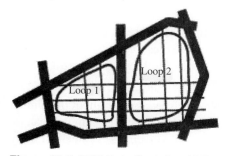

Figure 12.2 GDT Data Covering All Roads (Distorted with Corresponding Loops Indicated).

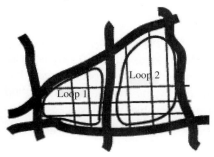

Figure 12.3 Merged Data.

Loops and Control Nodes

The sewing is accomplished by finding corresponding, closed, circular paths of travel along roads contained in both the NavTech and GDT data sets. A *NavTech loop* is defined as a path of segments in the NavTech data that makes the sharpest possible right-hand turn at every point, until it comes back to the point of origin. A *GDT loop* is defined as the closest match to the NavTech loop; it is comprised of roads in GDT data that are duplicated in both data sets. Kivera's software identifies the common road intersections occuring in both the NavTech loop and the GDT data sets along these loops. These are called *control nodes*. The links between the control nodes form the *loop path*.

Once a complete, closed loop path has been generated and matched, Kivera software then copies all the GDT links inside each loop and "rubber bands" their geographical position to compensate for the differences in geographcal positions of corresponding intersections. It then "glues" this shifted data on top of the NavTech data set, adding to, but not changing any elements from the original NavTech data set.

Some Specific Problems

Inconsistent Source Data   Of course, the process is not quite that simple. Because the original data was entered by humans, it is full of inconsistencies and other problems that make impossible a totally automated merge of the two data sets. Thus, at one critical stage, every single loop identified by the Kivera software requires inspection and possible alteration via a Kivera-proprietary user interface tool. This is necessary because inconsistencies between NavTech and GDT data are often not resolvable by algorithmic approaches. Approximately 20 percent of the loops generated and matched by the Kivera software require some human alteration to resolve discrepancies.

There are many cases in which the map representations of an area are totally different in the two data sets. Kivera has developed a manual process to resolve most of these cases and to allow all or most of the GDT data in these very confusing loops to be extracted. In some very rare cases, however, impossible areas are simply left out.

"Empties"   There are approximately 240,000 NavTech loops throughout the United States. Many of them are very small and simple and have no GDT data in them that can be clipped out and inserted into the NavTech loop. These are called *empties.* Typically, empty loops are generated inside of double-digitized highways in the NavTech data, between pairs of overpasses. Other varieties of empties include the inside circles and triangles created by the ramps surrounding an overpass. All 240,000, including approximately 140,000 empties, get passed to human inspector-adjusters.

Inspection Problems   Unfortunately, people are not perfect. Inspector-adjusters get tired and some understand the problems more completely than others. A second pass through the data by the more experienced and accurate members of the inspection team has proven to be necessary to catch errors that were undetected by the first pass. This second pass is assisted by a software program that locates loops with problems introduced by people and submits them along with English-language descriptions of the suspected problems to assist the person reviewing each loop. This tool reports both original data errors and human errors such as the following:

1. Control nodes that don't perfectly match
2. Inconsistencies like nonmatching road names and erroneous one-way streets
3. Severely divergent paths in the two data sets

ENDNOTES

1. © 2002 Brian Shenson and Kivera, Inc.

2. This section © 2002 Kivera, Inc.

Abbreviations

2G	Second-generation mobile network
3G	Third-generation mobile network
ACN	Automatic Crash Notification
A-GPS	Assisted GPS
AI	Artificial intelligence
AM	Amplitude modulation
ANSI	American National Standards Institute
AOA	Angle of arrival
API	Application programming interface
AuC	Authentication center
AWT	Abstract windowing toolkit
BREW	Binary Runtime Environment for Wireless
BSC	Base station controller
BSS	Base station subsystem
BTS	Base transceiver station
CCI	Common client interface
CDC	Connected device configuration
CDMA	Code Division Multiple Access
CDPD	Cellular digital packet data
CDR	Call detail record
CGI	Cell global identity
CLDC	Connected limited device configuration

COM	Component Object Model
COO	Cell of origin
CORBA	Common Object Request Broker Architecture
CRM	Customer relationship management
CSS	Cascading style sheets
CTIA	Cellular Telecommunications & Internet Association
DSSS	Direct Sequence Spread Spectrum
DTD	Document type definition
EIS	Enterprise information system
EJB	Enterprise JavaBean
E-OTD	Enhanced-observed time difference
FCC	Federal Communications Commission
FCD	Floating car data
FM	Frequency modulation
GGSN	Gateway GPRS support node
GI	Geospatial information
GIF	Graphics Interchange Format
GIS	Geographic information systems
GML	Geographic Markup Language
GMLC	Gateway mobile location center
GMT	Greenwich Mean Time
GPRS	General Packet Radio Service
GPS	Global positioning system
GSM	Global System for Mobile Communications
HIPAA	Health Information Portability and Accountability Act
HLR	Home location register
HTTP	Hypertext Transfer Protocol
HTTPS	Secure HTTP
IANA	Internet Assigned Numbers Authority
IDE	Interface Development Environment
IMEI	International Mobile Equipment Identifier
IN	Intelligent network
J2EE	Java 2 Enterprise Edition
J2ME	Java 2 MicroEdition
J2SE	Java 2 Standard Edition
JCA	Java Connector Architecture
JSP	Java Server Pages

JVM	Java virtual machine
KVM	Sun's K Java virtual machine
LCS	Location services
LIF	Location Interoperability Forum
LMU	Location measurement unit
ME	Mobile equipment
MeT	Mobile Electronic Transactions Standard
MFC	Microsoft Foundation Classes
MIDP	Mobile information device profile
MLC	Mobile location center
MLP	Mobile Location Protocol
MPC	Mobile positioning center
MS	Mobile station
MSC	Mobile switching center
MSID	Mobile station identifier
NSS	Network switching subsystem
OGC	Open GIS Consortium
OS	Operating system
P3P	Platform for Privacy Preferences
PDA	Personal digital assistant
POI	Point of interest
PRC	Palm OS Resource File Format
PSAP	Public safety answering points
PSDN	Public switched data network
PTD	Personal trusted device
RDS	Radio data services
RDS–TMC	Radio Data Services–Traffic Message Channel
RTD	Relative time difference
SA	GPS selective availability
SDE	Spatial data engine
SGSN	Serving GPRS support node
SIM	Subscriber identity module
SMLC	Serving mobile location center
SMS	Short Message Service
SNMP	Simple Network Management Protocol
SOAP	Simple Object Access Protocol
SRS	Spatial reference system

SSL	Secure Sockets Layer
SVG	Scalable vector graphics
TA	Timing advance
TDMA	Time Division Multiple Access
TDOA	Time difference of arrival
TMC	Traffic Message Channel
TOA	Time of arrival
UML	Unified Modeling Language
UMTS	Universal Mobile Telecommunications System
URI	Uniform resource identifier
URL	Uniform resource locator
UTM	Universal Transverse Mercator
VLR	Visitor location register
VML	Vector Markup Language
VXML	VoiceXML
W3C	World Wide Web Consortium
WAP	Wireless Application Protocol
WCEfA	Windows CE for Automotive
WGS	World Geodetic System
WML	Wireless Markup Language
WMLS	Wireless Markup Language Script
WSP	Wireless Session Protocol
WTLS	Wireless Transport Layer Security
XM	Digital satellite radio
XML	Extensible Markup Language
XSDL	XML Schema Definition Language
XSL	Extensible Stylesheet Language
XSLT	XSL Transformations

Geography Markup Language

The Open GIS Consortium (OGC) is an industry consortium focused on growing interoperability between technologies that involve spatial information and location. The OGC has the broadest goals among organizations developing location services standards. The OGC believes fundamentally that spatial and geographic information and location are being underutilized worldwide as tools to improve decision-making capabilities, economic productivity, and service delivery. Accordingly, the organization has as its core mission to "deliver spatial interface specifications that are openly available for global use."

The OGC approach to delivering on its mission is to (a) organize interoperability projects, (b) work toward consensus, (c) create a formal specification, (d) develop strategic business opportunities, (e) create strategic standards partnerships, and (f) promote demand for interoperable products. One of the most important initiatives of OGC is the Geography Markup Language.

GEOGRAPHY MARKUP LANGUAGE 2.0

Geography Markup Language (GML) is an XML format for transporting and storing geographic information, which includes both the spatial and nonspatial components of geographic features. As described by the OGC, this specification "defines the XML Schema syntax, mechanisms, and conventions that

- Provide an open, vendor-neutral framework for the definition of geospatial application schemas and objects;
- Allow profiles that support proper subsets of GML framework descriptive capabilities;
- Support the description of geospatial application schemas for specialized domains and information communities;

- Enable the creation and maintenance of linked geographic application schemas and datasets;
- Support the storage and transport of application schemas and data sets;
- Increase the ability of organizations to share geographic application schemas and the information they describe."

For updates to this specification, see *http://www.opengis.net/gml/*.

OpenGIS® IMPLEMENTATION SPECIFICATION, 20 FEBRUARY 2001

OGC Document Number: 01-029

This version:
> http://www.opengis.net/gml/01-029/GML2.html
> (Available as: PDF, zip archive of XHTML)

Latest version:
> http://www.opengis.net/gml/01-029/GML2.html

Previous versions:
> http://www.opengis.net/gml/00-029/GML.html

Editors:
> Simon Cox (CSIRO Exploration & Mining) <Simon.Cox@dem.csiro.au>
> Adrian Cuthbert (SpotOn MOBILE) <adrian@spotonmobile.com>
> Ron Lake (Galdos Systems, Inc.) <rlake@galdosinc.com>
> Richard Martell (Galdos Systems, Inc.) <rmartell@galdosinc.com>

Contributors:
> Simon Cox (CSIRO Exploration & Mining) <Simon.Cox@dem.csiro.au>
> Adrian Cuthbert (SpotOn MOBILE) <adrian@spotonmobile.com>
> Paul Daisey (U.S. Census Bureau) <pdaisey@geo.census.gov>
> John Davidson (OGC IP2000 Team) <georef@erols.com>
> Sandra Johnson (MapInfo Corporation) <sandra_johnson@mapinfo.com>
> Edric Keighan (Cubewerx Inc.) <ekeighan@cubewerx.com>
> Ron Lake (Galdos Systems Inc.) <rlake@galdosinc.com>
> Marwa Mabrouk (ESRI Ltd.) <mmabrouk@esri.com>
> Serge Margoulies (IONIC Software) <serge.margoulies@ionicsoft.com>
> Richard Martell (Galdos Systems, Inc.) <rmartell@galdosinc.com>
> Lou Reich (NASA/CSC) <louis.i.reich@gsfc.nasa.gov>
> Barry O'Rourke (Compusult Ltd.) <barry@compusult.nf.ca>
> Jayant Sharma (Oracle Corporation) <jsharma@us.oracle.com>
> Panagiotis (Peter) Vretanos (CubeWerx Inc.) <pvretano@cubewerx.com>

Abstract

The Geography Markup Language (GML) is an XML encoding for the transport and storage of geographic information, including both the spatial and non-spatial properties of geographic features. This specification defines the XML Schema syntax, mechanisms, and conventions that

- Provide an open, vendor-neutral framework for the definition of geospatial application schemas and objects;
- Allow profiles that support proper subsets of GML framework descriptive capabilities;
- Support the description of geospatial application schemas for specialized domains and information communities;
- Enable the creation and maintenance of linked geographic application schemas and datasets;
- Support the storage and transport of application schemas and data sets;
- Increase the ability of organizations to share geographic application schemas and the information they describe.

Implementers may decide to store geographic application schemas and information in GML, or they may decide to convert from some other storage format on demand and use GML only for schema and data transport.

Document status

This document is an OpenGIS® Implementation Specification.

XML instances which are compliant to this specification shall validate against a conforming application schema. A conforming application schema shall import the Geometry Schema (geometry.xsd), the Feature Schema (feature.xsd), and the XLinks schema (xlinks.xsd) as base schemas; furthermore, it shall be developed using the rules for the development of application schemas specified in section 5 of this document.

Sections 1 and 2 of this document present the background information and modeling concepts that are needed to understand GML. Section 3 presents the GML conceptual model which is independent of encoding. Section 4 presents material which discusses the encoding of the GML conceptual model using the XML Schema definition language (XSDL). This material is intended to demonstrate how to employ the normative GML geometry and feature schemas specified in Appendices A and B of this document. Section 5 of this document presents the rules for the development of conformant GML application schemas. Section 6 presents examples to illustrate techniques for constructing compliant GML application to model recurring geographic themes; these techniques are not normative but they do represent the collective experience of the editors of this document and are strongly recommended. Conforming profiles of this document shall be developed according to the rules specified in section 7. Appendix A presents the Geometry schema, Appendix B presents the Feature schema, and Appendix C presents the XLinks schema.

The OpenGIS Consortium (OGC) invites comments on this Implementation Specification—please submit them to gml.sig@opengis.org.

Table of Contents

1 Representing geographic features
   1.1 Introduction
   1.2 Feature and Geometry models
2 Overview of GML
   2.1 Design goals
   2.2 Schemas for geospatial data
   2.3 Graphical rendering
3 Conceptual framework
   3.1 Features and properties
   3.2 Geometric properties
   3.3 Application schemas
4 Encoding GML
   4.1 Introduction
   4.2 Encoding features without geometry
   4.3 Encoding geometry
   4.4 Encoding features with geometry
   4.5 Encoding feature collections
   4.6 Encoding feature associations
5 GML application schemas
   5.1 Introduction
   5.2 Rules for constructing application schemas (normative)
6 Worked examples of application schemas (non-normative)
7 Profiles of GML
Appendix A: The Geometry schema, v2.06 (normative)
Appendix B: The Feature schema, v2.06 (normative)
Appendix C: The XLinks schema, v2.01 (normative)
Appendix D: References
Appendix E: Revision history

1 REPRESENTING GEOGRAPHIC FEATURES

1.1 Introduction

This section introduces the key concepts required to understand how the Geography Markup Language (GML) models the world. It is based on the OGC Abstract Specification (available online: http://www.opengis.org/techno/specs.htm), which defines a geographic feature as "an abstraction of a real world phenomenon; it is a geographic feature if it is associated with a location relative to the Earth." Thus a digital representation of the real world can be thought of as a set of features. The state of a feature is defined by a set of properties, where each property can be thought of as a {name, type, value} triple. The number of properties a feature may have, together with their names and types, are determined by its type definition. Geographic features are those with properties that may be geometry-valued. A feature collection is a collection of features that can itself be regarded as a feature; as a consequence a feature collection has a feature type and thus may have distinct properties of its own, in addition to the features it contains.

This specification is concerned with what the OGC calls *simple features*: features whose geometric properties are restricted to 'simple' geometries for which coordinates are defined in two dimensions and the delineation of a curve is subject to linear interpolation. While this release of GML does permit coordinates to be specified in three dimensions, it currently provides no direct support for three-dimensional geometry constructs. The term 'simple features' was originally coined to describe the functionality defined in the set of OpenGIS® Implementation Specifications (available online: http://www.opengis.org/techno/specs.htm); GML follows the geometry model defined in these specifications. For example, the traditional 0, 1 and 2-dimensional geometries defined in a two-dimensional spatial reference system (SRS) are represented by points, line strings and polygons. In addition, the geometry model for simple features also allows geometries that are collections of other geometries (either homogeneous multi-point, multi-line string and multi-polygon collections, or heterogeneous geometry collections). In all cases the 'parent' geometry element is responsible for indicating in which SRS the measurements have been made.

How can GML be used to represent real-world phenomena? Suppose somebody wishes to build a digital representation of the city of Cambridge in England. The city could be represented as a feature collection where the individual features represent such things as rivers, roads and colleges; such a classification of real world phenomena determines the feature types that need to be defined. The choice of classification is related to the task to which the digital representation will ultimately be put. The River feature type might have a property called *name* whose value must be of the type 'string'. It is common practice to refer to the typed property; thus the River feature type is said to have a string property called *name*. Similarly, the Road feature type might have a string property called *classification* and an integer property called *number*. Properties with simple types (e.g. integer, string, float, boolean) are collectively referred to as simple properties.

The features required to represent Cambridge might have geometry-valued properties as well as simple properties. Just like other properties, geometric properties must be named. So the River feature type might have a geometric property called *centerLineOf* and the Road feature type might have a geometric property called *linearGeometry*. It is possible to be more precise about the type of geometry that can be used as a property value. Thus in the previous examples the geometric property could be specialised to be a line string property. Just as it is common to have multiple simple properties defined on a single feature type, so too a feature type may have multiple geometric properties.

1.2 Feature and Geometry models

The abstract feature model used by the Open GIS Consortium is shown in Figure 1.1 using Syntropy notation. While it is common practice in the Geospatial Information (GI) community to refer to the properties of a feature as attributes, this document refers to them as properties in order to avoid potential confusion with the attributes of XML elements.

Figure 1.1: The abstract feature model

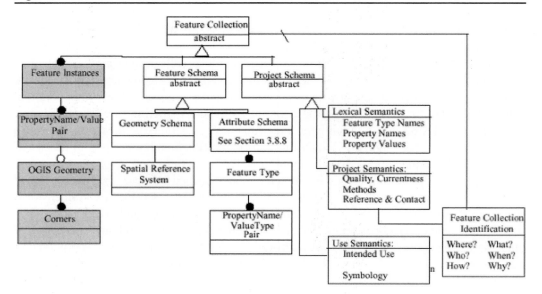

The 'Simple Features' model represents a simplification of the more general model described in the OpenGIS Abstract Specification, this simplification was the result of developing a number of implementation specifications. There are two major simplifications:

- Features are assumed to have either simple properties (booleans, integers, reals, strings) or geometric properties; and
- Geometries are assumed to be defined in two-dimensional SRS and use linear interpolation between coordinates.

A number of consequences follow from these simplifications. For example, simple features only provide support for vector data; and simple features are not sufficiently expressive to explicitly model topology. This version of GML addresses the first of these limitations in that it allows features to have complex or aggregate non-geometric properties. Such complex properties may themselves be composed of other complex and simple properties. Common examples include dates, times, and addresses. It is expected that future versions of GML will address the second of these limitations and provide more elaborate geometry models.

The geometry object model for simple features (Figure 1.2) has an (abstract) base Geometry class and associates each geometry object with an SRS that describes the coordinate space in which the object is defined. GML mirrors this class hierarchy but omits some intermediate (i.e. non-leaf) types such as Curve, Surface, MultiSurface, and MultiCurve.

Figure 1.2: The geometry model for simple features

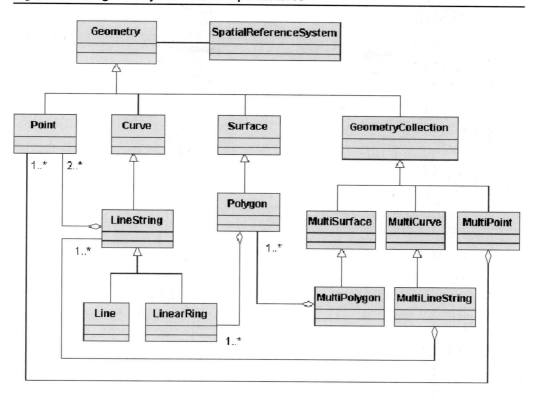

2   OVERVIEW OF GML

2.1   Design goals

GML was developed with a number of explicit design goals, a few of which overlap the objectives of XML itself:

- provide a means of encoding spatial information for both data transport and data storage, especially in a wide-area Internet context;
- be sufficiently extensible to support a wide variety of spatial tasks, from portrayal to analysis;
- establish the foundation for Internet GIS in an incremental and modular fashion;
- allow for the efficient encoding of geo-spatial geometry (e.g. data compression);
- provide easy-to-understand encodings of spatial information and spatial relationships, including those defined by the OGC Simple Features model;
- be able to separate spatial and non-spatial content from data presentation (graphic or otherwise);
- permit the easy integration of spatial and non-spatial data, especially for cases in which the non-spatial data is XML-encoded;

- be able to readily link spatial (geometric) elements to other spatial or non-spatial elements.
- provide a set common geographic modeling objects to enable interoperability of independently-developed applications.

GML is designed to support interoperability and does so through the provision of basic geometry tags (all systems that support GML use the same geometry tags), a common data model (features/properties), and a mechanism for creating and sharing application schemas. Most information communities will seek to enhance their interoperability by publishing their application schemas; interoperability may be further improved in some cases through the use of profiles as outlined in section 7.

2.2 Schemas for geospatial data

In general terms a schema defines the characteristics of a class of objects; in XML a schema also describes how data is marked up. GML strives to cater to a broad range of users, from neophytes to domain experts interested in modeling the semantics of geo-spatial information. Version 2.0 of GML is compliant with the XML Schema Candidate Recommendation published by the W3C in two parts on 24 October 2000 [XMLSchema1], [XMLSchema2]). GML has also been developed to be consistent with the XML Namespaces Recommendation [XMLName]. Namespaces are used to distinguish the definitions of features and properties defined in application-specific domains from one another, and from the core constructs defined in GML modules.

Geospatial feature types can be considered apart from their associated schemas. That is, a Road type (or class) exists independently of its schema definition whether it's expressed in terms of a DTD or an XML Schema. In GML 2.0 geospatial types are captured as element names, but these names assert the existence of model-level types separately from their XML Schema encodings. Consider the following example: a Road type is introduced in an application schema by declaring a global <Road> element of a type named RoadType (<element name="Road" type="RoadType"/>). We note the interplay of two perspectives: conceptual and implementation. Declaring that *width* is a property of a Road is a model-level assertion that says nothing about whether width is a floating point number with two decimal places or simply an integer—these are data and process characteristics at the implementation level.

GML 2.0 defines three base schemas for encoding spatial information. The Geometry schema (geometry.xsd) replaces the DTD that appeared in GML 1.0. This release provides an enhanced Feature schema that supports feature collections (as feature types) and includes common feature properties such as *fid* (a feature identifier), *name* and *description*. The XLink schema provides the XLink attributes to support linking functionality. Database implementations are required to provide an application schema that defines the schema expressions for the geographic features that they support, and these are derived from the definitions found in the Feature schema.

The XML schema definition language provides a superset of the facilities provided by the DTD validation mechanism defined in the XML 1.0 specification. XML Schema provides a rich set of primitive datatypes (e.g. string, boolean, float, month), and it allows the creation of built-in and user-defined datatypes. XML Schema offers several advantages when it comes to constraining GML encodings:

- it enables the intermingling of different vocabularies using namespaces;
- it permits finer control over the structure of the type definition hierarchy; and
- it confers extensibility and flexibility via derived types and substitution groups

2.3 Graphical rendering

GML has been designed to uphold the principle of separating content from presentation. GML provides mechanisms for the encoding of geographic feature data without regard to how the data may be presented to a human reader. Since GML is an XML application, it can be readily styled into a variety of presentation formats including vector and raster graphics, text, sound and voice. Generation of graphical output such as maps is one of the most common presentations of GML and this can be accomplished in a variety of ways including direct rendering by graphical applets or styling into an XML graphics technology (e.g. SVG [SVG] or X3D [VRML200x]). It should be noted that GML is not dependent on any particular XML graphical specification.

3 THE CONCEPTUAL FRAMEWORK

3.1 Features and properties

GML is an XML encoding for geographic features. In order to correctly interpret a GML document it is necessary to understand the conceptual model that underlies GML, which is described in the OGC Abstract Specification.

3.1.1 Object Model

A geographic feature is essentially a named list of properties. Some or all of these properties may be geospatial, describing the position and shape of the feature. Each feature has a *type*, which is equivalent to a *class* in object modeling terminology, such that the class-definition prescribes the named properties that a particular feature of that type is required to have. So a Road might be defined to have a name, a surface-construction, a destination, and a centreLine. The properties themselves are modeled in UML as *associations*, or as *attributes*, of the feature class. The feature property type is given by the *rolename* from an association, or by the *attribute name*. The *values* of the properties are themselves also instances of defined classes or types. So the Road name is a text-string, the surface-construction might be a text token selected from an enumerated list, the destination is another feature of type Town, and the centreLine is a LineString, which is a *geometry property.*

In UML it is partly a matter of taste whether a property is represented as an association or attribute, though it is common for a property with a complex or highly structured type to be modeled as an association, while simple properties are typically class attributes. If the value of a property only exists in the presence of the feature, such as the Road name, then it may use either a UML *composition association* or be represented as an attribute, as these two methods are functionally equivalent. However, if the value of a property is loosely bound to the object and the property value is an object that might exist independently of the feature, such as the Town that is the destination of a Road, then it must use a form of UML association called *aggregation* (see Figure 3.1).

Figure 3.1: Composition and aggregation relationships

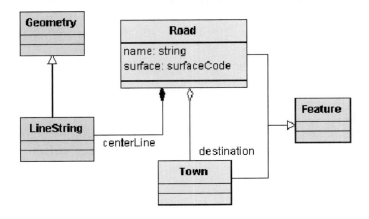

3.1.2 XML encoding of the object model

A feature is encoded as an XML element whose name is the feature type according to some classification. The feature instance contains feature properties, each as an XML element whose name is the property name. Each of these contains another element whose name is the type of the property value or instance; this produces a "layered" syntax in which properties and instances are interleaved.

GML adopts a uniform coding convention to help distinguish properties from instances: element names that represent *instances* of GML classes start with an uppercase letter (e.g. Polygon), while tags that represent *properties* start with a lowercase letter; all embedded words in the property name start with uppercase letters (e.g. centerLineOf).

3.1.3 Functional view of the object model

From a functional perspective we can consider a property as a function that maps a feature onto a property value. A property is characterised by the input feature type and the type of the value that is returned. For example, suppose the feature type House has a String property called *address* and a Polygon property called *extentOf*. Using functional notation we can then write:

address(House) = String
extentOf(House)= Polygon

This approach can also be applied to feature collections that have features as members:

featureMember(FeatureCollection) = Feature

3.2 Geometric properties

In general the definition of feature properties lies in the domain of application schemas. However, since the OGC abstract specification defines a small set of basic geometries, GML defines a set of geometric property elements to associate these geometries with features.

The GML Feature schema also provides descriptive names for the geometry properties, encoded as common English language terms. Overall, there are three levels of naming geometry properties in GML:

1. **Formal names** that denote geometry properties in a manner based on the type of geometry allowed as a property value
2. **Descriptive names** that provide a set of standardised synonyms or aliases for the formal names; these allow use of a more user-friendly set of terms.
3. **Application-specific names** chosen by users and defined in application schemas based on GML

The formal and descriptive names for the basic geometric properties are listed in Table 3.1; these names appear in the Feature schema to designate common geometric properties. The precise semantics of these geometry properties (e.g. "What does position of an object mean?" or "Are location and position synonymous?") is not specified.

Table 3.1: Basic geometric properties

Formal name	Descriptive name	Geometry type
boundedBy	-	Box
pointProperty	location, position, centerOf	Point
lineStringProperty	centerLineOf, edgeOf	LineString
polygonProperty	extentOf, coverage	Polygon
geometryProperty	-	*any*
multiPointProperty	multiLocation, multiPosition, multiCenterOf	MultiPoint
multiLineStringProperty	multiCenterLineOf, multiEdgeOf	MultiLineString
multiPolygonProperty	multiExtentOf, multiCoverage	MultiPolygon
multiGeometryProperty	-	MultiGeometry

There are no inherent restrictions in the type of geometry property a feature type may have. For example, a `RadioTower` feature type could have a *location* that returns a Point geometry to identify its location, and have another geometry property called *extentOf* that returns a Polygon geometry describing its physical structure. A geometric property can be modeled in UML as an association class. Figure 3.2 illustrates how the geometryProperty relation associates an abstract feature type with an abstract geometry type.

Figure 3.2: Geometric properties as instances of an association class

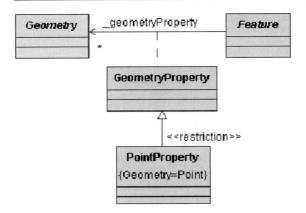

In Figure 3.2 we also see that a *pointProperty* is a concrete instance of `Geometry-Property` that links a feature instance with a <Point> instance. An important point needs to be emphasized here: GML uses property elements to carry the role name of an association; this preserves important semantic relationships that would otherwise be difficult—or impossible—to infer. Such a practice also helps to maintain congruence between a GML schema and its corresponding UML model (if one exists).

3.3 Application schemas

Three base XML Schema documents are provided by GML: feature.xsd which defines the general feature-property model, geometry.xsd which includes the detailed geometry components, and xlinks.xsd which provides the XLink attributes used to implement linking functionality. These schema documents alone do *not* provide a schema suitable for constraining data instances; rather, they provide base types and structures which may be used by an *application schema*. An application schema declares the actual feature types and property types of interest for a particular domain, using components from GML in standard ways. Broadly, these involve defining application-specific types which are derived from types in the standard GML schemas, or by directly including elements and types from the standard GML schemas.

The base GML schemas effectively provide a meta-schema, or a set of foundation classes, from which an application schema can be constructed. User-written application schemas may declare elements and/or define types to name and distinguish significant features and feature collections from each other; the methods used to accomplish this are presented in section 4. A set of (normative) guidelines and rules for developing an application schema which conforms with GML are given in section 5. By following these rules and deriving the types defined in an application schema from components in the base GML schemas, the application schema benefits from standardised constructs and is guaranteed to conform to the OGC Feature Model. A number of complete examples appear in section 6.

The GML Geometry schema is listed in Appendix A. The <import> element in the Geometry schema brings in the definitions and declarations contained in the XLinks schema. The GML Geometry schema includes type definitions for both abstract geometry elements, concrete (multi) point, line and polygon geometry elements, as well as complex type definitions for the underlying geometry types.

The GML Feature schema is listed in Appendix B. The <include> element in the Feature schema brings in the definitions and declarations contained in the Geometry schema; like the geometry schema, the Feature schema defines both abstract and concrete elements and types.

4 ENCODING GML

4.1 Introduction

This section describes how to encode geospatial (geographic) features in GML. The encoding of spatial features using GML 2.0 requires the use of two XML Schemas: the GML Feature Schema (feature.xsd) and the GML Geometry Schema (geometry.xsd); with these two simple schemas it is possible to encode a wide variety of geospatial information.

The remainder of this sub-section introduces the two XML Schemas using UML notation. The following sub-sections provide an introduction to encoding geospatial information in GML 2.0, broken down as follows:

- 4.2 Encoding a feature without geometry
- 4.3 Encoding geometry
- 4.4 Encoding a feature with geometry
- 4.5 Encoding collections of features
- 4.6 Encoding associations between features

First-time readers may wish to skip this sub-section and proceed directly to 4.2; starting from section 4.2 we present an extended set of examples which progessively introduce the components of GML 2.0. Individuals wishing to use GML as a simple means of structuring spatial information for transport are referred to Listing 6.6.

4.1.1 The Geometry schema

The Unified Modeling Language (UML) offers a fairly general means of visually representing the elements of an application schema; a class diagram presents a concise overview of defined types, and a package diagram depicts higher-level groupings of model elements. The base schemas can be viewed as distinct packages with the dependencies as illustrated in Figure 4.1.

Figure 4.1: Base schemas as packages

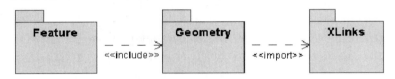

The GML Geometry schema includes type definitions for both abstract geometry elements, concrete (multi) point, line and polygon geometry elements, as well as complex type definitions for the underlying geometry types. Figure 4.2 is a UML representation of the Geometry schema; this diagram provides a bridge between the wide-ranging OGC Abstract Specification (Topic 1: *Feature Geometry*) and the GML Geometry schema—it includes many of the 'well-known' structures described in the abstract specification.

Figure 4.2: UML representation of the Geometry schema

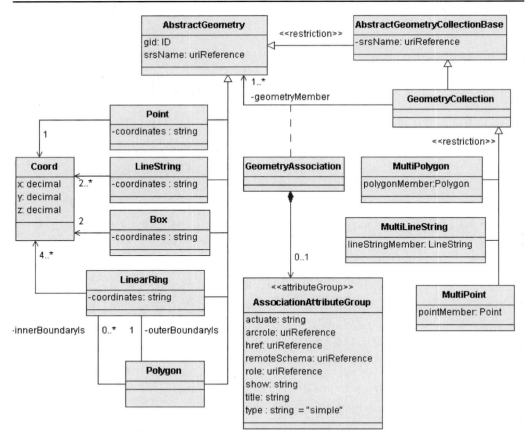

The <<restriction>> stereotype applied to a generalization relationship indicates that a subtype defined in the schema is derived by restriction from its supertype. For example, the `MultiLineString` class is a geometry collection in which a member must be a `LineString`. The complete GML Geometry schema appears in Appendix A; it is liberally documented with <annotation> elements. By convention, explicit named type definitions take the corresponding class name and append the 'Type' suffix (e.g. `LineString` becomes `LineStringType`). Type names are in mixed case with a leading capital; the names of geometric properties and attributes are in mixed case with a leading lower case character. The names of abstract elements are in mixed case with a leading underscore (e.g. _Feature) to highlight their abstract character.

The Geometry schema targets the 'gml' namespace. A namespace is a conceptual entity identified by a URI ("http://www.opengis.org/gml" for the core gml namespace) that denotes a collection of element names and type definitions that belong together—they comprise a cohesive vocabulary. User-defined schemas are required to declare their own target namespace as discussed in section 5.2.4.

4.1.2 The Feature schema

The Feature schema uses the <include> element to bring in the GML geometry constructs and make them available for use in defining feature types:

```
<include schemaLocation="geometry.xsd"/>
```

Figure 4.3 is a UML representation of the Feature schema. Note that a geometric property is modeled as an association class that links a feature with a geometry; concrete geometric property types such as `PointProperty` constrain the geometry to a particular type (e.g. `Point`).

The GML Feature schema is listed in Appendix B. The <include> element in the Feature schema brings in the definitions and declarations contained in the Geometry schema. Like the geometry schema, the Feature schema defines both abstract and concrete elements and types. User-written schemas may define elements and/or types to name and distinguish significant features and feature collections from each other.

Figure 4.3: UML representation of the Feature schema

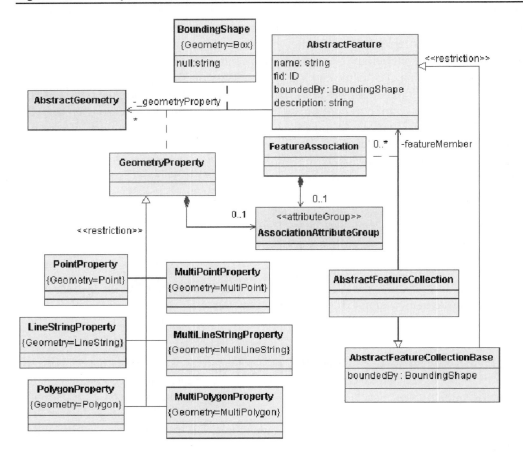

4.2 Encoding features without geometry

Although it is not anticipated that many features will be encoded in GML 2.0 without any
geometry properties, this section starts off with a simple example based on an aspatial feature. This is referred to as the 'Dean' example. There is a feature type called Dean that is
defined to have a string property called *familyName* and an integer property called *age*. In
addition a Dean feature can have zero or more string properties called *nickName*. Thus a
single instance of the Dean feature type might be encoded in XML as:

```
<Dean>
  <familyName>Smith</familyName>
  <age>42</age>
  <nickName>Smithy</nickName>
  <nickName>Bonehead</nickName>
</Dean>
```

Without any thought for GML, one could define the relevant XML Schema to support this as:

```
<element name="Dean" type="ex:DeanType" />
<complexType name="DeanType">
  <sequence>
    <element name="familyName" type="string"/>
    <element name="age" type="integer"/>
    <element name="nickName" type="string" minOccurs="0"
      maxOccurs="unbounded"/>
  </sequence>
</complexType>
```

However, using the Feature schema from GML, it is necessary to identify what plays the role of features (and their types) and what plays the role of properties. In the Dean example Dean is a feature type and *age* is a property. This is indicated in GML by the following:

```
<element name="Dean" type="ex:DeanType"
  substitutionGroup="gml:_Feature" />

<complexType name="DeanType">
  <complexContent>
    <extension base="gml:AbstractFeatureType">
      <sequence>
        <element name="familyName" type="string"/>
        <element name="age" type="integer"/>
        <element name="nickName" type="string" minOccurs="0"
          maxOccurs="unbounded"/>
      </sequence>
    </extension>
  </complexContent>
</complexType>
```

It should be noted that not only does the instance document validate against both of these XML Schema definitions but the content model of DeanType is the same in both XML Schema definitions. Using the Feature schema from GML enables the use of predefined attributes. For example features can be identified with the 'fid' attribute, and features can be described using a predefined *gml:description* property. These capabilities are inherited from gml:AbstractFeatureType (to be more precise Dean extends gml:AbstractFeatureType).

```
<Dean fid="D1123">
  <gml:description>A nice old chap</gml:description>
  <familyName>Smith</familyName>
  <age>42</age>
  <nickName>Smithy</nickName>
  <nickName>Bonehead</nickName>
</Dean>
```

4.3 Encoding geometry

This section describes how GML encodes basic geometry types, and it introduces the GML Geometry schema (geometry.xsd) that supports this encoding. The geometry schema corresponds quite closely to the geometry encoding embodied by the DTD put forth in the GML 1.0 document. The material in this section should be read by all prospective GML users.

In accord with the OGC Simple Features model, GML provides geometry elements corresponding to the following geometry classes:

- Point
- LineString
- LinearRing
- Polygon
- MultiPoint
- MultiLineString
- MultiPolygon
- MultiGeometry

In addition there are <coordinates> and <coord> elements for encoding coordinates, and a <Box> element for defining extents. The following sections describe in detail the encoding of each of these fundamental types of geometry elements.

4.3.1 Coordinates

The coordinates of any Geometry class instance are encoded either as a sequence of <coord> elements that encapsulate tuple components, or as a single string contained within a <coordinates> element. The advantage of using <coord> elements is that a validating XML parser can perform basic type checking and enforce constraints on the number of tuples that appear in a particular geometry instance. Both approaches can convey coordinates in one, two, or three dimensions. The relevant schema fragments can be found in the Geometry schema:

```
<element name="coord" type="gml:CoordType" />

<complexType name="CoordType">
  <sequence>
    <element name="X" type="decimal"/>
    <element name="Y" type="decimal" minOccurs="0"/>
    <element name="Z" type="decimal" minOccurs="0"/>
  </sequence>
</complexType>
```

An additional level of constraint restricts the number of tuples by data type. For example, a <Point> element contains exactly one coordinate tuple:

```
<Point srsName="http://www.opengis.net/gml/srs/epsg.xml#4326">
  <coord><X>5.0</X><Y>40.0</Y></coord>
</Point>
```

As an alternative, coordinates can also be conveyed by a single string. By default the coordinates in a tuple are separated by commas, and successive tuples are separated by a space character (#x20). While these delimiters are specified by several attributes, a user is free to define a localized coordinates list that is derived by restriction from `gml:Coordi-natesType`. An instance document could then employ the *xsi:type* attribute to substitute the localized coordinates list wherever a <coordinates> element is expected; such a subtype could employ other delimiters to reflect local usage.

It is expected that a specialized client application will extract and validate string content, as these functions will not be performed by a general XML parser. The formatting attributes will assume their default values if they are not specified for a particular instance; the <coordinates> element must conform to these XML Schema fragments:

```
<element name="coordinates" type="gml:CoordinatesType"/>

<complexType name="CoordinatesType">
  <simpleContent>
    <extension base="string">
      <attribute name="decimal" type="string" use="default" value="."/>
      <attribute name="cs" type="string" use="default" value=","/>
      <attribute name="ts" type="string" use="default" value="&#x20;"/>
    </extension>
  </simpleContent>
</complexType>
```

This would allow the Point example provided above to be encoded as:

```
<Point srsName="http://www.opengis.net/gml/srs/epsg.xml#4326">
  <coordinates>5.0,40.0</coordinates>
</Point>
```

4.3.2 Geometry elements

The coordinates for a geometry are defined within some spatial reference system (SRS), and all geometries must specify this SRS. GML 2.0 does not address the details of defining spatial reference systems. There is currently a proposed XML-based specification for handling coordinate reference systems and coordinate transformations [OGC00-040]. The *srsName* attribute of the geometry types can be used to test for the equality of SRS between different geometries. The *srsName* (since it is a URI reference) may be navigated to the definition of the SRS. It is expected that the pending SRS specification will be applicable to GML encodings, perhaps in the guise of a Geodesy module derived from that specification.

The optional *gid* attribute of the geometry types represents a unique identifier for geometry elements; this is an ID-type attribute whose value must be text string that conforms to all XML name rules (i.e. the first character cannot be a digit).

4.3.3 Primitive geometry elements

The **Point** element is used to encode instances of the Point geometry class. Each <Point> element encloses either a single <coord> element or a <coordinates> element containing exactly one coordinate tuple; the *srsName* attribute is optional since a Point element may be contained in other elements that specify a reference system. Similar considerations apply to the other geometry elements. The Point element, in common with other geometry types, also has an optional *gid* attribute that serves as an identifier. Here's an example:

```
<Point gid="P1" srsName="http://www.opengis.net/gml/srs/epsg.xml#4326">
  <coord><X>56.1</X><Y>0.45</Y></coord>
</Point>
```

The **Box** element is used to encode extents. Each <Box> element encloses either a sequence of two <coord> elements or a <coordinates> element containing exactly two coordinate tuples; the first of these is constructed from the minimum values measured along all axes, and the second is constructed from the maximum values measured along all axes. A value for the *srsName* attribute should be provided, since a Box cannot be contained by other geometry classes. A Box instance looks like this:

```
<Box srsName="http://www.opengis.net/gml/srs/epsg.xml#4326">
  <coord><X>0.0</X><Y>0.0</Y></coord>
  <coord><X>100.0</X><Y>100.0</Y></coord>
</Box>
```

A **LineString** is a piece-wise linear path defined by a list of coordinates that are assumed to be connected by straight line segments. A closed path is indicated by having coincident first and last coordinates. At least two coordinates are required. Here's an example of a LineString instance:

```
<LineString srsName="http://www.opengis.net/gml/srs/epsg.xml#4326">
  <coord><X>0.0</X><Y>0.0</Y></coord>
  <coord><X>20.0</X><Y>35.0</Y></coord>
  <coord><X>100.0</X><Y>100.0</Y></coord>
</LineString>
```

A **LinearRing** is a closed, simple piece-wise linear path which is defined by a list of coordinates that are assumed to be connected by straight line segments. The last coordinate must be coincident with the first coordinate and at least four coordinates are required (the three to define a ring plus the fourth duplicated one). Since a LinearRing is used in the construction of Polygons (which specify their own SRS), the *srsName* attribute is not needed.

A **Polygon** is a connected surface. Any pair of points in the polygon can be connected to one another by a path. The boundary of the Polygon is a set of LinearRings. We distinguish the outer (exterior) boundary and the inner (interior) boundaries; the LinearRings of the interior boundary cannot cross one another and cannot be contained within one another. There must be at most one exterior boundary and zero or more interior boundary elements. The ordering of LinearRings and whether they form clockwise or anti-clockwise paths is not important. A following example of a Polygon instance has two inner boundaries and uses coordinate strings:

```
<Polygon gid="_98217" srsName="http://www.opengis.net/gml/srs/
  epsg.xml#4326">
  <outerBoundaryIs>
    <LinearRing>
      <coordinates>0.0,0.0 100.0,0.0 100.0,100.0 0.0,100.0 0.0,0.0
        </coordinates>
    </LinearRing>
  </outerBoundaryIs>
  <innerBoundaryIs>
    <LinearRing>
      <coordinates>10.0,10.0 10.0,40.0 40.0,40.0 40.0,10.0 10.0,10.0
        </coordinates>
    </LinearRing>
  </innerBoundaryIs>
  <innerBoundaryIs>
    <LinearRing>
      <coordinates>60.0,60.0 60.0,90.0 90.0,90.0 90.0,60.0 60.0,60.0
        </coordinates>
    </LinearRing>
  </innerBoundaryIs>
</Polygon>
```

4.3.4 Geometry collections

There are a number of homogeneous geometry collections that are predefined in the Geometry schema. A **MultiPoint** is a collection of Points; a **MultiLineString** is a collection of LineStrings; and a **MultiPolygon** is a collection of Polygons. All of these collections each use an appropriate membership property to contain elements. It should be noted that the *srsName* attribute can only occur on the outermost GeometryCollection and must not appear as an attribute of any of the enclosed geometry elements. Here's an example of a MultiLineString instance with three members:

```
<MultiLineString srsName="http://www.opengis.net/gml/srs/
  epsg.xml#4326">
  <lineStringMember>
    <LineString>
      <coord><X>56.1</X><Y>0.45</Y></coord>
      <coord><X>67.23</X><Y>0.98</Y></coord>
    </LineString>
  </lineStringMember>
  <lineStringMember>
    <LineString>
      <coord><X>46.71</X><Y>9.25</Y></coord>
      <coord><X>56.88</X><Y>10.44</Y></coord>
    </LineString>
  </lineStringMember>
```

```
  <lineStringMember>
    <LineString>
      <coord><X>324.1</X><Y>219.7</Y></coord>
      <coord><X>0.45</X><Y>4.56</Y></coord>
    </LineString>
  </lineStringMember>
</MultiLineString>
```

In addition the Geometry schema defines a heterogeneous geometry collection represented by the **MultiGeometry** element that provides a container for arbitrary geometry elements; it might contain any of the primitive geometry elements such as Points, LineStrings, Polygons, MultiPoints, MultiLineStrings, MultiPolygons and even other GeometryCollections. The MultiGeometry element has a generic *geometryMember* property which returns the next geometry element in the collection. An example of a heterogeneous MultiGeometry instance appears below.

```
<MultiGeometry gid="c731" srsName="http://www.opengis.net/gml/srs/
  epsg.xml#4326">
  <geometryMember>
    <Point gid="P6776">
      <coord><X>50.0</X><Y>50.0</Y></coord>
    </Point>
  </geometryMember>

  <geometryMember>
    <LineString gid="L21216">
      <coord><X>0.0</X><Y>0.0</Y></coord>
      <coord><X>0.0</X><Y>50.0</Y></coord>
      <coord><X>100.0</X><Y>50.0</Y></coord>
    </LineString>
  </geometryMember>

  <geometryMember>
    <Polygon gid="_877789">
      <outerBoundaryIs>
        <LinearRing>
          <coordinates>0.0,0.0 100.0,0.0 50.0,100.0 0.0,0.0
            </coordinates>
        </LinearRing>
      </outerBoundaryIs>
    </Polygon>
  </geometryMember>
</MultiGeometry>
```

4.4 Encoding features with geometry

GML 2.0 provides a pre-defined set of geometry properties that can be used to relate geometries of particular types to features. Consider the case where the `DeanType` feature definition has a point property called *location*, which is one of the pre-defined descriptive names that can substitute for the formal name *pointProperty*.

```
<Dean>
  <familyName>Smith</familyName>
  <age>42</age>
  <nickName>Smithy</nickName>
  <nickName>Bonehead</nickName>
  <gml:location>
    <gml:Point>
      <gml:coord><gml:X>1.0</gml:X><gml:Y>1.0</gml:Y></gml:coord>
    </gml:Point>
  <gml:location>
</Dean>
```

which is based on the following application schema fragment:

```
<element name="Dean" type="ex:DeanType"
substitutionGroup="gml:_Feature"/>

<complexType name="DeanType">
  <complexContent>
    <extension base="gml:AbstractFeatureType">
      <sequence>
        <element name="familyName" type="string"/>
        <element name="age" type="integer"/>
        <element name="nickName" type="string" minOccurs="0"
          maxOccurs="unbounded"/>
        <element ref="gml:location"/>
      </sequence>
    </extension>
  </complexContent>
</complexType>
```

Alternatively one can define geometry properties specific to an application schema. For example one might wish to name the property specifying the location of the `Dean` instance as *deanLocation*:

```
<element name="Dean" type="ex:DeanType"
substitutionGroup="gml:_Feature"/>

<complexType name="DeanType">
  <complexContent>
    <extension base="gml:AbstractFeatureType">
      <sequence>
        <element name="familyName" type="string"/>
        <element name="age" type="integer"/>
        <element name="nickName" type="string" minOccurs="0"
          maxOccurs="unbounded"/>
        <element name="deanLocation" type="gml:PointPropertyType"/>
      </sequence>
    </extension>
  </complexContent>
</complexType>
```

In this example `gml:PointPropertyType` is available as a useful pre-defined property type for a feature to employ in just the same way that strings and integers are. The local declaration of the <deanLocation> element basically establishes an alias for <gml:point-Property> as a subelement of `Dean`.

The exclusive use of globally-scoped element declarations reflects a different authoring style that 'pools' all elements in the same symbol space (see section 2.5 of the XML Schema specification, Part 1 for further details); this style also allows us to assign elements to a substitution group such that designated elements can substitute for a particular *head element*, which must be declared as a global element. The *deanLocation* property would be declared globally and referenced in a type definition as shown below:

```
<element name="Dean" type="ex:DeanType"
substitutionGroup="gml:_Feature"/>
<element name="deanLocation" type="gml:PointPropertyType"
  substitutionGroup="gml:pointProperty"/>

<complexType name="DeanType">
  <complexContent>
    <extension base="gml:AbstractFeatureType">
      <sequence>
        <element name="familyName" type="string"/>
        <element name="age" type="integer"/>
        <element name="nickName" type="string" minOccurs="0"
          maxOccurs="unbounded"/>
        <element ref="ex:deanLocation" />
      </sequence>
    </extension>
  </complexContent>
</complexType>
```

4.5 Encoding feature collections

GML 2.0 provides support for building feature collections. An element in an application schema that plays the role of a feature collection must derive from `gml:Abstract-FeatureCollectionType` and declare that it can substitute for the (abstract) <gml:_FeatureCollection> element. A feature collection can use the *featureMember* property to show containment of other features and/or feature collections.

Consider the 'Cambridge' example (described in Section 5) where a `CityModel` feature collection contains `Road` and `River` feature members. In this modification to the example the features are contained within the `CityModel` using the generic *gml:featureMember* property that instantiates the `gml:FeatureAssociationType`:

```
<CityModel fid="Cm1456">
  <dateCreated>Feb 2000</dateCreated>
  <gml:featureMember>
    <River fid="Rv567">....</River>
  </gml:featureMember>

  <gml:featureMember>
    <River fid="Rv568">....</River>
  </gml:featureMember>

  <gml:featureMember>
    <Road fid="Rd812">....</Road>
  </gml:featureMember>
</CityModel>
```

With the following associated application schema fragments:

```
<element name="CityModel" type="ex:CityModelType"
  substitutionGroup="gml:_FeatureCollection"/>
<element name="River" type="ex:RiverType" substitutionGroup=
  "gml:_Feature"/>
<element name="Road" type="ex:RoadType" substitutionGroup=
  "gml:_Feature"/>

<complexType name="CityModelType">
  <complexContent>
    <extension base="gml:AbstractFeatureCollectionType">
      <sequence>
        <element name="dateCreated" type="month"/>
      </sequence>
    </extension>
  </complexContent>
</complexType>

<complexType name="RiverType">
  <complexContent>
    <extension base="gml:AbstractFeatureType">
      <sequence>....</sequence>
    </extension>
  </complexContent>
</complexType>

<complexType name="RoadType">
  <complexContent>
    <extension base="gml:AbstractFeatureType">
      <sequence>.....</sequence>
    </extension>
  </complexContent>
</complexType>
```

GML 2.0 provides a mechanism for feature identification. All GML features have an optional 'fid' attribute of type ID inherited from `gml:AbstractFeatureType`; this means that features in the same GML document cannot share a 'fid' value. The 'fid' attribute and simple link elements constructed using XLink attributes provide a means of unambiguously referencing specific features within a GML document.

Within a feature collection, a <featureMember> element can either contain a single feature or point to a feature that is stored remotely (including elsewhere in the same document). A simple link element can be constructed by including a specific set of XLink attributes. The XML Linking Language (XLink) is currently a Proposed Recommendation of the World Wide Web Consortium [XLink]. XLink allows elements to be inserted into XML documents so as to create sophisticated links between resources; such links can be used to reference remote properties.

A simple link element can be used to implement pointer functionality, and this functionality has been built into various GML 2.0 elements by including the `gml:AssociationAttributeGroup` in these constructs:

- gml:FeatureAssociationType,
- the geometry collection types, and
- all of the pre-defined geometry property types.

As an example, we can modify the `CityModel` fragment shown above to include a remote `River` members *without* making any changes to the application schema.

```
<CityModel fid="Cm1456">
  <dateCreated>Feb 2000</dateCreated>

  <gml:featureMember xlink:type="simple"
    xlink:href="http://www.myfavoritesite.com/rivers.xml#Rv567"/>

  <gml:featureMember xlink:type="simple"
    xlink:href="http://www.myfavoritesite.com/rivers.xml#Rv568"/>

  <gml:featureMember>
    <Road fid="Rd812">....</Road>
  </gml:featureMember>
</CityModel>
```

It should be noted that the *featureMember* property can both point to a remote feature *and* contain a feature. It is not possible in XML Schema to preclude this practice using a purely grammar-based validation approach. The GML 2.0 specification regards a *featureMember* in this state to be undefined.

The most basic syntax for a simple link is as follows:

```
<propertyName xlink:type="simple"
  xlink:title="Description of target instance"
  xlink:href="http://www.myfavoritesite.com/locations.xml#identifier" />
```

where the xlink:title attribute is optional. The xlink:href attribute must point to an object whose type matches that of the value type of the property. It is up to the application to validate that this is the case, an XML parser would not place any constraints on the element linked to by the href attribute.

To enhance clarity a new attribute defined in the 'gml' namespace is introduced to supplement the basic XLink attributes: *remoteSchema*. The *remoteSchema* attribute is a URI reference that uses XPointer syntax to identify the GML schema fragment that constrains the resource pointed to by the locator attribute; this additional attribute is included in the `gml:AssociationAttributeGroup` and so is already available to *featureMember* properties so that they can be expressed like this (assuming "RiverType" is the value of some identifier):

```
<gml:featureMember xlink:type="simple"
  gml:remoteSchema="http://www.myfavoritesite.com/types.xsd#RiverType"
  xlink:href="http://www.myfavoritesite.com/rivers#Rv567"/>
```

XLink attributes can **only** be placed on property elements, and there are no constraints on the values of the xlink:title attribute. Simple XLinks also allow the use of the xlink:role attribute. However this is most commonly a reflection of the property (for example the *featureMember* role name) that is using the link.

The XLink specification requires that the xlink:href attribute point to the resource participating in the link by providing a Uniform Resource Identifier (URI). For example, such a URI may constitute a request to a remote server such as an OGC Web Feature Server (WFS). It might be noted that in response to a 'GETFEATURE' request a WFS will return the GML description of a feature, provided the feature identifier (the value of its 'fid' attribute) is known:

```
xlink:href="http://www.myfavoritesite.com/wfs?WFSVER=0.0.12
  &amp;REQUEST=GETFEATURE
  &amp;FEATUREID=Rv567"
```

The value of the xlink:href **must** be a valid URI per IETF RFC 2396 [RFC2396] and RFC 2732 [RFC2732]; as a consequence, certain characters in the URI string must be escaped according to the URI encoding rules set forth in the XLink specification and in the aforementioned IETF documents.

It might be noted that the WFS 'GETFEATURE' request returns a single feature. If the *gml:remoteSchema* attribute is being used, then it should point to the definition of the relevant feature type. Alternatively an XLink can be used to encode an entire query request to a WFS (required character references do not appear in the query string for clarity):

```
xlink:href="http://www.myfavoritesite.com/wfs?
  WFSVER=0.0.12&
  REQUEST=GETFEATURE&
  TYPENAME=INWATERA_1M&
  FILTER='<Filter>
            <Not>
              <Disjoint>
                <PropertyName>INWATERA_1M.WKB_GEOM</PropertyName>
                <Polygon srsName="http://www.opengis.net/gml/srs/
                  epsg.xml#4326">
                <outerboundaryIs>
                  <LinearRing>
                    <coordinates>
                      -150,50 -150,60 -125,60 -125,50 -150,50
                    </coordinates>
                  </LinearRing>
                </outerboundaryIs>
              </Polygon>
            </Disjoint>
          </Not>
        </Filter>'"
```

In this case the WFS may 'manufacture' a generic feature collection to hold the results of the query. But since feature collections also have a feature type, the *gml:remote-Schema* attribute (if used) should point to the feature type definition of this feature collection, not the types of the features returned by the query.

4.6 Encoding feature associations

The essential purpose of XML document structure is to describe data and the relationships between components within it. With GML it is possible to denote relationships either by containment (binary relationships only) or by linking, but there is no *a priori* reason for preferring one style over another. The examples presented so far have emphasized containment, generally using the property name (i.e. the role name) as a 'wrapper' to make the logical structure explicit.

The GML data model is based on three basic constructs: Feature, Feature Type (i.e. Class), and Property. Both Classes and Properties are resources that are independently defined in GML. A property is defined by specifying the range and domain of the property, each of which must be defined GML types. A GML feature is a Class on which are defined a number of properties; zero, one, or more of these properties are geometry-valued. A *Resource* is anything that has an identity [RFC2396]; this can include electronic documents, a person, a service or a geographic feature.

In the previous section specialized links were used to allow a feature collection in a GML document to contain features external to the document. Had the features been present in the GML document, there would have been no compelling reason to use a link, since the 'containment' relationship could have been indicated by using nesting in the GML document. However there are many relationships that cannot be encoded using containment.

Consider the 'adjacency' relationship between three `LandParcel` features. In total there are three relationships representing the three pairings of features. Using the nesting approach encouraged so far, one might be tempted to encode this in the following, albeit messy, way using an *adjacentTo* property (we will present a much cleaner form later):

```
<LandParcel fid="Lp2034">
  <area>2345</area>
  <gml:extentOf>...</extentOf>
  <adjacentTo>
    <LandParcel fid="Lp2035">
      <area>9812</area>
      <gml:extentOf>...</extentOf>
      <adjacentTo xlink:type="simple" xlink:href="#Lp2034"/>
      <adjacentTo>
        <LandParcel fid="Lp2036">
          <area>8345</area>
          <gml:extentOf>...</extentOf>
          <adjacentTo xlink:type="simple" xlink:href="#Lp2034"/>
          <adjacentTo xlink:type="simple" xlink:href="#Lp2035"/>
        </LandParcel>
      </adjacentTo>
    </LandParcel>
  </adjacentTo>
  <adjacentTo xlink:type="simple" xlink:href="#Lp2036"/>
</LandParcel>
```

In the above fragment the links are being used to identify features within the same GML document. In this example the adjacency relationship is binary (it connects pairs of features) and bi-directional (it can be navigated in both directions). This has been achieved by using two *adjacentTo* properties (each with a simple XLink) to represent each relationship. The relationship itself has no identity in this encoding, and it is not possible to record properties on the relationship. Relationships with these characteristics are sometimes referred to as 'lightweight' relationships.

In the above encoding the *adjacentTo* property sometimes 'nests' the related feature and sometimes 'points' to it. A more symmetrical version of the above would be:

```
<LandParcel fid="Lp2034">
  <area>2345</area>
  <gml:extentOf>...</extentOf>
  <adjacentTo xlink:type="simple" xlink:href="#Lp2035"/>
  <adjacentTo xlink:type="simple" xlink:href="#Lp2036"/>
</LandParcel>
....
<LandParcel fid="Lp2035">
  <area>9812</area>
  <gml:extentOf>...</extentOf>
  <adjacentTo xlink:type="simple" xlink:href="#Lp2034"/>
  <adjacentTo xlink:type="simple" xlink:href="#Lp2036"/>
</LandParcel>
```

```
....
<LandParcel fid="Lp2036">
  <area>8345</area>
  <gml:extentOf>...</extentOf>
  <adjacentTo xlink:type="simple" xlink:href="#Lp2034"/>
  <adjacentTo xlink:type="simple" xlink:href="#Lp2035"/>
</LandParcel>
```

Here the ellipses represent the necessary containment constructs holding this set of LandParcels in some root feature collection. However both examples conform to the same application schema:

```
<element name="LandParcel" type="ex:LandParcelType"
  substitutionGroup="gml:_Feature"/>
<element name="adjacentTo" type="ex:AdjacentToType"
  substitutionGroup="gml:featureMember" />

<complexType name="LandParcelType">
  <complexContent>
    <extension base="gml:AbstractFeatureType">
      <sequence>
        <element name="area" type="integer"/>
        <element ref="gml:extentOf"/>
        <element ref="ex:adjacentTo" minOccurs="0"
          maxOccurs="unbounded"/>
      </sequence>
    </extension>
  </complexContent>
</complexType>

<complexType name="AdjacentToType">
  <complexContent>
    <restriction base="gml:FeatureAssociationType">
      <sequence>
        <element ref="ex:LandParcel"/>
      </sequence>
    </restriction>
  </complexContent>
</complexType>
```

A 'heavyweight' relationship provides identity for the relationship and an opportunity to have properties on the relationship. To extend the adjacency example, one might wish to have a heavyweight adjacency relationship represented by an AdjacentPair feature type with the length of the common border recorded as a *commonBoundaryLength* property. This might be encoded in GML as:

```
<LandParcel fid="Lp2034">
  <area>2345</area>
  <gml:extentOf>...</extentOf>
</LandParcel>
....
<LandParcel fid="Lp2035">
  <area>9812</area>
  <gml:extentOf>...</extentOf>
</LandParcel>
....
<AdjacentPair fid="Ad1465">
  <commonBoundaryLength>231</commonBoundaryLength>
  <adjacentTo xlink:type="simple" xlink:href="#Lp2034"/>
  <adjacentTo xlink:type="simple" xlink:href="#Lp2035"/>
</Adjacent>
```

Note that in this example the `AdjacentPair` instances could exist in a separate GML document from the `LandParcel` features since the *adjacentTo* properties can point to features outside of the GML document. Whether the instances are in a single or multiple documents, they all conform to the same application schema:

```
<element name="LandParcel" type="ex:LandParcelType"
  substitutionGroup="gml:_Feature"/>
<element name="adjacentTo" type="ex:AdjacentToType"
  substitutionGroup="gml:featureMember"/>
<element name="AdjacentPair" type="ex:AdjacentPairType"
  substitutionGroup="gml:_Feature"/>

<complexType name="LandParcelType">
  <complexContent>
    <extension base="gml:AbstractFeatureType">
      <sequence>
        <element name="area" type="integer"/>
        <element ref="gml:extentOf"/>
        <!-- note adjacentTo has been removed -->
      </sequence>
    </extension>
  </complexContent>
</complexType>

<complexType name="AdjacentToType">
  <complexContent>
    <restriction base="gml:FeatureAssociationType">
      <sequence>
        <element ref="ex:LandParcel"/>
      </sequence>
    </restriction>
  </complexContent>
</complexType>
```

```
<complexType name="AdjacentPairType">
  <complexContent>
    <extension base="gml:AbstractFeatureType">
      <sequence>
        <element name="commonBoundaryLength" type="integer"/>
        <element ref="ex:adjacentTo" minOccurs="2" maxOccurs="2"/>
      </sequence>
    </extension>
  </complexContent>
</complexType>
```

The use of extended-type XLink elements offers a more concise means of handling n-ary associations and support for creating third-party linkbases. However, such topics are beyond the scope of this document and interested readers are encouraged to consult the XLink specification for details.

5 GML APPLICATION SCHEMAS

5.1 Introduction

The base schemas (Geometry, Feature, XLink) can be viewed as the components of an application framework for developing schemas or sets of schemas that pertain to a particular domain (e.g. Forestry), jurisdiction (e.g. France), or information community. Furthermore, such application schemas may be developed in a more horizontal fashion to support many information communities.

There are some basic conformance requirements that every application schema must satisfy. Specifically, a conforming GML application schema must heed the following general requirements:

1. an application schema must adhere to the detailed schema development rules described in Section 5.2
2. an application schema must not change the name, definition, or data type of mandatory GML elements.
3. abstract type definitions may be freely extended or restricted.
4. the application schema must be made available to anyone receiving data structured according to that schema.
5. the relevant schemas must specify a target namespace that *must not* be http:// www.opengis.net/gml (i.e. the 'gml' namespace).

A set of logically-related GML schemas, which we term the GML Framework, is depicted in Figure 5.1. The GML schemas provide basic constructs for handling geospatial data in a modular manner. A more specialized application framework containing component application schemas would typically be created for a particular theme or domain, but may also be quite horizontal in nature.

Figure 5.1: GML as a core framework

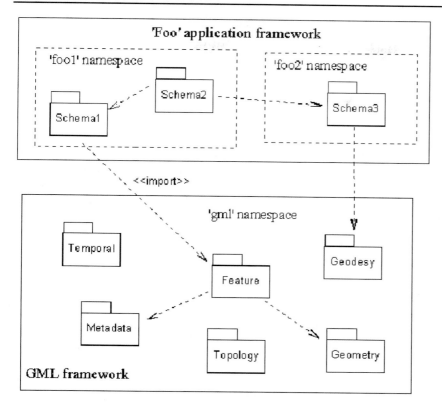

5.2 Rules for constructing application schemas (normative)

The following rules must be adhered to when constructing GML application schemas.

5.2.1 Defining new feature types

Developers of application schemas can create their own feature or feature collection types, but they must ensure that these concrete feature and feature collection types are sub-typed (either directly or indirectly) from the corresponding GML types: `gml:Abstract-FeatureType` or `gml:AbstractFeatureCollectionType`.

```
<complexType name="MyFeature1Type">
  <complexContent>
    <extension base="gml:AbstractFeatureType">
      <sequence>
        <!-- additional child elements inserted here -->
      </sequence>
    </extension>
  </complexContent>
</complexType>
```

```
<complexType name="MyFeature2Type">
  <complexContent>
    <extension base="foo:MyFeature1Type">
      <sequence>
        <!-- additional child elements inserted here -->
      </sequence>
    </extension>
  </complexContent>
</complexType>
```

5.2.2 Defining new geometry types

Authors may create their own geometry types if GML lacks the desired construct. To do this, authors must ensure that these concrete geometry and geometry collection types are subtyped (either directly or indirectly) from the corresponding GML types: `AbstractGeometryType` or `GeometryCollectionType`:

```
<complexType name="MyGeometry1Type">
  <complexContent>
    <extension base="gml:AbstractGeometryType">
      <sequence>
        <!-- additional child elements inserted here -->
      </sequence>
    </extension>
  </complexContent>
</complexType>
```

Any user-defined geometry subtypes shall inherit the elements and attributes of the base GML geometry types without restriction, but may extend these base types to meet application requirements, such as providing a finer degree of interoperability with legacy systems and data sets.

5.2.3 Defining new geometry properties

Authors may create their own geometry properties that encapsulate geometry types they have defined according to subsection 5.2.2; they must ensure that these properties are subtyped (either directly or indirectly) from `gml:GeometryPropertyType` and that they do not change the cardinality of the target instance, which must be a bonafide geometry construct:

```
<complexType name="MyGeometry1PropertyType">
  <complexContent>
    <restriction base="gml:GeometryPropertyType">
      <sequence minOccurs="0">
        <element ref="foo:MyGeometry1Type" />
      </sequence>
      <attributeGroup ref="gml:AssociationAttributeGroup"/>
    </restriction>
  </complexContent>
</complexType>
```

5.2.4 Declaring a target namespace

Authors must declare a target namespace for their schemas. All elements declared in the schema, along with their type definitions, will reside in this namespace. Validation will not succeed if the instance document does not reside in the schema's target namespace. Note that it is not a requirement that URIs actually point to anything concrete, such as a schema document; namespaces are basically just a mechanism to keep element names distinct, thereby preventing namespace 'collisions'.

To use namespaces, elements are given qualified names (QName) that consist of two parts: the *prefix* is mapped to a URI reference and signifies the namespace to which the element belongs; the *local part* conforms to the usual NCName production rules from the W3C Namespaces Recommendation:

```
NCName ::= (Letter | '_') (NCNameChar)*
NCNameChar ::= Letter | Digit | '.' | '-' | '_' | CombiningChar |
  Extender
```

In each worked example the namespace for all elements is explicitly indicated in order to show how vocabularies from different namespaces can intermingle in a single instance document. The use of fully qualified names is specified by setting the value of the *element-FormDefault* attribute of <schema> to "qualified":

```
<schema  targetNamespace="http://www.bar.net/foo"
         xmlns="http://www.w3.org/2000/10/XMLSchema"
         xmlns:gml="http://www.opengis.net/gml"
         xmlns:foo="http://www.bar.net/foo"
         elementFormDefault="qualified"
         version="0.1">

  <!-- import constructs from the GML Feature and Geometry schemas -->
  <import namespace="http://www.opengis.net/gml"
    schemaLocation="feature.xsd"/>
  . . .
</schema>
```

5.2.5 Importing schemas

A conforming instance document can employ constructs from multiple namespaces, as indicated in Figure 5.2. Schema-A in the 'foo' namespace plugs into the GML framework via the Feature schema (which also brings along the geometry constructs). The Feature schema resides in the 'gml' namespace along with the Geometry schema, so it uses the *include* mechanism. However, since Schema-A targets a different namespace, it must employ the *import* mechanism to use the core GML constructs.

Figure 5.2: Using schemas from multiple namespaces

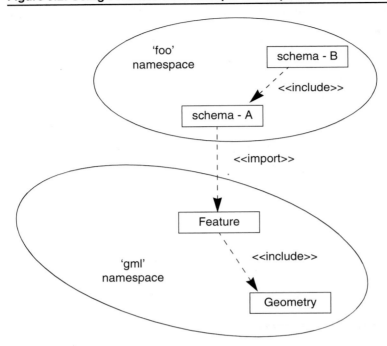

Authors must bear in mind that the 'import' mechanism is the only means whereby an application schema can bring in GML constructs and make use of them. Since a conforming application schema cannot target the core GML namespace ("http://www.opengis.net/gml"), other mechanisms such as 'include' and 'redefine' cannot be employed.

5.2.6 Using substitution groups

Global elements that define substitution groups shall be used for both class (e.g. feature, geometry etc.) and properties that are defined in GML application schemas, and can substitute for the corresponding GML elements wherever these are expected. Declare an element globally and specify a suitable substitution group if the element is required to substitute for another (possibly abstract) element; substitution groups thus enable type promotion (i.e. treating a specific type as a more general supertype). The following global declaration ensures that if foo:CircleType is a defined geometry type, then a <Circle> element can appear wherever the (abstract) gml:_Geometry element is expected:

```
<schema . . .>
  <element name="Circle" type="foo:CircleType"
    substitutionGroup="gml:_Geometry" />
  . . .
</schema>
```

Identical elements declared in more than one complex type definition in a schema should reference a global element. If the <Circle> element is declared globally in the 'foo' namespace (as shown above), it is referenced from within a type definition as follows:

```
<complexType name="MyHubPropertyType">
  <complexContent>
    <restriction base="gml:GeometryPropertyType">
      <sequence minOccurs="0">
        <element ref="foo:Circle" />
      </sequence>
      <attributeGroup ref="gml:AssociationAttributeGroup"/>
    </restriction>
  </complexContent>
</complexType>
```

5.2.7 Declaring additional properties

Schema authors can use the built-in geometric properties or derive their own when necessary as shown in subsection 5.2.3. GML provides a number of predefined geometric properties: *location*, *centerLineOf*, *extentOf*, and so on. Authors can also apply a different name to a base type and use it instead:

```
<element name="crashSite" type="gml:PointPropertyType"
  substitutionGroup="gml:pointProperty" />
```

A feature may also have properties that possess a more complex content model (like geometry properties or feature members). It's important to keep in mind that complex properties represent binary relationships, and that property elements carry the role name of that association. In general, the value of a property element must be one or more instances (either local or remote) of some defined simple or complex type.

As an example, consider the *inspectedBy* property of a waste handling facility that associates a <WasteFacility> instance with an <Inspector> instance:

```
<complexType name="WasteFacilityType">
  <complexContent>
    <extension base="gml:AbstractFeatureType">
      <sequence>
        <!-- other properties -->
        <element name="inspectedBy" type="foo:InspectedByType" />
      </sequence>
    </extension>
  </complexContent>
</complexType>

<complexType name="InspectedByType">
  <sequence minOccurs="0">
    <element ref="foo:Inspector"/>
  </sequence>
  <attributeGroup ref="gml:AssociationAttributeGroup" />
</complexType>
```

The above fragment indicates that `foo:InspectionType` includes the simple-type XLink attributes (that are bundled up in `gml:AssociationAttributeGroup`) to take advantage of the optional pointer functionality; this can be useful if the same inspector assesses multiple facilities and is the target of multiple references:

```
<WasteFacility>
  . . .
  <inspectedBy xlink:type="simple" xlink:href="inspectors.xml#n124" />
</WasteFacility>
```

The definition of the `Inspector` type is not shown here, but it must appear in some application schema. Furthermore, an inspector need not be a feature instance—perhaps an inspector is a `Person` that possesses properties such as an 'oid' identifier. Without the *href* and other XLink attributes a valid instance might then look like the following:

```
<WasteFacility>
  . . .
  <inspectedBy>
    <Inspector oid="n124">. . .</Inspector>
  </inspectedBy>
</WasteFacility>
```

5.2.8 Defining new feature association types

Developers of application schemas can create their own feature association types that must derive from `gml:FeatureAssociationType`. The target instance must be a bonafide GML feature, and it may appear once (explicitly minOccurs="0", implicitly max-Occurs="1"):

```
<complexType name="MyFeatureAssociationType">
  <complexContent>
    <restriction base="gml:FeatureAssociationType">
      <sequence minOccurs="0">
        <element ref="foo:MyFeature1Type" />
      </sequence>
      <attributeGroup ref="gml:AssociationAttributeGroup"/>
    </restriction>
  </complexContent>
</complexType>
```

In many cases it may be desirable to allow instances of only certain feature types as members of a feature collection. A *Feature Filter* as described below shall be applied to ensure that only properly labeled features are valid members.

Feature Filter

Intent:

Restrict membership in a feature collection to permit only instances of specified feature types as allowable members.

Also Known As:

Barbarians at the Gate

Motivation:

Feature collections that extend `gml:AbstractFeatureCollectionType` are somewhat 'promiscuous' in that they will accept any concrete GML feature as an allowable member. A "Feature Filter" can be applied to ensure that only properly labeled features are valid members.

Implementation:

1. Declare a set of abstract elements to 'label' allowable members in a feature collection:
```
<element name="_SchoolFeature" type="gml:AbstractFeatureType"
    substitutionGroup="gml:_Feature" abstract="true"/>
```
2. Define a filter by restricting `gml:FeatureAssociationType`:
```
<complexType name="SchoolMemberType">
  <complexContent>
    <restriction base="gml:FeatureAssociationType">
      <sequence minOccurs="0">
        <element ref="ex:_SchoolFeature"/>
      </sequence>
      <attributeGroup ref="gml:AssociationAttributeGroup"/>
    </restriction>
  </complexContent>
</complexType>
```
3. Label allowable features as they are declared globally:
```
<element name="School" type="ex:SchoolType"
    substitutionGroup="ex:_SchoolFeature"/>
```

6 WORKED EXAMPLES OF APPLICATION SCHEMAS (NON-NORMATIVE)

This section presents several examples to illustrate the construction of application schema that employ the base GML schemas. All examples in this document have been validated using the XSV parser (version 1.166/1.77, 2000-09-28), a Python implementation that supports the current XML Schema Candidate Recommendation; several other parsers (both commercial and open-source implementations) now also solidly support the current specification. An online validation service that uses XSV is available at this URL: <http://www.w3.org/2000/09/webdata/xsv>; while this checking service requires that all documents be web-accessible, it is also possible to download the source files and use the parser offline.

Two examples are briefly summarized here and are explored in some depth using UML class diagrams, corresponding GML application schemas, and sample instance documents. One example is based on a simple model of the city of Cambridge, and is described more fully in Box 1.1 below.

Box 1.1: The Cambridge example

This example has a single feature collection of type `CityModel` and contains two features using a containment relationship called 'cityMember'. The feature collection has a string property called *dateCreated* with the value 'Feb 2000' and a geometric property called *boundedBy* with a 'Box' value. The Box geometry (which represents the 'bounding box' of the feature collection) is expressed in the SRS identified by the value of the *srsName* attribute: this URI reference points to a fragment in another XML document that contains information about the reference system.

The first feature member is an instance of `RiverType` with the name "Cam" and description "The river that runs through Cambridge"; it has a geometric property called *centerLineOf* with a LineString value. The LineString geometry is expressed in the same SRS used by the bounding box.

The second feature member is an instance of `RoadType` with description "M11". It has a string property called *classification* with value "motorway" and an integer property called *number* with value "11". The road has a geometric property called *linearGeometry* with a LineString value; this LineString geometry is also expressed in the same SRS used by the bounding box.

In the 'Cambridge' example the first feature member uses only standard property names defined by GML, whereas the second feature member uses application-specific property names. Thus this example will demonstrate how GML is capable of being used in a custom application model, but it is not intended to provide examples of how the various types of geometry are encoded.

A second example will be used to illustrate how GML can represent a hierarchy of feature collections; this will be referred to as the 'Schools' example and it is summarized in Box 1.2.

Box 1.2: The Schools example

In this example we have a root feature collection of type `StateType` that contains two features collections (instances of `SchoolDistrictType`) using the pre-defined containment relationship 'featureMember'. The State collection also has a *studentPopulation* property. Each of the `SchoolDistrict` collections contains two School or College features using the containment relationship 'schoolMember'.

A `SchoolDistrict` feature has a string property called *name* and a polygon property called *extentOf*. A `School` feature has a string property called *address* and a point property called *location*. A `College` feature also has a string property called *address* plus a point property called *pointProperty*.

Figure 6.1 is a UML diagram for the Cambridge example. As shown, allowable city members must be `Road` or `River` instances; a `Mountain` instance is not a valid member of the feature collection.

Figure 6.1: UML diagram for the Cambridge example

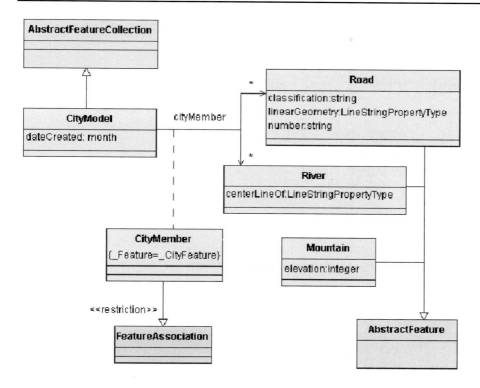

Listing 6.1 is a custom city schema for the Cambridge example. The explicit reference to "city.xsd" in the root element of the instance document in Listing 6.2 (i.e. the value of the *xsi:schemaLocation* attribute) is not required, but in this case it provides a hint to the validating parser regarding the location of a relevant schema document.

Listing 6.1: city.xsd   **View source**

```xml
<?xml version="1.0" encoding="UTF-8"?>
<!-- File: city.xsd   -->
<schema targetNamespace="http://www.opengis.net/examples"
        xmlns:ex="http://www.opengis.net/examples"
        xmlns:xlink="http://www.w3.org/1999/xlink"
        xmlns:gml="http://www.opengis.net/gml"
        xmlns="http://www.w3.org/2000/10/XMLSchema"
        elementFormDefault="qualified"
        version="2.03">
```

```xml
<annotation>
  <appinfo>city.xsd v2.03 2001-02</appinfo>
  <documentation xml:lang="en">
    GML schema for the Cambridge example
  </documentation>
</annotation>

<!-- import constructs from the GML Feature and Geometry schemas -->
<import namespace="http://www.opengis.net/gml"
  schemaLocation="feature.xsd"/>

<!-- ============================================================
      global element declarations
     ============================================================ -->

<element name="CityModel" type="ex:CityModelType"
    substitutionGroup="gml:_FeatureCollection" />
<element name="cityMember" type="ex:CityMemberType"
  substitutionGroup="gml:featureMember"/>
<element name="Road" type="ex:RoadType"
  substitutionGroup="ex:_CityFeature"/>
<element name="River" type="ex:RiverType"
  substitutionGroup="ex:_CityFeature"/>
<element name="Mountain" type="ex:MountainType"
  substitutionGroup="gml:_Feature"/>

<!-- a label for restricting membership in the CityModel collection -->
<element name="_CityFeature" type="gml:AbstractFeatureType"
  abstract="true"
  substitutionGroup="gml:_Feature"/>

<!-- ============================================================
      type definitions for city model
     ============================================================ -->

<complexType name="CityModelType">
  <complexContent>
    <extension base="gml:AbstractFeatureCollectionType">
      <sequence>
        <element name="dateCreated" type="month"/>
      </sequence>
    </extension>
  </complexContent>
</complexType>

<complexType name="CityMemberType">
  <annotation>
    <documentation>
      A cityMember is restricted to those features (or feature
      collections)that are declared equivalent to ex:_CityFeature.
    </documentation>
```

```
      </annotation>
      <complexContent>
        <restriction base="gml:FeatureAssociationType">
          <sequence minOccurs="0">
            <element ref="ex:_CityFeature"/>
          </sequence>
          <attributeGroup ref="gml:AssociationAttributeGroup"/>
        </restriction>
      </complexContent>
    </complexType>

    <complexType name="RiverType">
      <complexContent>
        <extension base="gml:AbstractFeatureType">
          <sequence>
            <element ref="gml:centerLineOf"/>
          </sequence>
        </extension>
      </complexContent>
    </complexType>

    <complexType name="RoadType">
      <complexContent>
        <extension base="gml:AbstractFeatureType">
          <sequence>
            <element name="linearGeometry"
              type="gml:LineStringPropertyType"/>
            <element name="classification" type="string"/>
            <element name="number" type="string"/>
          </sequence>
        </extension>
      </complexContent>
    </complexType>

    <!-- this is just here to demonstrate feature member restriction -->
    <complexType name="MountainType">
      <complexContent>
        <extension base="gml:AbstractFeatureType">
          <sequence>
            <element name="elevation" type="integer"/>
          </sequence>
        </extension>
      </complexContent>
    </complexType>
</schema>
```

Note that the application schema targets the 'ex' namespace; it imports the GML feature and geometry constructs from the 'gml' namespace. The <boundedBy> element is defined in the Feature schema; the <name> and <description> elements are also defined there. The <CityModel> element is an instance of the user-defined ex:CityModelType

type that is derived by extension from `gml:AbstractFeatureCollectionType`. The types `ex:RiverType` and `ex:RoadType` are both derived by extension from `gml:AbstractFeatureType`, which is defined in the GML Feature schema; these derivations assure that the application schema conforms with the GML implementation specification of the OGC Simple Feature model.

Listing 6.2 is a simple schema-valid instance document that conforms to city.xsd. A few words of explanation about the <Mountain> feature are in order! If this particular cityMember is uncommented in Listing 6.2, it will raise a validation error because even though the mountain is a well-formed GML feature, it is not recognized as a valid *city* feature. Note that in city.xsd the <Road> and <River> features are declared equivalent to ex:_CityFeature using the *substitutionGroup* attribute; this abstract element functions as a *label* that restricts membership in the <CityModel> feature collection—only features so labeled are allowable members, as defined by `CityMemberType`. This technique demonstrates the application of the "Feature Filter" (see 5.2.7) that restricts membership in GML feature collections.

One <cityMember> element in Listing 6.2 functions as a simple link by employing several XLink attributes; in effect we have a pointer entitled "Trinity Lane". Any <featureMember> element may behave as a simple link that references a remote resource. The link can point to a document fragment using an *xpointer scheme* that identifies a location, point, or range in the target document [XPointer]. In this case the value of the *href* attribute for the remote member contains an HTTP query string that can retrieve the feature instance; the *remoteSchema* attribute points to a schema fragment that constrains the instance: namely, the complex type definition in city.xsd that bears the name "RoadType".

Listing 6.2: cambridge.xml   **View source**

```
<?xml version="1.0" encoding="UTF-8"?>
<!-- File: cambridge.xml -->
<CityModel xmlns="http://www.opengis.net/examples"
           xmlns:gml="http://www.opengis.net/gml"
           xmlns:xlink="http://www.w3.org/1999/xlink"
           xmlns:xsi="http://www.w3.org/2000/10/XMLSchema-instance"
           xsi:schemaLocation="http://www.opengis.net/
             examples city.xsd">

  <gml:name>Cambridge</gml:name>
  <gml:boundedBy>
    <gml:Box srsName="http://www.opengis.net/gml/srs/epsg.xml#4326">
      <gml:coord><gml:X>0.0</gml:X><gml:Y>0.0</gml:Y></gml:coord>
      <gml:coord><gml:X>100.0</gml:X><gml:Y>100.0</gml:Y></gml:coord>
    </gml:Box>
  </gml:boundedBy>

  <cityMember>
    <River>
      <gml:description>The river that runs through Cambridge.
        </gml:description>
      <gml:name>Cam</gml:name>
```

```
      <gml:centerLineOf>
        <gml:LineString srsName="http://www.opengis.net/gml/srs/
          epsg.xml#4326">
          <gml:coord><gml:X>0</gml:X><gml:Y>50</gml:Y></gml:coord>
          <gml:coord><gml:X>70</gml:X><gml:Y>60</gml:Y></gml:coord>
          <gml:coord><gml:X>100</gml:X><gml:Y>50</gml:Y></gml:coord>
        </gml:LineString>
      </gml:centerLineOf>
    </River>
  </cityMember>

  <cityMember>
    <Road>
      <gml:name>M11</gml:name>
        <linearGeometry>
          <gml:LineString srsName="http://www.opengis.net/gml/srs/
            epsg.xml#4326">
            <gml:coord><gml:X>0</gml:X><gml:Y>5.0</gml:Y></gml:coord>
            <gml:coord><gml:X>20.6</gml:X><gml:Y>10.7</gml:Y>
              </gml:coord>
            <gml:coord><gml:X>80.5</gml:X><gml:Y>60.9</gml:Y>
              </gml:coord>
          </gml:LineString>
        </linearGeometry>
      <classification>motorway</classification>
      <number>11</number>
    </Road>
  </cityMember>

  <cityMember xlink:type="simple" xlink:title="Trinity Lane"
    xlink:href="http://www.foo.net/cgi-bin/wfs?FeatureID=C10239"
    gml:remoteSchema="city.xsd#xpointer
      (//complexType[@name='RoadType'])"/>

  <!-- a mountain doesn't belong here! Uncomment this cityMember and see
       the parser complain!
  <cityMember>
    <Mountain>
      <gml:description>World's highest mountain is in Nepal!
        </gml:description>
      <gml:name>Everest</gml:name>
      <elevation>8850</elevation>
    </Mountain>
  </cityMember>
  -->

  <dateCreated>2000-11</dateCreated>
</CityModel>
```

Figure 6.2 is a UML diagram for the Schools example. The `SchoolDistrictclass` is associated with the `State` class via the *featureMember* relationship, and instances of the `School` or `College` classes are members of the SchoolDistrict collection.

Figure 6.2: UML diagram for the Schools example

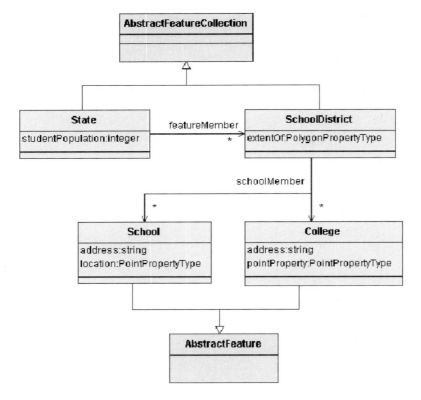

Listing 6.3 is an application schema for the Schools example. The purpose of this example is to demonstrate that feature collections may indeed contain other feature collections. To keep things fairly simple no attempt has been made to restrict membership in any of the collections; this means that a valid instance document could contain **any** GML feature within the <State> and <SchoolDistrict> collections, not just those pertaining to educational institutions. Sub-section 5.2.7 describes a design pattern for restricting collection membership.

Listing 6.3: schools.xsd   **View source**

```xml
<?xml version="1.0" encoding="UTF-8"?>
<!-- File: schools.xsd  -->
<schema targetNamespace="http://www.opengis.net/examples"
       xmlns:ex="http://www.opengis.net/examples"
       xmlns:xlink="http://www.w3.org/1999/xlink"
```

```
              xmlns:gml="http://www.opengis.net/gml"
              xmlns="http://www.w3.org/2000/10/XMLSchema"
              elementFormDefault="qualified" version="2.01">

  <annotation>
    <appinfo>schools.xsd v2.01 2001-02</appinfo>
    <documentation xml:lang="en">
      GML schema for Schools example.
    </documentation>
  </annotation>

  <!-- import constructs from the GML Feature and Geometry schemas -->
  <import namespace="http://www.opengis.net/gml"
    schemaLocation="feature.xsd"/>

  <!-- =============================================================
       global element declarations
       ============================================================= -->

  <element name="State" type="ex:StateType"
    substitutionGroup="gml:_FeatureCollection"/>
  <element name="SchoolDistrict" type="ex:SchoolDistrictType"
    substitutionGroup="gml:_FeatureCollection"/>
  <element name="schoolMember" type="gml:FeatureAssociationType"
    substitutionGroup="gml:featureMember"/>
  <element name="School" type="ex:SchoolType"
    substitutionGroup="gml:_Feature"/>
  <element name="College" type="ex:CollegeType"
    substitutionGroup="gml:_Feature"/>
  <element name="address" type="string"/>

  <!-- =============================================================
       type definitions for state educational institutions
       ============================================================= -->

  <complexType name="StateType">
    <complexContent>
      <extension base="gml:AbstractFeatureCollectionType">
        <sequence>
          <element name="studentPopulation" type="integer"/>
        </sequence>
      </extension>
    </complexContent>
  </complexType>

  <complexType name="SchoolDistrictType">
    <complexContent>
      <extension base="gml:AbstractFeatureCollectionType">
        <sequence>
          <element ref="gml:extentOf"/>
        </sequence>
```

```
        </extension>
      </complexContent>
    </complexType>

    <complexType name="SchoolType">
      <complexContent>
        <extension base="gml:AbstractFeatureType">
          <sequence>
            <element ref="ex:address"/>
            <element ref="gml:location"/>
          </sequence>
        </extension>
      </complexContent>
    </complexType>

    <complexType name="CollegeType">
      <complexContent>
        <extension base="gml:AbstractFeatureType">
          <sequence>
            <element ref="ex:address"/>
            <element ref="gml:pointProperty" />
          </sequence>
        </extension>
      </complexContent>
    </complexType>
</schema>
```

A few interesting things are happening in this example. The root <State> element is an instance of `ex:StateType`, which is derived from the abstract `gml:AbstractFeature-CollectionType` defined in the GML Feature schema. One of the child elements, <SchoolDistrict>, is also a feature collection; in effect we have a feature collection containing a feature collection as one of its members. Listing 6.4 is a conforming instance document. Refer to Box 1.2 for a summary of the example.

Listing 6.4: schools.xml   **View source**

```
<?xml version="1.0" encoding="UTF-8"?>
<!-- File: schools.xml -->
<State xmlns="http://www.opengis.net/examples"
       xmlns:gml="http://www.opengis.net/gml"
       xmlns:xlink="http://www.w3.org/1999/xlink"
       xmlns:xsi="http://www.w3.org/2000/10/XMLSchema-instance"
       xsi:schemaLocation="http://www.opengis.net/
          examples schools.xsd">

  <gml:description>
    Educational institutions with student populations exceeding 500.
  </gml:description>
  <gml:name>School districts in the North Region.</gml:name>
```

```
<gml:boundedBy>
  <gml:Box srsName="http://www.opengis.net/gml/srs/epsg.xml#4326">
    <gml:coord><gml:X>0</gml:X><gml:Y>0</gml:Y></gml:coord>
    <gml:coord><gml:X>50</gml:X><gml:Y>50</gml:Y></gml:coord>
  </gml:Box>
</gml:boundedBy>

<gml:featureMember>
  <SchoolDistrict>
    <gml:name>District 28</gml:name>
    <gml:boundedBy>
      <gml:Box srsName="http://www.opengis.net/gml/srs/
        epsg.xml#4326">
        <gml:coord><gml:X>0</gml:X><gml:Y>0</gml:Y></gml:coord>
        <gml:coord><gml:X>50</gml:X><gml:Y>40</gml:Y></gml:coord>
      </gml:Box>
    </gml:boundedBy>

    <schoolMember>
      <School>
        <gml:name>Alpha</gml:name>
        <address>100 Cypress Ave.</address>
        <gml:location>
          <gml:Point srsName="http://www.opengis.net/gml/srs/
            epsg.xml#4326">
            <gml:coord><gml:X>20.0</gml:X><gml:Y>5.0</gml:Y>
              </gml:coord>
          </gml:Point>
        </gml:location>
      </School>
    </schoolMember>

    <schoolMember>
      <School>
        <gml:name>Beta</gml:name>
        <address>1673 Balsam St.</address>
        <gml:location>
          <gml:Point srsName="http://www.opengis.net/gml/srs/
            epsg.xml#4326">
            <gml:coord><gml:X>40.0</gml:X><gml:Y>5.0</gml:Y>
              </gml:coord>
          </gml:Point>
        </gml:location>
      </School>
    </schoolMember>

    <gml:extentOf>
      <gml:Polygon srsName="http://www.opengis.net/gml/srs/
        epsg.xml#4326">
        <gml:outerBoundaryIs>
          <gml:LinearRing>
```

```
        <gml:coord><gml:X>0</gml:X><gml:Y>0</gml:Y></gml:coord>
        <gml:coord><gml:X>50</gml:X><gml:Y>0</gml:Y></gml:coord>
        <gml:coord><gml:X>50</gml:X><gml:Y>40</gml:Y></gml:coord>
        <gml:coord><gml:X>0</gml:X><gml:Y>0</gml:Y></gml:coord>
      </gml:LinearRing>
    </gml:outerBoundaryIs>
  </gml:Polygon>
 </gml:extentOf>
</SchoolDistrict>
</gml:featureMember>

<gml:featureMember>
  <SchoolDistrict>
    <gml:name>District 32</gml:name>
    <gml:boundedBy>
      <gml:Box srsName="http://www.opengis.net/gml/srs/
        epsg.xml#4326">
      <gml:coord><gml:X>0</gml:X><gml:Y>0</gml:Y></gml:coord>
      <gml:coord><gml:X>30</gml:X><gml:Y>50</gml:Y></gml:coord>
      </gml:Box>
    </gml:boundedBy>

    <schoolMember>
      <School>
        <gml:name>Gamma</gml:name>
        <address>651 Sequoia Ave.</address>
        <gml:location>
          <gml:Point srsName="http://www.opengis.net/gml/srs/
            epsg.xml#4326">
          <gml:coord><gml:X>5.0</gml:X><gml:Y>20.0</gml:Y>
            </gml:coord>
          </gml:Point>
        </gml:location>
      </School>
    </schoolMember>

    <schoolMember xlink:type="simple" xlink:href="http:abc.com">
      <College>
        <gml:name>Delta</gml:name>
        <address>260 University Blvd.</address>
        <gml:pointProperty>
          <gml:Point srsName="http://www.opengis.net/gml/srs/
            epsg.xml#4326">
          <gml:coord><gml:X>5.0</gml:X><gml:Y>40.0</gml:Y>
            </gml:coord>
          </gml:Point>
        </gml:pointProperty>
      </College>
    </schoolMember>
```

```
    <schoolMember xlink:type="simple" xlink:title=
      "Epsilon High School"
      xlink:href="http:www.state.gov/schools/cgi-bin/
        wfs?schoolID=hs736"
      gml:remoteSchema="schools.xsd#xpointer
        (//complexType[@name='SchoolType'])"/>

    <gml:extentOf>
      <gml:Polygon srsName="http://www.opengis.net/gml/srs/
        epsg.xml#4326">
        <gml:outerBoundaryIs>
          <gml:LinearRing>
            <gml:coord><gml:X>0</gml:X><gml:Y>0</gml:Y></gml:coord>
            <gml:coord><gml:X>40</gml:X><gml:Y>50</gml:Y></gml:coord>
            <gml:coord><gml:X>50</gml:X><gml:Y>50</gml:Y></gml:coord>
            <gml:coord><gml:X>0</gml:X><gml:Y>0</gml:Y></gml:coord>
          </gml:LinearRing>
        </gml:outerBoundaryIs>
      </gml:Polygon>
    </gml:extentOf>
  </SchoolDistrict>
  </gml:featureMember>
  <studentPopulation>392620</studentPopulation>
</State>
```

Note the use of <coord> elements to convey coordinate values; the XML parser constrains the number of tuples according to geometry type. For example, a <Point> element has exactly one coordinate tuple, and a <LinearRing> has at least four.

The final example illustrates the construction of a "horizontal" application schema in that it might be applied to a range of application domains. This example is simple data interchange schema that facilitates the exchange of application-level data structures (i.e. features and/or feature collections); such a schema provides a generic means of transferring instances of simple features.

Listing 6.5 is a sample instance document that conforms to the schema in Listing 6.6. A feature may include any of the predefined simple geometric properties (e.g. those that return Points, Polygons, or LineStrings). Non-spatial properties must reflect one of the atomic datatypes of XML Schema.

Listing 6.5: gmlpacket.xml   **View source**

```
<?xml version="1.0" encoding="UTF-8"?>
<!-- File: gmlpacket.xml -->
<gmlPacket xmlns="http://www.opengis.net/examples/packet"
           xmlns:gml="http://www.opengis.net/gml"
           xmlns:xsi="http://www.w3.org/2000/10/XMLSchema-instance"
           xsi:schemaLocation=
              "http://www.opengis.net/examples/packet  gmlpacket.xsd">
```

```
<gml:description>Road network elements</gml:description>
<gml:boundedBy>
  <gml:null>unknown</gml:null>
</gml:boundedBy>

<packetMember>
  <StaticFeature fid="Highway-99" featureType="Road">
    <gml:centerLineOf>
      <gml:LineString>
        <gml:coordinates>...</gml:coordinates>
      </gml:LineString>
    </gml:centerLineOf>
    <property>
      <propertyName>numLanes</propertyName>
      <value dataType="integer">2</value>
    </property>
    <property>
      <propertyName>surfaceType</propertyName>
      <value dataType="string">asphalt</value>
    </property>
  </StaticFeature>
</packetMember>

<Metadata>
  <title>Vancouver-Squamish corridor</title>
</Metadata>
</gmlPacket>
```

Listing 6.6 is the 'gmlpacket' schema. The <gmlPacket> element is the root feature collection; this schema restricts allowable feature members to instances of `pak:Static-FeatureType`. None of the type definitions in the gmlpacket schema can be extended or restricted in any manner, and this schema cannot serve as the basis for any other application schema (i.e. it cannot be imported or included into another schema).

Listing 6.6: gmlpacket.xsd   <u>**View source**</u>

```
<?xml version="1.0" encoding="UTF-8"?>
<!-- File: gmlpacket.xsd  -->
<schema targetNamespace="http://www.opengis.net/examples/packet"
        xmlns:pak="http://www.opengis.net/examples/packet"
        xmlns:gml="http://www.opengis.net/gml"
        xmlns="http://www.w3.org/2000/10/XMLSchema"
        elementFormDefault="qualified"
        finalDefault="#all" version="0.5">

  <annotation>
    <appinfo>gmlpacket.xsd v0.5 2001-02</appinfo>
    <documentation xml:lang="en">
      GML schema for simple data transfer.
```

```
        </documentation>
      </annotation>

      <!-- import constructs from the GML Feature and Geometry schemas -->
      <import namespace="http://www.opengis.net/gml"
        schemaLocation="feature.xsd"/>


      <element name="gmlPacket" type="pak:GMLPacketType"/>
      <element name="packetMember" type="pak:packetMemberType"/>
      <element name="StaticFeature" type="pak:StaticFeatureType"/>


      <complexType name="GMLPacketBaseType">
        <complexContent>
          <restriction base="gml:AbstractFeatureCollectionType">
            <sequence>
              <element ref="gml:description" minOccurs="0"/>
              <element ref="gml:name" minOccurs="0"/>
              <element ref="gml:boundedBy"/>
              <element ref="pak:packetMember" minOccurs="0"
                maxOccurs="unbounded"/>
            </sequence>
            <attribute name="fid" type="ID" use="optional"/>
          </restriction>
        </complexContent>
      </complexType>


      <complexType name="GMLPacketType">
        <complexContent>
          <extension base="pak:GMLPacketBaseType">
            <sequence>
              <element name="Metadata" type="pak:MetadataType"
                minOccurs="0"/>
            </sequence>
          </extension>
        </complexContent>
      </complexType>


      <complexType name="packetMemberType">
        <complexContent>
          <restriction base="gml:FeatureAssociationType">
            <sequence minOccurs="0">
              <element ref="pak:StaticFeature"/>
            </sequence>
            <attributeGroup ref="gml:AssociationAttributeGroup"/>
          </restriction>
        </complexContent>
      </complexType>


      <complexType name="MetadataType">
        <sequence>
          <element name="origin" type="string" minOccurs="0"/>
```

```xml
        <element name="title" type="string" minOccurs="0"/>
        <element name="abstract" type="string" minOccurs="0"/>
      </sequence>
    </complexType>

    <complexType name="StaticFeatureType">
      <complexContent>
        <extension base="gml:AbstractFeatureType">
          <sequence>
            <element ref="gml:pointProperty" minOccurs="0"/>
            <element ref="gml:polygonProperty" minOccurs="0"/>
            <element ref="gml:lineStringProperty" minOccurs="0"/>
            <element name="property" type="pak:PropertyType"
              minOccurs="0" maxOccurs="unbounded"/>
          </sequence>
          <attribute name="featureType" type="string" use="required"/>
        </extension>
      </complexContent>
    </complexType>

    <complexType name="PropertyType">
      <sequence>
        <element name="propertyName" type="string"/>
        <element name="value">
          <complexType>
            <simpleContent>
              <extension base="string">
                <attribute name="dataType" type="pak:DataType"
                  use="required"/>
              </extension>
            </simpleContent>
          </complexType>
        </element>
      </sequence>
    </complexType>

    <simpleType name="DataType">
      <restriction base="string">
        <enumeration value="string"/>
        <enumeration value="integer"/>
        <enumeration value="long"/>
        <enumeration value="decimal"/>
        <enumeration value="boolean"/>
        <enumeration value="time"/>
        <enumeration value="date"/>
      </restriction>
    </simpleType>
  </schema>
```

7   PROFILES OF GML

GML is a fairly complex specification that is rich in functionality. In general an implementation need not exploit the entire specification, but may employ a subset of constructs corresponding to specific relevant requirements. A profile of GML could be defined to enhance interoperability and to curtail ambiguity by allowing only a specific subset of GML; different application schemas could conform to such a profile in order to take advantage of any interoperability or performance advantages that it offers in comparison with full-blown GML. Such profiles could be defined inside other OGC specifications.

There may be cases where reduced functionality is acceptable, or where processing requirements compel use of a logical subset of GML. For example, applications that don't need to handle XLink attributes in any form can adhere to a specific profile that excludes them; the constraint in this case would be to not use links. Other cases might include defining constraints on the level of nesting allowed inside tags (i.e. tree depth), or only allowing features with homogeneous properties as members of a feature collection. In many cases such constraints can be enforced via new schemas; others may be enforced through procedural agreements within an information community.

Here are the guidelines for developing a profile:

1. A profile of GML is a restriction of the basic descriptive capability of GML.
2. Any such profile must be fully compliant with the GML 2.0 specification.
3. GML abstract type definitions may be freely extended or restricted.
4. A profile *must not* change the name, definition, or data type of mandatory GML elements.
5. The schema or schemas that define a profile must be made available to any application schemas which will conform to the profile.
6. The relevant schema or schemas that define a profile must reside in a specified namespace that *must not* be `http://www.opengis.net/gml` (i.e. the core 'gml' namespace)

APPENDIX A: THE GEOMETRY SCHEMA, V2.06 (NORMATIVE)

```
<?xml version="1.0" encoding="UTF-8"?>
<!--  File: geometry.xsd  -->
<schema targetNamespace="http://www.opengis.net/gml"
        xmlns="http://www.w3.org/2000/10/XMLSchema"
        xmlns:gml="http://www.opengis.net/gml"
        xmlns:xlink="http://www.w3.org/1999/xlink"
        elementFormDefault="qualified"
        version="2.06">

  <annotation>
    <appinfo>geometry.xsd v2.06 2001-02</appinfo>
    <documentation xml:lang="en">
      GML Geometry schema. Copyright (c) 2001 OGC, All Rights Reserved.
    </documentation>
  </annotation>
```

```
<!-- bring in the XLink attributes -->
<import namespace="http://www.w3.org/1999/xlink"
  schemaLocation="xlinks.xsd"/>


<!-- ============================================================
     global declarations
     ============================================================ -->

<element name="_Geometry" type="gml:AbstractGeometryType"
  abstract="true"/>
<element name="_GeometryCollection" type="gml:GeometryCollectionType"
  abstract="true"/>
<element name="geometryMember" type="gml:GeometryAssociationType"/>

<!-- primitive geometry elements -->
<element name="Point" type="gml:PointType" substitutionGroup=
  "gml:_Geometry"/>
<element name="LineString" type="gml:LineStringType"
  substitutionGroup="gml:_Geometry"/>
<element name="LinearRing" type="gml:LinearRingType"
  substitutionGroup="gml:_Geometry"/>
<element name="Polygon" type="gml:PolygonType"
  substitutionGroup="gml:_Geometry"/>
<element name="Box" type="gml:BoxType"/>


<!-- aggregate geometry elements -->
<element name="MultiGeometry" type="gml:GeometryCollectionType"/>
<element name="MultiPoint" type="gml:MultiPointType"
  substitutionGroup="gml:_Geometry"/>
<element name="MultiLineString" type="gml:MultiLineStringType"
  substitutionGroup="gml:_Geometry"/>
<element name="MultiPolygon" type="gml:MultiPolygonType"
  substitutionGroup="gml:_Geometry"/>


<!-- coordinate elements -->
<element name="coord" type="gml:CoordType"/>
<element name="coordinates" type="gml:CoordinatesType"/>


<!-- this attribute gives the location where an element is defined -->
<attribute name="remoteSchema" type="uriReference" />


<!-- ============================================================
     abstract supertypes
     ============================================================ -->

<complexType name="AbstractGeometryType" abstract="true">
  <annotation>
    <documentation>
      All geometry elements are derived from this abstract supertype;
      a geometry element may have an identifying attribute ('gid').
      It may be associated with a spatial reference system.
```

```
        </documentation>
      </annotation>
      <attribute name="gid" type="ID" use="optional"/>
      <attribute name="srsName" type="uriReference" use="optional"/>
  </complexType>


  <complexType name="AbstractGeometryCollectionBaseType"
      abstract="true">
      <annotation>
        <documentation>
          This abstract base type for geometry collections just makes the
          srsName attribute mandatory.
        </documentation>
      </annotation>
      <complexContent>
        <restriction base="gml:AbstractGeometryType">
          <attribute name="gid" type="ID" use="optional"/>
          <attribute name="srsName" type="uriReference" use="required"/>
        </restriction>
      </complexContent>
  </complexType>


  <attributeGroup name="AssociationAttributeGroup">
      <annotation>
        <documentation>
          These attributes can be attached to any element, thus allowing
          it to act as a pointer. The 'remoteSchema' attribute allows
          an element that carries link attributes to indicate that the
          element is declared in a remote schema rather than by the
          schema that constrains the current document instance.
        </documentation>
      </annotation>
      <attributeGroup ref="xlink:simpleLink"/>
      <attribute ref="gml:remoteSchema" use="optional"/>
  </attributeGroup>

  <complexType name="GeometryAssociationType">
      <annotation>
        <documentation>
          An instance of this type (e.g. a geometryMember) can either
          enclose or point to a primitive geometry element. When serving
          as a simple link that references a remote geometry instance,
          the value of the gml:remoteSchema attribute can be used to
          locate a schema fragment that constrains the target instance.
        </documentation>
      </annotation>
      <sequence>
        <element ref="gml:_Geometry" minOccurs="0"/>
      </sequence>
      <attributeGroup ref="gml:AssociationAttributeGroup"/>
  </complexType>
```

```
<!-- ============================================================
     primitive geometry types
     ============================================================ -->

<complexType name="PointType">
  <annotation>
    <documentation>
      A Point is defined by a single coordinate tuple.
    </documentation>
  </annotation>
  <complexContent>
    <extension base="gml:AbstractGeometryType">
      <sequence>
        <choice>
          <element ref="gml:coord"/>
          <element ref="gml:coordinates"/>
        </choice>
      </sequence>
    </extension>
  </complexContent>
</complexType>

<complexType name="LineStringType">
  <annotation>
    <documentation>
      A LineString is defined by two or more coordinate tuples, with
      linear interpolation between them.
    </documentation>
  </annotation>
  <complexContent>
    <extension base="gml:AbstractGeometryType">
      <sequence>
        <choice>
          <element ref="gml:coord" minOccurs="2" maxOccurs=
            "unbounded"/>
          <element ref="gml:coordinates"/>
        </choice>
      </sequence>
    </extension>
  </complexContent>
</complexType>

<complexType name="LinearRingType">
  <annotation>
    <documentation>
      A LinearRing is defined by four or more coordinate tuples, with
      linear interpolation between them; the first and last
      coordinates must be coincident.
    </documentation>
  </annotation>
```

```
        <complexContent>
          <extension base="gml:AbstractGeometryType">
            <sequence>
              <choice>
                <element ref="gml:coord" minOccurs="4" maxOccurs=
                  "unbounded"/>
                <element ref="gml:coordinates"/>
              </choice>
            </sequence>
          </extension>
        </complexContent>
      </complexType>

      <complexType name="BoxType">
        <annotation>
          <documentation>
            The Box structure defines an extent using a pair of
            coordinate tuples.
          </documentation>
        </annotation>
        <complexContent>
          <extension base="gml:AbstractGeometryType">
            <sequence>
              <choice>
                <element ref="gml:coord" minOccurs="2" maxOccurs="2"/>
                <element ref="gml:coordinates"/>
              </choice>
            </sequence>
          </extension>
        </complexContent>
      </complexType>

      <complexType name="PolygonType">
        <annotation>
          <documentation>
            A Polygon is defined by an outer boundary and zero or more
            inner boundaries which are in turn defined by LinearRings.
          </documentation>
        </annotation>
        <complexContent>
          <extension base="gml:AbstractGeometryType">
            <sequence>
              <element name="outerBoundaryIs">
                <complexType>
                  <sequence>
                    <element ref="gml:LinearRing"/>
                  </sequence>
                </complexType>
              </element>
              <element name="innerBoundaryIs" minOccurs="0" maxOccurs=
                "unbounded">
```

```
            <complexType>
              <sequence>
                <element ref="gml:LinearRing"/>
              </sequence>
            </complexType>
          </element>
        </sequence>
      </extension>
    </complexContent>
  </complexType>


  <!-- =============================================================
       aggregate geometry types
       ============================================================= -->

  <complexType name="GeometryCollectionType">
    <annotation>
      <documentation>
        A geometry collection must include one or more geometries,
        referenced through geometryMember elements. User-defined
        geometry collections that accept GML geometry classes as
        members must instantiate--or derive from--this type.
      </documentation>
    </annotation>
    <complexContent>
      <extension base="gml:AbstractGeometryCollectionBaseType">
        <sequence>
          <element ref="gml:geometryMember" maxOccurs="unbounded"/>
        </sequence>
      </extension>
    </complexContent>
  </complexType>

  <complexType name="MultiPointType">
    <annotation>
      <documentation>
        A MultiPoint is defined by one or more Points, referenced
        through pointMember elements.
      </documentation>
    </annotation>
    <complexContent>
      <restriction base="gml:GeometryCollectionType">
        <sequence>
          <element name="pointMember" maxOccurs="unbounded">
            <complexType>
              <sequence>
                <element ref="gml:Point"/>
              </sequence>
            </complexType>
          </element>
        </sequence>
```

```
      </restriction>
    </complexContent>
  </complexType>


  <complexType name="MultiLineStringType">
    <annotation>
      <documentation>
        A MultiLineString is defined by one or more LineStrings,
        referenced through lineStringMember elements.
      </documentation>
    </annotation>
    <complexContent>
      <restriction base="gml:GeometryCollectionType">
        <sequence>
          <element name="lineStringMember" maxOccurs="unbounded">
            <complexType>
              <sequence>
                <element ref="gml:LineString"/>
              </sequence>
            </complexType>
          </element>
        </sequence>
      </restriction>
    </complexContent>
  </complexType>


  <complexType name="MultiPolygonType">
    <annotation>
      <documentation>
        A MultiPolygon is defined by one or more Polygons, referenced
        through polygonMember elements.
      </documentation>
    </annotation>
    <complexContent>
      <restriction base="gml:GeometryCollectionType">
        <sequence>
          <element name="polygonMember" maxOccurs="unbounded">
            <complexType>
              <sequence>
                <element ref="gml:Polygon"/>
              </sequence>
            </complexType>
          </element>
        </sequence>
      </restriction>
    </complexContent>
  </complexType>
```

```
<!-- ============================================================
     There are two ways to represent coordinates: (1) as a sequence
     of <coord> elements that encapsulate tuples, or (2) using a
     single <coordinates> string.
     ============================================================ -->

<complexType name="CoordType">
  <annotation>
    <documentation>
      Represents a coordinate tuple in one, two, or three dimensions.
    </documentation>
  </annotation>
  <sequence>
    <element name="X" type="decimal"/>
    <element name="Y" type="decimal" minOccurs="0"/>
    <element name="Z" type="decimal" minOccurs="0"/>
  </sequence>
</complexType>

<complexType name="CoordinatesType">
  <annotation>
    <documentation>
      Coordinates can be included in a single string, but there is no
      facility for validating string content. The value of the 'cs'
      attribute is the separator for coordinate values, and the value
      of the 'ts' attribute gives the tuple separator (a single space
      by default); the default values may be changed to reflect local
      usage.
    </documentation>
  </annotation>
  <simpleContent>
    <extension base="string">
      <attribute name="decimal" type="string" use="default"
        value="."/>
      <attribute name="cs" type="string" use="default" value=","/>
      <attribute name="ts" type="string" use="default"
        value="&#x20;"/>
    </extension>
  </simpleContent>
</complexType>
</schema>
```

APPENDIX B: THE FEATURE SCHEMA, V2.06 (NORMATIVE)

```
<?xml version="1.0" encoding="UTF-8"?>
<!-- File: feature.xsd  -->
<schema targetNamespace="http://www.opengis.net/gml"
        xmlns:gml="http://www.opengis.net/gml"
        xmlns="http://www.w3.org/2000/10/XMLSchema"
        elementFormDefault="qualified"
        version="2.06">
```

```
<annotation>
  <appinfo>feature.xsd v2.06 2001-02</appinfo>
  <documentation xml:lang="en">
    GML Feature schema. Copyright (c) 2001 OGC, All Rights Reserved.
  </documentation>
</annotation>

<!-- include constructs from the GML Geometry schema -->
<include schemaLocation="geometry.xsd"/>

<!-- ============================================================
      global declarations
============================================================ -->

<element name="_Feature" type="gml:AbstractFeatureType" abstract=
  "true"/>
<element name="_FeatureCollection" type=
  "gml:AbstractFeatureCollectionType"
  abstract="true" substitutionGroup="gml:_Feature"/>
<element name="featureMember" type="gml:FeatureAssociationType"/>

<!-- some basic geometric properties of features -->
<element name="_geometryProperty" type="gml:GeometryPropertyType"
  abstract="true"/>
<element name="geometryProperty" type="gml:GeometryPropertyType" />
<element name="boundedBy" type="gml:BoundingShapeType"/>

<element name="pointProperty" type="gml:PointPropertyType"
  substitutionGroup="gml:_geometryProperty"/>
<element name="polygonProperty" type="gml:PolygonPropertyType"
  substitutionGroup="gml:_geometryProperty"/>
<element name="lineStringProperty" type="gml:LineStringPropertyType"
  substitutionGroup="gml:_geometryProperty"/>
<element name="multiPointProperty" type="gml:MultiPointPropertyType"
  substitutionGroup="gml:_geometryProperty"/>
<element name="multiLineStringProperty" type=
  "gml:MultiLineStringPropertyType"
  substitutionGroup="gml:_geometryProperty"/>
<element name="multiPolygonProperty" type=
  "gml:MultiPolygonPropertyType"
  substitutionGroup="gml:_geometryProperty"/>
<element name="multiGeometryProperty" type=
  "gml:MultiGeometryPropertyType"
  substitutionGroup="gml:_geometryProperty"/>

<!-- common aliases for geometry properties -->
<element name="location" type="gml:PointPropertyType"
  substitutionGroup="gml:pointProperty"/>
<element name="centerOf" type="gml:PointPropertyType"
  substitutionGroup="gml:pointProperty"/>
<element name="position" type="gml:PointPropertyType"
```

```
                    substitutionGroup="gml:pointProperty"/>
<element name="extentOf" type="gml:PolygonPropertyType"
    substitutionGroup="gml:polygonProperty"/>
<element name="coverage" type="gml:PolygonPropertyType"
    substitutionGroup="gml:polygonProperty"/>
<element name="edgeOf" type="gml:LineStringPropertyType"
    substitutionGroup="gml:lineStringProperty"/>
<element name="centerLineOf" type="gml:LineStringPropertyType"
    substitutionGroup="gml:lineStringProperty"/>
<element name="multiLocation" type="gml:MultiPointPropertyType"
    substitutionGroup="gml:multiPointProperty"/>
<element name="multiCenterOf" type="gml:MultiPointPropertyType"
    substitutionGroup="gml:multiPointProperty"/>
<element name="multiPosition" type="gml:MultiPointPropertyType"
    substitutionGroup="gml:multiPointProperty"/>
<element name="multiCenterLineOf" type=
    "gml:MultiLineStringPropertyType"
    substitutionGroup="gml:multiLineStringProperty"/>
<element name="multiEdgeOf" type="gml:MultiLineStringPropertyType"
    substitutionGroup="gml:multiLineStringProperty"/>
<element name="multiCoverage" type="gml:MultiPolygonPropertyType"
    substitutionGroup="gml:multiPolygonProperty"/>
<element name="multiExtentOf" type="gml:MultiPolygonPropertyType"
    substitutionGroup="gml:multiPolygonProperty"/>

<!-- common feature descriptors -->
<element name="description" type="string"/>
<element name="name" type="string"/>

<!-- ============================================================
     abstract supertypes
     ============================================================ -->

<complexType name="AbstractFeatureType" abstract="true">
  <annotation>
    <documentation>
      An abstract feature provides a set of common properties. A
      concrete feature type must derive from this type and specify
      additional properties in an application schema. A feature may
      optionally possess an identifying attribute ('fid').
    </documentation>
  </annotation>
  <sequence>
    <element ref="gml:description" minOccurs="0"/>
    <element ref="gml:name" minOccurs="0"/>
    <element ref="gml:boundedBy" minOccurs="0"/>
    <!-- additional properties must be specified in an application
      schema -->
  </sequence>
  <attribute name="fid" type="ID" use="optional"/>
</complexType>
```

```
<complexType name="AbstractFeatureCollectionBaseType" abstract=
  "true">
  <annotation>
    <documentation>
      This abstract base type just makes the boundedBy element
      mandatory for a feature collection.
    </documentation>
  </annotation>
  <complexContent>
    <restriction base="gml:AbstractFeatureType">
      <sequence>
        <element ref="gml:description" minOccurs="0"/>
        <element ref="gml:name" minOccurs="0"/>
        <element ref="gml:boundedBy"/>
      </sequence>
      <attribute name="fid" type="ID" use="optional"/>
    </restriction>
  </complexContent>
</complexType>

<complexType name="AbstractFeatureCollectionType" abstract="true">
  <annotation>
    <documentation>
      A feature collection contains zero or more featureMember
      elements.
    </documentation>
  </annotation>
  <complexContent>
    <extension base="gml:AbstractFeatureCollectionBaseType">
      <sequence>
        <element ref="gml:featureMember" minOccurs="0" maxOccurs=
          "unbounded"/>
      </sequence>
    </extension>
  </complexContent>
</complexType>

<complexType name="GeometryPropertyType">
  <annotation>
    <documentation>
      A simple geometry property encapsulates a geometry element.
      Alternatively, it can function as a pointer (simple-type link)
      that refers to a remote geometry element.
    </documentation>
  </annotation>
  <sequence minOccurs="0">
    <element ref="gml:_Geometry"/>
  </sequence>
  <attributeGroup ref="gml:AssociationAttributeGroup"/>
</complexType>
```

```xml
<complexType name="FeatureAssociationType">
  <annotation>
    <documentation>
      An instance of this type (e.g. a featureMember) can either
      enclose or point to a feature (or feature collection); this
      type can be restricted in an application schema to allow only
      specified features as valid participants in the association.
      When serving as a simple link that references a remote feature
      instance, the value of the gml:remoteSchema attribute can be
      used to locate a schema fragment that constrains the target
      instance.
    </documentation>
  </annotation>
  <sequence minOccurs="0">
    <element ref="gml:_Feature"/>
  </sequence>
  <attributeGroup ref="gml:AssociationAttributeGroup"/>
</complexType>

<complexType name="BoundingShapeType">
  <annotation>
    <documentation>
      Bounding shapes--a Box or a null element are currently allowed.
    </documentation>
  </annotation>
  <sequence>
    <choice>
      <element ref="gml:Box"/>
      <element name="null" type="gml:NullType" />
    </choice>
  </sequence>
</complexType>

<!-- ============================================================
     geometry properties
     ============================================================ -->

<complexType name="PointPropertyType">
  <annotation>
    <documentation>
      Encapsulates a single point to represent position, location, or
      centerOf properties.
    </documentation>
  </annotation>
  <complexContent>
    <restriction base="gml:GeometryPropertyType">
      <sequence minOccurs="0">
        <element ref="gml:Point"/>
      </sequence>
      <attributeGroup ref="gml:AssociationAttributeGroup"/>
    </restriction>
```

```
        </complexContent>
      </complexType>

      <complexType name="PolygonPropertyType">
        <annotation>
          <documentation>
            Encapsulates a single polygon to represent coverage or extentOf
            properties.
          </documentation>
        </annotation>
        <complexContent>
          <restriction base="gml:GeometryPropertyType">
            <sequence minOccurs="0">
              <element ref="gml:Polygon"/>
            </sequence>
            <attributeGroup ref="gml:AssociationAttributeGroup"/>
          </restriction>
        </complexContent>
      </complexType>

      <complexType name="LineStringPropertyType">
        <annotation>
          <documentation>
            Encapsulates a single LineString to represent centerLineOf or
            edgeOf properties.
          </documentation>
        </annotation>
        <complexContent>
          <restriction base="gml:GeometryPropertyType">
            <sequence minOccurs="0">
              <element ref="gml:LineString"/>
            </sequence>
            <attributeGroup ref="gml:AssociationAttributeGroup"/>
          </restriction>
        </complexContent>
      </complexType>

      <complexType name="MultiPointPropertyType">
        <annotation>
          <documentation>
            Encapsulates a MultiPoint element to represent the following
            discontiguous geometric properties: multiLocation,
            multiPosition, multiCenterOf.
          </documentation>
        </annotation>
        <complexContent>
          <restriction base="gml:GeometryPropertyType">
            <sequence minOccurs="0">
              <element ref="gml:MultiPoint"/>
            </sequence>
            <attributeGroup ref="gml:AssociationAttributeGroup"/>
```

```
      </restriction>
    </complexContent>
  </complexType>

  <complexType name="MultiLineStringPropertyType">
    <annotation>
      <documentation>
        Encapsulates a MultiLineString element to represent the
        following discontiguous geometric properties: multiEdgeOf,
        multiCenterLineOf.
      </documentation>
    </annotation>
    <complexContent>
      <restriction base="gml:GeometryPropertyType">
        <sequence minOccurs="0">
          <element ref="gml:MultiLineString"/>
        </sequence>
        <attributeGroup ref="gml:AssociationAttributeGroup"/>
      </restriction>
    </complexContent>
  </complexType>

  <complexType name="MultiPolygonPropertyType">
    <annotation>
      <documentation>
        Encapsulates a MultiPolygon to represent the following
        discontiguous geometric properties: multiCoverage,
        multiExtentOf.
      </documentation>
    </annotation>
    <complexContent>
      <restriction base="gml:GeometryPropertyType">
        <sequence minOccurs="0">
          <element ref="gml:MultiPolygon"/>
        </sequence>
        <attributeGroup ref="gml:AssociationAttributeGroup"/>
      </restriction>
    </complexContent>
  </complexType>

  <complexType name="MultiGeometryPropertyType">
    <annotation>
      <documentation>Encapsulates a MultiGeometry element.
        </documentation>
    </annotation>
    <complexContent>
      <restriction base="gml:GeometryPropertyType">
        <sequence minOccurs="0">
          <element ref="gml:MultiGeometry"/>
        </sequence>
        <attributeGroup ref="gml:AssociationAttributeGroup"/>
```

```
      </restriction>
    </complexContent>
  </complexType>

  <simpleType name="NullType">
    <annotation>
      <documentation>
        If a bounding shape is not provided for a feature collection,
        explain why. Allowable values are:
        innapplicable - the features do not have geometry
        unknown - the boundingBox cannot be computed
        unavailable - there may be a boundingBox but it is not divulged
        missing - there are no features
      </documentation>
    </annotation>
    <restriction base="string">
      <enumeration value="inapplicable"/>
      <enumeration value="unknown"/>
      <enumeration value="unavailable"/>
      <enumeration value="missing"/>
    </restriction>
  </simpleType>
</schema>
```

APPENDIX C: THE XLINKS SCHEMA (NORMATIVE)

At the time that GML 2.0 was finalised, the World Wide Web Consortium (W3C) had not produced a normative schema to support its XLink recommendation. As an interim measure, this schema has been produced by the editors of GML 2.0 to provide the XLink attributes for general use; pending the provision of a definitive schema by the W3C, this schema shall be considered a normative component of GML 2.0.

```
<?xml version="1.0" encoding="UTF-8"?>
<!-- File: xlinks.xsd  -->
<schema targetNamespace="http://www.w3.org/1999/xlink"
        xmlns="http://www.w3.org/2000/10/XMLSchema"
        xmlns:xlink="http://www.w3.org/1999/xlink"
        version="2.01">

  <annotation>
    <appinfo>xlinks.xsd v2.01 2001-02</appinfo>
    <documentation xml:lang="en">
      This schema provides the XLink attributes for general use.
  </documentation>
  </annotation>

  <!-- ==============================================================
       global declarations
       ============================================================== -->
```

```
<!-- locator attribute -->
<attribute name="href" type="uriReference" />

<!-- semantic attributes -->
<attribute name="role" type="uriReference" />
<attribute name="arcrole" type="uriReference" />
<attribute name="title" type="string" />

<!-- behavior attributes -->
<attribute name="show">
  <annotation>
    <documentation>
      The 'show' attribute is used to communicate the desired
      presentation of the ending resource on traversal from the
      starting resource; it's value should be treated as follows:
      new - load ending resource in a new window, frame, pane, or
            other presentation context
      replace - load the resource in the same window, frame, pane,
                or other presentation context
      embed - load ending resource in place of the presentation of
              the starting resource
      other - behavior is unconstrained; examine other markup in the
              link for hints
      none - behavior is unconstrained
    </documentation>
  </annotation>
  <simpleType>
    <restriction base="string">
      <enumeration value="new"/>
      <enumeration value="replace"/>
      <enumeration value="embed"/>
      <enumeration value="other"/>
      <enumeration value="none"/>
    </restriction>
  </simpleType>
</attribute>

<attribute name="actuate">
  <annotation>
    <documentation>
      The 'actuate' attribute is used to communicate the desired
      timing of traversal from the starting resource to the ending
      resource; it's value should be treated as follows:
      onLoad - traverse to the ending resource immediately on
               loading the starting resource
      onRequest - traverse from the starting resource to the ending
                  resource only on a post-loading event triggered
                  for this purpose
      other - behavior is unconstrained; examine other markup in
              link for hints
      none - behavior is unconstrained
    </documentation>
```

```
    </annotation>
    <simpleType>
      <restriction base="string">
        <enumeration value="onLoad"/>
        <enumeration value="onRequest"/>
        <enumeration value="other"/>
        <enumeration value="none"/>
      </restriction>
    </simpleType>
  </attribute>

  <!-- traversal attributes -->
  <attribute name="label" type="string" />
  <attribute name="from" type="string" />
  <attribute name="to" type="string" />

  <!-- ============================================================
       Attributes grouped by XLink type, as specified in the W3C
       Proposed Recommendation (dated 2000-12-20)
       ============================================================ -->

  <attributeGroup name="simpleLink">
    <attribute name="type" type="string" use="fixed" value="simple"
      form="qualified"/>
    <attribute ref="xlink:href" use="optional"/>
    <attribute ref="xlink:role" use="optional"/>
    <attribute ref="xlink:arcrole" use="optional"/>
    <attribute ref="xlink:title" use="optional"/>
    <attribute ref="xlink:show" use="optional"/>
    <attribute ref="xlink:actuate" use="optional"/>
  </attributeGroup>

  <attributeGroup name="extendedLink">
    <attribute name="type" type="string" use="fixed" value="extended"
      form="qualified"/>
    <attribute ref="xlink:role" use="optional"/>
    <attribute ref="xlink:title" use="optional"/>
  </attributeGroup>

  <attributeGroup name="locatorLink">
    <attribute name="type" type="string" use="fixed" value="locator"
      form="qualified"/>
    <attribute ref="xlink:href" use="required"/>
    <attribute ref="xlink:role" use="optional"/>
    <attribute ref="xlink:title" use="optional"/>
    <attribute ref="xlink:label" use="optional"/>
  </attributeGroup>

  <attributeGroup name="arcLink">
    <attribute name="type" type="string" use="fixed" value="arc"
      form="qualified"/>
```

```
      <attribute ref="xlink:arcrole" use="optional"/>
      <attribute ref="xlink:title" use="optional"/>
      <attribute ref="xlink:show" use="optional"/>
      <attribute ref="xlink:actuate" use="optional"/>
      <attribute ref="xlink:from" use="optional"/>
      <attribute ref="xlink:to" use="optional"/>
   </attributeGroup>

   <attributeGroup name="resourceLink">
      <attribute name="type" type="string" use="fixed" value="resource"
        form="qualified"/>
      <attribute ref="xlink:role" use="optional"/>
      <attribute ref="xlink:title" use="optional"/>
      <attribute ref="xlink:label" use="optional"/>
   </attributeGroup>

   <attributeGroup name="titleLink">
      <attribute name="type" type="string" use="fixed" value="title"
        form="qualified"/>
   </attributeGroup>

   <attributeGroup name="emptyLink">
      <attribute name="type" type="string" use="fixed" value="none"
        form="qualified"/>
      </attributeGroup>
</schema>
```

APPENDIX D: REFERENCES

[OGC00-0040]

Whiteside, A, and J. Bobbit. 2000. *Recommended Definition Data for Coordinate Reference Systems and Coordinate Transformations*. OGC Project Document 00-040r7.

[RFC2396]

Uniform Resource Identifiers (URI): Generic Syntax. (August 1998). Available [Online]: <ftp://www.ietf.org/rfc/rfc2396.txt>

[RFC2732]

Format for Literal IPv6 Addresses in URLs. (December 1999). Available [Online]: <http://www.ietf.org/rfc/rfc2732.txt>

[SVG]

Scalable Vector Graphics (SVG) 1.0 Specification. W3C Candidate Recommendation (2 November 2000). Available [Online]: <http://www.w3.org/TR/2000/CR-SVG-20001102/index.html>

[VRML200x]

The Virtual Reality Modeling Language. Draft International Standard ISO/IEC 14772:200x. Available [Online]: <http://www.web3d.org/TaskGroups/x3d/specification/>

[XLink]

XML Linking Language (XLink) Version 1.0. W3C Proposed Recommendation (20 December 2000). Available [Online]: <http://www.w3.org./TR/xlink/>

[XMLName]

Namespaces in XML. W3C Recommendation (14 January 1999). Available [Online]: <http://www.w3.org/TR/1999/REC-xml-names-19990114/>

[XMLSchema-1]

XML Schema Part 1: Structures. W3C Candidate Recommendation (24 October 2000). Available [Online]: <http://www.w3.org/TR/xmlschema-1/>

[XMLSchema-2]

XML Schema Part 2: Datatypes. W3C Candidate Recommendation (24 October 2000). Available [Online]: <http://www.w3.org/TR/xmlschema-2/>

[XPointer]

XML Pointer Language (XPointer) Version 1.0. W3C Candidate Recommendation (7 June 2000). Available [Online]: <http://www.w3.org./TR/xptr/>

APPENDIX E: REVISION HISTORY

REC-GML2 (Recommendation, 2001-02-20):

- extensive restructuring of content
- fixed usage of 'type' attribute in xlinks utility schema
- the XLink schema is now normative
- feature collections can now be empty (per OGC Abstract Specification)
- renamed FeatureMemberType and GeometryMemberType to FeatureAssociationType and GeometryAssociationType, respectively
- updated UML diagrams
- added UML diagrams for the 'Cambridge' and 'Schools' examples
- expanded discussion about profiles and application frameworks (sec. 7)

PR-GML2 (Proposed Recommendation, 2001-01-15):

- Moved discussion of XLink attributes to section 4.4 and deleted the GeoLinks schema
- Removed material on temporal data (sec. 6) to a separate GML Module
- purged RDF remnants
- changed the names of identifying attributes ('fid' for features, 'gid' for geometry elements)
- added gml:AssociationType to support pointers to remote feature properties
- miscellaneous editorial changes

CR-GML2 (Candidate Recommendation, 2000-12-22):

- changed srsName attribute to type="uriReference"
- the featureMember element and all basic geometric properties can also function as pointers to remote resources (i.e. as simple-type XLink elements)
- added a brief introduction to section 5 that clarifies the different ways GML can be used to express associations (i.e. containment vs. linking)
- added an example to section 5.4 demonstrating the use of a linkbase to store instances of feature relationships (Listings 5.7 - 5.9)
- inserted comments concerning feature identification (sec. 4.2)
- miscellaneous corrections: invalid feature identifiers; invalid *xlink:role* attributes
- coordinate strings may now contain 1D, 2D, or 3D tuples
- added a description of the Feature Filter design pattern (sec. 4.4.6)
- removed the gml:label attribute from the GeoLinks schema

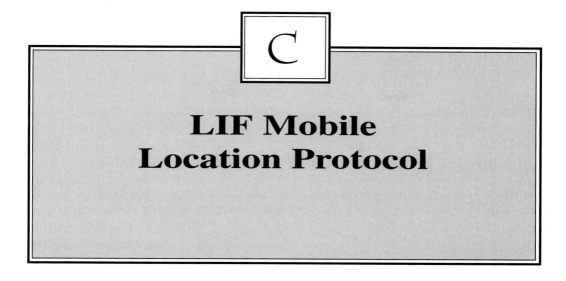

LIF Mobile
Location Protocol

The Location Interoperability Forum (LIF) is an initiative established jointly by Ericsson, Motorola, and Nokia to define and promote a common and ubiquitous solution for determining the location of mobile devices across different wireless network boundaries. The members of LIF include a mix of network operators, equipment manufacturers, and service providers responsible for deploying equipment for this solution. LIF has developed a standards-based solution to locate mobile devices called the Mobile Location Protocol. Issues considered in developing the Mobile Location Protocol specification include the following:

- How to ensure end user privacy and location information security
- How to locate legacy phones (with no existing positioning capability)
- How to offer location services to roaming users
- How to support applications developed internally, by third-party developers and by service providers
- Availability of location capable terminals
- Interoperability of equipment and applications provided by multiple vendors
- Provisioning, billing, and revenue-sharing models

For updates to this specification, see *http://www.locationforum.org*. Note that the contents of Appendix C are reproduced as they appear in the LIF Web site.

LIF TS 101 SPECIFICATION
VERSION 3.0.0, 6 JUNE 2002

Mobile Location Protocol Specification

Abstract

The purpose of this specification is to define a simple and secure access method that allows Internet applications to query location information from a wireless network, irrespective of its underlying air interface technologies and positioning methods.

This specification covers the core of a Mobile Location Protocol that can be used by a location-based application to request MS location information from a location server (GMLC/MPC or other entity in the wireless network).

This specification has been prepared by LIF to provide a simple and secure API (Application Programmer's Interface) to the location server, but that also could be used for other kinds of location servers and entities in the wireless network.

The API is based on existing and well-known Internet technologies as HTTP, SSL/TLS and XML, in order to facilitate the development of location-based applications.

Contents

1	Revision History	5
2	Introduction	6
2.1	Abbreviations	6
2.2	Notational Conventions and Generic Grammar	6
3	General	8
3.1	Overview	8
3.2	MLP structure	8
3.3	MLP extension mechanism	12
4	Mobile Location Service Definitions	14
4.1	Transport Protocol Layer Definitions	14
4.2	Element Layer Definitions	14
4.3	Service Layer Definitions	18
5	Elements and attributes in DTD	35
5.1	add_info	35
5.2	alt	35
5.3	alt_acc	35
5.4	angle	36
5.5	angularUnit	36
5.6	Box	37
5.7	cc	38
5.8	cellid	38
5.9	CircularArcArea	39
5.10	CircularArea	40
5.11	code	40
5.12	codeSpace	41
5.13	codeword	41
5.14	distanceUnit	41
5.15	direction	42
5.16	edition	42
5.17	EllipticalArea	43
5.18	eme_event	43
5.19	esrd	44
5.20	esrk	45
5.21	GeometryCollection	46
5.22	hor_acc	46
5.23	id	47
5.24	inRadius	47
5.25	interval	47
5.26	lac	48
5.27	lev_conf	48
5.28	LinearRing	49
5.29	LineString	50
5.30	ll_acc	51
5.31	lmsi	51
5.32	loc_type	51
5.33	max_loc_age	52
5.34	mcc	52

5.35	mnc	53
5.36	ms_action	53
5.37	msid	54
5.38	MultiLineString	55
5.39	MultiPoint	56
5.40	MultiPolygons	56
5.41	ndc	57
5.42	nmr	57
5.43	radius	57
5.44	startAngle	58
5.45	stopAngle	58
5.46	Point	58
5.47	Polygon	59
5.48	prio	60
5.49	pwd	60
5.50	outRadius	61
5.51	req_id	61
5.52	resp_req	61
5.53	resp_timer	62
5.54	result	62
5.55	semiMajor	63
5.56	semiMinor	63
5.57	serviceid	64
5.58	requestmode	64
5.59	session	65
5.60	sessionid	65
5.61	speed	66
5.62	start_time	66
5.63	stop_time	67
5.64	subclient	68
5.65	ta	68
5.66	time	69
5.67	time_remaining	69
5.68	trl_pos	70
5.69	url	70
5.70	vlrno	71
5.71	vmscno	71
5.72	X	71
5.73	Y	72
5.74	Z	72
5.75	Service attributes	73
6	Result codes	74
6.1	Result codes	74
7	References	76
7.1	References (Normative)	76
7.2	References (Informative)	76
8	Appendix A (informative): Adaptation to 3GPP LCS	78
8.1	Version mapping between 3GPP TS23.271 and this specification	78
8.2	The terminology mapping table with 3GPP LCS Specifications	78
8.3	The corresponding terms used for the location procedures in 3GPP LCS Definition	79

8.4   Error Mapping (informative)...80
9   Appendix B - HTTP Mapping..81
9.1   Location Services using HTTP..81
9.2   Request and Response Encapsulation 82
10   Appendix C: Geographic Information 86
10.1   Coordinate Reference systems (Informative)..............................86
10.2   Coordinate Reference System Transformations (Informative) 88
10.3   Methodology for defining CRSs and transformations in this protocol
         (Normative)...88
10.4   Supported coordinate systems and datum..................................88
10.5   Shapes representing a geographical position 89

1   Revision History

1.0	23-Jan-2001	Sanjiv Bhatt, Motorola	Motorola, Nokia, Ericsson contribution to LIF
1.1	26-Jan-2001	Sanjiv Bhatt, Motorola	Updated after review in MLP adhoc committee in LIF #2 meeting
1.1.1	5-Nov-2001	Sanjiv Bhatt, Motorola	Updated after SIG#6 meeting
1.1.2	17-Nov-2001	Sanjiv Bhatt, Motorola	Updated after SIG#7 meeting
2.0.0	20-Nov-2001	Sanjiv Bhatt, Motorola	Final version (public release)
2.1.0	10-Mar-2002	Sanjiv Bhatt, Motorola	Updated after SIG#8 meeting
2.2.0	02-Apr-2002	Sanjiv Bhatt, Motorola	Updated after SIG#9 meeting
2.2.1	16-Apr-2002	Sanjiv Bhatt, Motorola	Updated before public review
2.3.0	15-May-2002	Sanjiv Bhatt, Motorola	Updated after SIG#10 meeting
3.0.0	06-Jun-2002	Sanjiv Bhatt, Motorola	Final changes

2   Introduction

The Mobile Location Protocol (MLP) is an application-level protocol for getting the position of mobile stations (mobile phones, wireless personal digital assistants, etc.) independent of underlying network technology. The MLP serves as the interface between a Location Server and a Location Services (LCS) Client. This specification defines the core set of operations that a Location Server should be able to perform.

2.1   Abbreviations

ANSI	American National Standards Institute
DTD	Document Type Definition
GMLC	Gateway Mobile Location Center
GMT	Greenwich Mean Time
HTTP	Hypertext Transfer Protocol
HTTPS	HTTP Secure
LCS	Location Services
MLC	Mobile Location Center
MLP	Mobile Location Protocol
MPC	Mobile Positioning Center
MS	Mobile Station
MSID	Mobile Station Identifier
SSL	Secure Socket Layer
TLS	Transport Layer Security
URI	Uniform Resource Identifier
URL	Uniform Resource Locator
UTM	Universal Transverse Mercator
WGS	World Geodetic System
XML	Extensible Markup Language

2.2   Notational Conventions and Generic Grammar

The following rules are used throughout this specification to describe basic parsing constructs. ANSI X3.4-1986 defines the US-ASCII coded character set, see ref. [5]

CR	= <US-ASCII CR, carriage return (13)>
LF	= <US-ASCII LF, linefeed (10)>
SP	= <US-ASCII SP, space (32)>

A set of characters enclosed in brackets ([]) is a one-character expression that matches any of the characters in that set. E.g., "[lcs]" matches either an "l", "c", or "s". A range of characters is indicated with a dash. E.g., "[a-z]" matches any lower-case letter.

The one-character expression can be followed by an interval operator, for
example [a-zA-Z]{min,max} in which case the one-character expression is
repeated at least min and at most max times. E.g., "[a-zA-Z]{2,4}" matches for
example the strings "at", "Good", and "biG".

DTD Syntax Notation

The table below describes the special characters and separators used in the
DTDs defining the different services.

Character	Meaning
+	One or more occurrence
*	Zero or more occurrences
?	Optional
()	A group of expressions to be matched together
|	OR...as in, "this or that"
,	Strictly ordered. Like an AND

3   General

3.1   Overview

The Mobile Location Protocol (MLP) is an application-level protocol for querying the position of mobile stations independent of underlying network technology. The MLP serves as the interface between a Location Server and a location-based application.

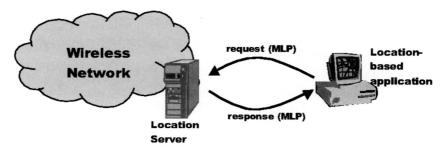

Possible realizations of a Location Server are the GMLC, which is the location server defined in GSM and UMTS, and the MPC, which is defined in ANSI standards. Since the location server should be seen as a logical entity, other implementations are possible.

In the most scenarios (except where explicitly mentioned) an LCS client initiates the dialogue by sending a query to the location server and the server responds to the query.

3.2   MLP structure

In our heterogeneous world, different devices may support different means of communication. A ubiquitous protocol for location services should support different transport mechanisms.

In MLP, the transport protocol is separated from the XML content. The following diagram shows a layered view of MLP.

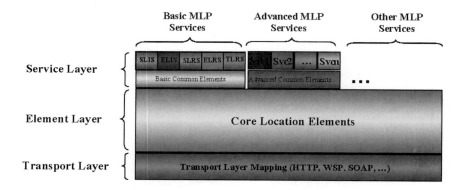

On the lowest level, the transport protocol defines how XML content is transported. Possible MLP transport protocols include HTTP, WSP, SOAP and others.

The Element Layer defines all common elements used by the services in the service layer. Currently MLP defines the following set of DTDs making up the element layer of MLP:

MLP_ID.DTD	Identify Element Definitions
MLP_FUNC.DTD	Function Element Definitions
MLP_LOC.DTD	Location Element Definitions
MLP_SHAPE.DTD	Shape Element Definitions
MLP_QOP.DTD	Quality of Position Element Definitions
MLP_GSM_NET.DTD	GSM Network Parameters Element Definitions
MLP_CTXT	Context Element Definitions

The Service Layer defines the actual services offered by the MLP framework. Basic MLP Services are based on location services defined by 3GPP, and are defined by this specification. The "Advanced MLP Services" and "Other MLP Services" are additional services that either will be specified in other specifications or are specified by other fora that conform to the MLP framework.

Note: The boxes representing services in the Service Layer may contain more than one message. e.g. SLIS (Standard Location Immediate Service) consists of **slir** (Standard Location Immediate Request), **slia** (Standard Location Immediate Answer) and **slirep** (Standard Location Immediate Report) messages. Messages for each service are listed in the table below.

The Service Layer is divided into two sub-layers. The topmost defines the services mentioned in the previous paragraph. The lower sub-layer holds common elements which are specific for that group of services. If an element is common to more than one group of services then that element is defined in the element layer. The present specification specifies no element sub-layer.

There are a number of different possible types of location services. Each implementation of location server can select which services it wants/needs to support. The services are described in the table below.

Service	Description
Standard Location Immediate Service	This is a standard query service with support for a large set of parameters. This service is used when a (single) location response is required immediately (within a set time) or the request may be served by several asynchronous location responses (until a predefined timeout limit is reached). This service consists of the following messages: • Standard Location Immediate Request • Standard Location Immediate Answer • Standard Location Immediate Report
Emergency Location Immediate Service	This is a service used especially for querying of the location of a mobile subscriber that has initiated an emergency call. The response to this service is required immediately (within a set time). This service consists of the following messages: • Emergency Location Immediate Request • Emergency Location Immediate Answer
Standard Location Reporting Service	This is a service that is used when a mobile subscriber wants an LCS Client to receive the MS location. The position is sent to the LCS Client from the location server. Which application and its address are specified by MS or defined in the location server. This service consists of the following message: • Standard Location Report
Emergency Location Reporting Service	This is a service that is used when the wireless network automatically initiates the positioning at an emergency call. The position and related data is then sent to the emergency application from the location server. Which application and its address are defined in the location server. This service consists of the following message: • Emergency Location Report
Triggered Location Reporting Service	This is a service used when the mobile subscriber's location should be reported at a specific time interval or on the occurrence of a specific event. This service consists of the following messages: • Triggered Location Reporting Request • Triggered Location Reporting Answer • Triggered Location Report • Triggered Location Reporting Stop Request • Triggered Location Reporting Stop Answer

3.3   MLP extension mechanism

The MLP specification has been designed with extensibility in mind. Examples of design principles employed to achieve this include:

- Separate DTDs for definitions that are common to all messages, e.g. client address and shapes, so they can be re-used.

- Message extension mechanism allowing the addition of new messages (specific for the HTTP mapping). This mechanism works by specifying an entity parameter, '%extension;', referring to an extension DTD. The extension DTD MUST contain another entity parameter, '%extension.message', containing the definition of the extension as a string together with the actual parameters being added

- Parameter extension mechanism allows the addition of new parameters to existing messages. This mechanism works by specifying an entity parameter, '%extension;', referring to an extension DTD. The extension DTD MUST contain another entity parameter, '%extension.param', containing the definition of the extension as a string together with the actual messages being added.

Each extension parameters should have a vendor specific prefix in order to guarantee their uniqueness.

In order to use the extension, the extension DTD has to be explicitly referenced in the XML document.

The Location Server may ignore any extension that is not recognized and process the message as if the extension is not available.

Example 1: Message extension

```
<!—truckco_MLP_extension -->

<!ENTITY        % extension.message      "| truckco_message">

<!ELEMENT       truckco_message          (truckco_data)>
<!ATTLIST       truckco_message
                ver  CDATA                #FIXED "x.y.z">
```

```
<?xml version = "1.0" ?>
<!DOCTYPE svc_init SYSTEM "MLP_SVC_INIT_300.DTD " [
  <!ENTITY % extension SYSTEM
      "http://www.truckco.com/truckco_MLP_extension.dtd">
  %extension;
]>
<svc_init ver="3.0.0">
  <hdr ver="3.0.0">
    ...
  </hdr>
  <truckco_message ver="x.y.z">
    <truckco_data>
      ...
```

```
      </truckco_data>
    </truckco_message>
</svc_init>
```

Example 2: Parameter extension

```
<!--truckco_MLP_extension -->

<!ENTITY      % extension.param      "truckco_extension">

<!ELEMENT     trucko_extension       (#PCDATA) >
```

```
<?xml version = "1.0" ?>
<!DOCTYPE svc_init SYSTEM "MLP_SVC_INIT_300.DTD" [
  <!ENTITY % extension SYSTEM
      "http://www.truckco.com/truckco_MLP_extension.dtd">
  %extension;
]>
<svc_init ver="3.0.0">
  <hdr ver="3.0.0">
   ...
  </hdr>
  <slir ver="3.0.0">
   ...
    <truckco_extension>
     ...
    </truckco_extension>
  </slir>
</svc_init>
```

4   Mobile Location Service Definitions

4.1   Transport Protocol Layer Definitions

MLP can be implemented using various transport mechanism as stated in section 3.2. The following mappings are specified for MLP:

Mapping	Section
HTTP	Appendix B - HTTP Mapping

4.2   Element Layer Definitions

4.2.1   Identity Element Definitions

```
<!-- MLP_ID -->

<!ELEMENT    msid                                        (#PCDATA)>
<!ATTLIST    msid
             type (MSISDN | IMSI | IMEI | MIN | MDN |    "MSISDN"
             EME MSID | ASID | OPE ID | IPV4 | IPV6 |
             SESSID)
             enc (ASC | CRP)                             "ASC">
<!ELEMENT    msid_range                                  (start_msid, stop_msid)>
<!ELEMENT    msids                                       (((msid, codeword?, session?)
                                                         | (msid_range, codeword*))+)>
<!ELEMENT    codeword                                    (#PCDATA)>
<!ELEMENT    esrd                                        (#PCDATA)>
<!ATTLIST    esrd
             type (NA)                                   "NA">
<!ELEMENT    esrk                                        (#PCDATA)>
<!ATTLIST    esrk
             type (NA)                                   "NA">
<!ELEMENT    session                                     (#PCDATA)>
<!ATTLIST    session
             type (APN | DIAL)                           #REQUIRED>
<!ELEMENT    start_msid                                  (msid)>
<!ELEMENT    stop_msid                                   (msid)>
```

Note: The type attributes of the msid elements that form the start_msid and stop_msid elements must be the same.

4.2.2   Function Element Definitions

```
<!-- MLP_FUNC -->

<!ELEMENT    eme_event                                   (eme_pos+)>
<!ATTLIST    eme_event
             eme_trigger (EME_ORG |                      #REQUIRED>
             EME_REL)
<!ELEMENT    tlrr_event        .                         ( ms_action)>
<!ELEMENT    ms_action                                   EMPTY>
<!ATTLIST    ms_action
             type (MS_AVAIL)                             #REQUIRED>
<!ELEMENT    interval                                    (#PCDATA)>
<!ELEMENT    loc_type                                    EMPTY>
<!ATTLIST    loc_type
             type (CURRENT | LAST                        "CURRENT">
             |CURRENT_OR_LAST | INITIAL)
<!ELEMENT    prio                                        EMPTY>
<!ATTLIST    prio
             type (NORMAL | HIGH)                        "NORMAL">
<!ELEMENT    pushaddr                                    (url, id?, pwd?)>
<!ELEMENT    req_id                                      (#PCDATA)>
```

```
<!ELEMENT    start_time                        (#PCDATA)>
<!ATTLIST    start_time
             utc_off CDATA                     "0000">
<!ELEMENT    stop_time                         (#PCDATA)>
<!ATTLIST    stop_time
             utc_off CDATA                     "0000">
<!ELEMENT    url                               (#PCDATA)>
<!ELEMENT    time_remaining                    (#PCDATA)>
```

4.2.3   Location Element Definitions

```
<!-- MLP_LOC -->
<!ELEMENT    pos                               (msid, (pd | poserr), gsm_net_param?)>
<!ELEMENT    eme_pos                           (msid, (pd | poserr), esrd?, esrk?)>
<!ELEMENT    trl_pos                           (msid, (pd | poserr))>
<!ATTLIST    trl_pos
             trl_trigger (PERIODIC |           #REQUIRED>
             MS_AVAIL)
<!ELEMENT    pd                                (time, shape, (alt, alt_acc?)?, speed?,
                                               direction?, lev_conf?)>
<!ELEMENT    poserr                            (result, add_info?, time)>
<!ELEMENT    add_info                          (#PCDATA)>
<!ELEMENT    result                            (#PCDATA)>
<!ATTLIST    result
             resid CDATA                       #REQUIRED>
<!ELEMENT    time                              (#PCDATA)>
<!ATTLIST    time
             utc off CDATA                     "0000">
<!ELEMENT    alt                               (#PCDATA)>
<!ELEMENT    alt_acc                           (#PCDATA)>
<!ELEMENT    direction                         (#PCDATA)>
<!ELEMENT    speed                             (#PCDATA)>
<!ELEMENT    lev_conf                          (#PCDATA)>
<!ELEMENT    geo_info                          (CoordinateReferenceSystem)>
<!ELEMENT    CoordinateReferenceSystem         (Identifier)>
<!ELEMENT    Identifier                        (code, codeSpace, edition)>
<!ELEMENT    code                              (#PCDATA)>
<!ELEMENT    codeSpace                         (#PCDATA)>
<!ELEMENT    edition                           (#PCDATA)>
```

Examples of geo_info encoding.

The encoding for WGS84 is:

```
<CoordinateReferenceSystem>
    <Identifier>
        <code>4326</code>
        <codeSpace>EPSG</codeSpace>
        <edition>6.1</edition>
    </Identifier>
</CoordinateReferenceSystem>
```

The encoding for the Transverse Mercator coordinate system based on the OSGB1936 is:

```
<CoordinateReferenceSystem>
    <Identifier>
        <code>27700</code>
        <codeSpace>EPSG</codeSpace>
        <edition>6.1</edition>
    </Identifier>
</CoordinateReferenceSystem>
```

Note that the GML V2.1.1 Implementation Specification is limited to use of only well-known CRSs, so this XML is currently abbreviated by a single attribute name and value:

```
srsName="http://www.opengis.net/gml/srs/epsg.xml#4326"
```

4.2.4   Shape Element Definitions

```
<!-- MLP_SHAPE -->
<!ELEMENT    shape                    (Point | LineString | Polygon | Box | CircularArea |
                                      CircularArcArea | EllipticalArea | GeometryCollection |
                                      MultiLineString | MultiPoint | MultiPolygon)>
<!ELEMENT    distanceUnit             (#PCDATA)>
<!ELEMENT    angularUnit              (#PCDATA)>
<!ELEMENT    angle                    (#PCDATA)>
<!ELEMENT    coord                    (X, Y?, Z?)>
<!ELEMENT    X                        (#PCDATA)>
<!ELEMENT    Y                        (#PCDATA)>
<!ELEMENT    Z                        (#PCDATA)>
<!ELEMENT    Point                    (coord)>
<!ATTLIST    Point
             gid ID                   #IMPLIED
             srsName CDATA            #IMPLIED>
<!ELEMENT    LineString               (coord, coord+)>
<!ATTLIST    LineString
             gid ID                   #IMPLIED
             srsName CDATA            #IMPLIED>
<!ELEMENT    Box                      (coord, coord)>
<!ATTLIST    Box
             gid ID                   #IMPLIED
             srsName CDATA            #IMPLIED>
<!ELEMENT    LinearRing               (coord, coord, coord, coord*)>
<!ATTLIST    LinearRing
             gid ID                   #IMPLIED
             srsName CDATA            #IMPLIED>
<!ELEMENT    Polygon                  (outerBoundaryIs, innerBoundaryIs*)>
<!ATTLIST    Polygon
             gid ID                   #IMPLIED
             srsName CDATA            #IMPLIED>
<!ELEMENT    outerBoundaryIs          (LinearRing)>
<!ELEMENT    innerBoundaryIs          (LinearRing)>
<!ELEMENT    CircularArcArea          (coord, inRadius, outRadius, startAngle, stopAngle,
                                      angularUnit?, distanceUnit?)>
<!ATTLIST    CircularArcArea
             gid ID                   #IMPLIED
             srsName CDATA            #IMPLIED>
<!ELEMENT    CircularArea             (coord, radius, distanceUnit?)>
<!ATTLIST    CircularArea
             gid ID                   #IMPLIED
             srsName CDATA            #IMPLIED>
<!ELEMENT    EllipticalArea           (coord, angle, semiMajor, semiMinor, angularUnit,
                                      distanceUnit?)>
<!ATTLIST    EllipticalArea
             gid ID                   #IMPLIED
             srsName CDATA            #IMPLIED>
<!ELEMENT    inRadius                 (#PCDATA)>
<!ELEMENT    outRadius                (#PCDATA)>
<!ELEMENT    radius                   (#PCDATA)>
<!ELEMENT    semiMajor                (#PCDATA)>
<!ELEMENT    semiMinor                (#PCDATA)>
<!ELEMENT    startAngle               (#PCDATA)>
<!ELEMENT    stopAngle                (#PCDATA)>
<!ELEMENT    GeometryCollection       (shape+)>
<!ATTLIST    GeometryCollection
             gid ID                   #IMPLIED
             srsName CDATA            #IMPLIED>
<!ELEMENT    MultiLineString          (LineString+)>
<!ATTLIST    MultiLineString
             gid ID                   #IMPLIED
             srsName CDATA            #IMPLIED>
<!ELEMENT    MultiPoint               (Point+)>
<!ATTLIST    MultiPoint
             gid ID                   #IMPLIED
             srsName CDATA            #IMPLIED>
<!ELEMENT    MultiPolygon             ((Polygon| Box | CircularArea | CircularArcArea |
                                      EllipticalArea)+)>
<!ATTLIST    MultiPolygon
             gid ID                   #IMPLIED
             srsName CDATA            #IMPLIED>
```

4.2.5   Quality of Position Element Definitions

```
<!-- MLP_QOP -->
<!ELEMENT    eqop                        (resp_req?, resp_timer?, (ll_acc | hor_acc)?,
                                         alt_acc?, max_loc_age?)>
<!ELEMENT    qop                         ((ll_acc | hor_acc)?, alt_acc?)>
<!ELEMENT    ll_acc                      (#PCDATA)>
<!ELEMENT    hor_acc                     (#PCDATA)>
<!ELEMENT    max_loc_age                 (#PCDATA)>
<!ELEMENT    resp_req                    EMPTY>
<!ATTLIST    resp_req
             type (NO_DELAY | LOW_DELAY | "DELAY_TOL">
             DELAY_TOL)
<!ELEMENT    resp_timer                  (#PCDATA)>
```

4.2.6   Network Parameters Element Definitions

```
<!-- MLP_GSM_NET -->
<!ELEMENT    gsm_net_param               (cgi?, neid?, nmr?, ta?, lmsi?)>
<!ELEMENT    cgi                         (mcc, mnc, lac, cellid)>
<!ELEMENT    neid                        (vmscid | vlrid | (vmscid, vlrid))>
<!ELEMENT    vmscid                      (cc?, ndc?, vmscno)>
<!ELEMENT    vlrid                       (cc?, ndc?, vlrno)>
<!ELEMENT    nmr                         (#PCDATA)>
<!ELEMENT    mcc                         (#PCDATA)>
<!ELEMENT    mnc                         (#PCDATA)>
<!ELEMENT    ndc                         (#PCDATA)>
<!ELEMENT    cc                          (#PCDATA)>
<!ELEMENT    vmscno                      (#PCDATA)>
<!ELEMENT    vlrno                       (#PCDATA)>
<!ELEMENT    lac                         (#PCDATA)>
<!ELEMENT    cellid                      (#PCDATA)>
<!ELEMENT    ta                          (#PCDATA)>
<!ELEMENT    lmsi                        (#PCDATA)>
```

Note: The above table corresponds to GSM specific network element
             identifiers and network parameters. This information may be
             considered operator sensitive

4.2.7   Context Element Definitions

```
<!-- MLP_CTXT -->
<!ELEMENT    client                      (id, pwd?, serviceid?, requestmode?)>
<!ELEMENT    sessionid                   (#PCDATA)>
<!ELEMENT    id                          (#PCDATA)>
<!ELEMENT    requestor                   (id, serviceid?)>
<!ELEMENT    pwd                         (#PCDATA)>
<!ELEMENT    serviceid                   (#PCDATA)>
<!ELEMENT    requestmode                 EMPTY>
<!ATTLIST    requestmode
             type (ACTIVE | PASSIVE)     "PASSIVE">
<!ELEMENT    subclient                   (id, pwd?, serviceid?)>
<!ATTLIST    subclient
             last_client (YES | NO)      "NO">
```

4.3   Service Layer Definitions

Each message may have two main parts, namely a context or header part and a body part. The body part consists of the request/answer and is described in sections 4.3.2 - 4.3.7. The context or header part consists of the information that identifies the client as defined in section 4.3.1.

4.3.1   Header Components

The **subclient** elements, if present, identify the ASPs, resellers and portals in the chain of service providers between the network and the end-user. The distinction between **client** and **subclient** elements is that the **client** element identifies the provider of the service that the Location Server has the initial relationship with, whereas the **subclient** elements identify the chain of other service providers up to the end-user. The final service provider in the chain is identified as such (last_client="YES"). On the other hand **requestor** is indicating the initiator of the location request, so in this context besides an ASP it could also be an MS subscriber who is asking the position of another target MS. The identity of the **requestor** may be an MSISDN or any other identifier identifying the initiator of the location request.

The **sessionid** element is used to represent the current session between the LCS Client and the Location Server. It may be used to replace the **id** and **pwd** elements, used in the context by the LCS Client to "login" to the Location Server, for the transactions that make up a session. For the first transaction of the session the LCS Client will need to "login" as usual. The Location Server may optionally return the **sessionid** in the response to this first transaction. If the Location Server does not return a **sessionid** the LCS Client will need to continue to "login" for subsequent transactions. The LCS Client can opt to ignore the **sessionid** if desired and continue to "login" for subsequent transactions.

The Location Server will decide the policy to be used to determine how the **sessionid** will be created and maintained. For example, the Location Server may determine the session as being just the transactions pertaining to a single service/MSID combination – this being restrictive and hence secure whilst still being useable, or the Location Server may allow the session to apply to a number of transactions between the Location Server and LCS Client. The Location Server may also allow the **sessionid** to be used for a particular period of time. The Location Server may also decide to return a different **sessionid** on each response, which the LCS Client will then use on the next transaction of the session.

The **sessionid** cannot be used instead of the **req_id** as this latter id refers to a set of reports that have been requested to be delivered from the Location Server to the LCS Client and do not form part of an existing LCS Client to Location Server connection. These reports are delivered by the Location Server "logging in" to the LCS Client for each one and the use of a **sessionid**, here would allow the security of the LCS Client to be breached.

4.3.1.1   Context DTD

```
<!-- MLP_HDR -->

<!ELEMENT      hdr                    ((client | sessionid | (client , sessionid)), subclient*,
                                      requestor?)>
<!ATTLIST      hdr
               ver CDATA              #FIXED "3.0.0">
```

Example 1: ASP as Initiator

```
<hdr ver="3.0.0">
  <client>
    <id>theasp</id>
    <pwd>thepwd</pwd>
    <serviceid>0005</serviceid>
    <requestmode type="PASSIVE"/>
  </client>
  <subclient last_client="YES">
    <id>thelastasp</id>
    <serviceid>0007</serviceid>
  </subclient>
  <requestor>
    <id>theoriginalasp</id>
    <serviceid>0003</serviceid>
  </requestor>
</hdr>
```

Example 2: MS as Initiator

```
<hdr ver="3.0.0">
  <client>
    <id>theasp</id>
    <pwd>thepwd</pwd>
    <serviceid>0005</serviceid>
    <requestmode type="ACTIVE"/>
  </client>
  <requestor>
    <id>461018765710</id>
  </requestor>
</hdr>
```

4.3.2   Standard Location Immediate Service

This is a standard service for requesting the location of one or more Mobile Subscribers. The service is used when a location response is required immediately (within a set time).

When a lot of positioning reports are requested, it may take an unacceptably long time to get the all responses from the network. If the Location Server supports it the LCS Client can define how to receive the location responses, either at a time with the response of the request, or individually using one or more connections initiated by the Location Server.

The extended service supports a number of different formats for describing the location of the mobile subscriber. It has also support for requesting a certain Quality of Service, Type of location and priority.

The service consists of the following messages:
• Standard Location Immediate Request
• Standard Location Immediate Answer
• Standard Location Immediate Report

The following message flow encapsulates this service:

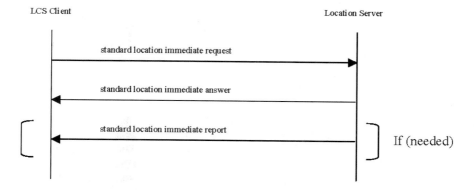

4.3.2.1   Standard Location Immediate Request DTD

```
<!-- MLP_SLIR -->

<!ENTITY      % extension.param        "">

<!ELEMENT     slir                     ((msids | (msid, codeword?, gsm_net_param)+),
                                       eqop?, geo_info?, loc_type?, prio?, pushaddr?
                                       %extension.param;)>
<!ATTLIST     slir
              ver CDATA                #FIXED "3.0.0"
              res_type (SYNC | ASYNC)  "SYNC">
```

Example

```
<slir ver="3.0.0" res_type="SYNC">
  <msids>
    <msid type="IPV4">93.10.0.250</msid>
    <msid_range>
      <start_msid>
        <msid>461018765710</msid>
      </start_msid>
      <stop_msid>
        <msid>461018765712</msid>
      </stop_msid>
    </msid_range>
    <msid type="ASID">441728922342</msid>
    <msid_range>
      <start_msid>
        <msid>461018765720</msid>
      </start_msid>
      <stop_msid>
        <msid>461018765728</msid>
      </stop_msid>
    </msid_range>
  </msids>
  <eqop>
    <resp_req type="LOW_DELAY" />
    <hor_acc>1000</hor_acc>
  </eqop>
  <geo_info>
    <CoordinateReferenceSystem>
      <Identifier>
        <code>4004</code>
        <codeSpace>EPSG</codeSpace>
        <edition>6.1</edition>
      </Identifier>
    </CoordinateReferenceSystem>
  </geo_info>
  <loc_type type="CURRENT_OR_LAST" />
  <prio type="HIGH" />
</slir>
```

4.3.2.2   Standard Location Immediate Answer DTD

```
<!-- MLP_SLIA -->

<!ENTITY     % extension.param          "">

<!ELEMENT    slia                       ((pos+ | req_id | (result, add_info?))
                                        %extension.param;)>

<!ATTLIST    slia
             ver CDATA                  #FIXED "3.0.0">
```

Example 1: Successful positioning of multiple subscribers

```
<slia ver="3.0.0" >
  <pos>
    <msid>461011334411</msid>
    <pd>
      <time utc_off="+0200">20020623134453</time>
      <shape>
              <CircularArea srsName="www.epsg.org#4004">
          <coord>
            <X>301628.312</X>
```

```
                        <Y>451533.431</Y>
                      </coord>
                      <radius>240</radius>
                </CircularArea>
              </shape>
            </pd>
          </pos>
          <pos>
            <msid>461018765710</msid>
            <pd>
              <time utc_off="+0300">20020623134454</time>
              <shape>
                <CircularArea srsName="www.epsg.org#4004">
                  <coord>
                    <X>301228.302</X>
                    <Y>865633.863</Y>
                  </coord>
                  <radius>570</radius>
                </CircularArea>
              </shape>
            </pd>
          </pos>
          <pos>
            <msid>461018765711</msid>
            <pd>
              <time utc_off="+0300">20020623110205</time>
              <shape>
                        <CircularArea srsName="www.epsg.org#4004">
                  <coord>
                    <X>781234.322</X>
                    <Y>762162.823</Y>
                  </coord>
                  <radius>15</radius>
                </CircularArea>
              </shape>
            </pd>
          </pos>
          <pos>
            <msid>461018765712</msid>
            <poserr>
              <result resid="10">QOP NOT ATTAINABLE</result>
              <time>20020623134454</time>
            </poserr>
          </pos>
        </slia>
```

Example 2: Service not supported

```
<slia ver="3.0.0" >
  <result resid="108">SERVICE NOT SUPPORTED</result>
  <add_info>'slir' is not supported by the location server</add_info>
</slia>
```

4.3.2.3   Standard Location Immediate Report DTD

```
<!-- MLP_SLIREP -->

<!ENTITY      % extension.param           "">

<!ELEMENT     slirep                      (req_id, pos+ %extension.param;)>
<!ATTLIST     slirep
              ver CDATA                   #FIXED "3.0.0">
```

Example

```
<slirep ver="3.0.0">
  <req_id>25267</req_id>
  <pos>
    <msid type="IPV6">10:A1:45::23:B7:89</msid>
    <pd>
      <time utc_off="+0300">20020813010423</time>
      <shape>
        <CircularArea srsName="www.epsg.org#4326">
          <coord>
            <X>35 03 28.244N</X>
            <Y>135 47 08.711E</Y>
          </coord>
          <radius>15</radius>
        </CircularArea>
      </shape>
    </pd>
  </pos>
</slirep>
```

4.3.3 Emergency Location Immediate Service

The emergency location immediate service is used to retrieve the position of a
mobile subscriber that is involved in an emergency call or have initiated an
emergency service in some other way.

The service consists of the following messages:
- Emergency Location Immediate Request
- Emergency Location Immediate Answer

The following message flow encapsulates this service:

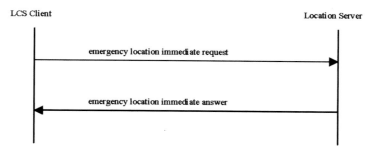

4.3.3.1 Emergency Location Immediate Request DTD

```
<!-- MLP_EME_LIR -->

<!ENTITY      % extension.param        "">

<!ELEMENT     eme_lir                   ((msids | (msid, gsm_net_param)+), qop?,
                                        geo_info?, loc_type? %extension.param;)>
<!ATTLIST     eme_lir
              ver CDATA                 #FIXED "3.0.0">
```

Example

```
<eme_lir ver="3.0.0">
  <msids>
    <msid type="EME_MSID">520002-51-431172-6-06</msid>
  </msids>
  <geo_info>
    <CoordinateReferenceSystem>
      <Identifier>
        <code>4004</code>
        <codeSpace>EPSG</codeSpace>
        <edition>6.1</edition>
      </Identifier>
    </CoordinateReferenceSystem>
  </geo_info>
  <loc_type type="CURRENT_OR_LAST" />
</eme_lir>
```

4.3.3.2   Emergency Location Immediate Answer DTD

```
<!-- MLP_EME_LIA -->

<!ENTITY      % extension.param      "">

<!ELEMENT     eme_lia                ((eme_pos+ | (result, add_info?)) %extension.param;)>
<!ATTLIST     eme_lia
              ver CDATA              #FIXED "3.0.0">
```

Example

```
<eme_lia ver="3.0.0">
  <eme_pos>
    <msid type="EME_MSID">520002-51-431172-6-06</msid>
    <pd>
      <time utc_off="+0300">20020623134453</time>
      <shape>
        <CircularArea srsName="www.epsg.org#4004">
          <coord>
            <X>N301628.312</X>
            <Y>-451533.431</Y>
          </coord>
          <radius>15</radius>
        </CircularArea>
      </shape>
    </pd>
    <esrk>7839298236</esrk>
  </eme_pos>
</eme_lia>
```

4.3.4   Standard Location Reporting Service

When a mobile subscriber wants an LCS client to receive the MS location a standard location report is generated. The LCS Client that the location report should be sent to is specified by MS or defined within the Location Server.

The service consists of the following message:
• Standard Location Report

The following message flow encapsulates this service:

4.3.4.1   Standard Location Report DTD

```
<!-- MLP_SLREP -->
<!ENTITY      % extension.param           "">

<!ELEMENT     slrep                       (pos+ %extension.param;)>
<!ATTLIST     slrep
              ver CDATA                   #FIXED "3.0.0">
```

Example

```
<slrep ver="3.0.0">
  <pos>
    <msid>461011678298</msid>
    <pd>
      <time>20020813010423</time>
      <shape>
        <CircularArea srsName="www.epsg.org#4004">
          <coord>
            <X>301628.312</X>
            <Y>451533.431</Y>
          </coord>
          <radius>15</radius>
        </CircularArea>
      </shape>
    </pd>
  </pos>
</slrep>
```

4.3.5   Emergency Location Reporting Service

If the wireless network spontaneously initiates a positioning when a user initiates or releases an emergency call, an emergency location report is generated. The application(s) that the emergency location report should be sent to is defined within the location server. Data as required geographical format and address to application is also defined within the location server.

The service consists of the following message:
- Emergency Location Report

The following message flow encapsulates this service:

4.3.5.1   Emergency Location Report DTD

```
<!-- MLP_EMEREP -->

<!ENTITY          % extension.param        "">

<!ELEMENT         emerep                   (eme_event %extension.param;)>
<!ATTLIST         emerep
                  ver CDATA                #FIXED "3.0.0">
```

Example

```
<emerep ver="3.0.0">
  <eme_event eme_trigger="EME_ORG">
    <eme_pos>
      <msid>461011678298</msid>
      <pd>
        <time utc_off="+0300">20020623010003</time>
        <shape>
          <CircularArea srsName="www.epsg.org#4004">
            <coord>
              <X>301628.312</X>
              <Y>451533.431</Y>
            </coord>
            <radius>15</radius>
          </CircularArea>
        </shape>
      </pd>
    </eme_pos>
  </eme_event>
</emerep>
```

4.3.6   **Triggered Location Reporting Service**

The triggered location reporting service is used when an application wants the position of the list of MS to be tracked. The triggers could be:

- The periodicity time defined by an interval

- An MS action, defined as the event "UE available" in 3GPP TS 23.271 rel. 4 [ref. 11].

The report will be triggered when one of the pre-defined MS's actions occurred or the time interval elapses. The service consists of the following messages:
- Triggered Location Reporting Request
- Triggered Location Reporting Answer
- Triggered Location Report
- Triggered Location Reporting Stop Request
- Triggered Location Reporting Stop Answer

The following message flow encapsulates this service:

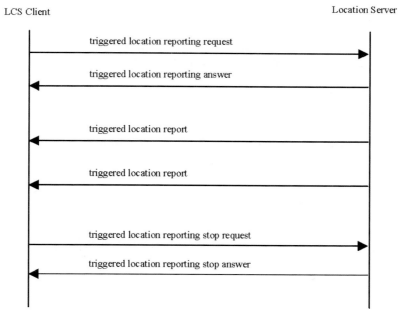

> **Note**: _It is the intention that Triggered services will support entering or leaving an area in future releases. An area may be defined as a specified geographical area, a city or locale, a country or a network. Other triggers that may be supported are specific events not yet defined, such a subscriber being in proximity to a friend in a FriendFinder application. Other events are FFS within 3GPP and are targeted for rel. 6._

4.3.6.1   Triggered Location Reporting Request DTD

```
<!-- MLP_TLRR -->

<!ENTITY        % extension.param        "">

<!ELEMENT       tlrr                     (msids, interval?, start_time?, stop_time?,
                                         tlrr_event?, qop?, geo_info?, pushaddr?,
                                         loc_type?, prio? %extension.param;)>

<!ATTLIST       tlrr
                ver CDATA                #FIXED "3.0.0">
```

The following rules apply to the use of 'start_time', 'stop_time', 'interval' and 'tlrr_event':

- TLRR with 'interval' is interpreted as a request for periodic location reports, and TLRR with 'tlrr_event' is interpreted as a request for a location report on the occurrence of a specific event. 'interval' and 'tlrr_event' can be combined. When neither 'interval' nor 'tlrr_event' is specified in TLRR, Location Server MUST reject the request with an error indication '106' to the client.

- If no START_TIME is specified reporting starts immediately.

- If no STOP_TIME is specified the reporting will occur until explicitly canceled with 'Triggered Location Stop Request' or a time out occurs (depending on system configuration). Timeout may be reported to the LCS client by 'time_remaining' in triggered location report.

- If START_TIME is 'older' than current time the Location Server MUST reject the request with an error indication '110' to the client.

- If STOP_TIME is 'older' than current time the Location Server MUST reject the request with an error indication '110' to the client.

- If STOP_TIME is earlier than START_TIME the implementation MUST reject the request with an error indication '110' to the client.

- If STOP_TIME is equal to START_TIME the Location Server MUST return a single location report to the client at the specified time. Any interval specified MUST be ignored.

Example 1: TLRR for periodic location reports during a specified period

```
<tlrr ver="3.0.0">
  <msids>
    <msid>461011678298</msid>
  </msids>
  <interval>00003000</interval>
  <start_time utc_off="+0300">20021003112700</start_time>
  <stop_time utc_off="+0300">20021003152700</stop_time>
  <qop>
    <hor_acc>100</hor_acc>
  </qop>
  <geo_info>
    <CoordinateReferenceSystem>
      <Identifier>
        <code>4326</code>
        <codeSpace>EPSG</codeSpace>
        <edition>6.1</edition>
      </Identifier>
    </CoordinateReferenceSystem>
  </geo_info>
  <pushaddr>
    <url>http://location.application.com</url>
  </pushaddr>
  <loc_type type="CURRENT" />
  <prio type="HIGH" />
</tlrr>
```

Example 2: TLRR for single location report at a specified time. 'stop_time' is specified equal to 'start_time'.

```
<tlrr ver="3.0.0">
  <msids>
    <msid>461011678298</msid>
  </msids>
  <interval>00003000</interval>
  <start_time utc_off="+0300">20021003112700</start_time>
  <stop_time utc_off="+0300">20021003112700</stop_time>
  <qop>
    <hor_acc>100</hor_acc>
  </qop>
  <geo_info>
    <CoordinateReferenceSystem>
      <Identifier>
        <code>4004</code>
        <codeSpace>EPSG</codeSpace>
        <edition>6.1</edition>
      </Identifier>
    </CoordinateReferenceSystem>
  </geo_info>
  <pushaddr>
    <url>http://location.application.com</url>
  </pushaddr>
  <loc_type type="CURRENT" />
  <prio type="HIGH" />
</tlrr>
```

Example 3: TLRR for a location report on the occurrence of a MS_AVAIL event after a specified time.

```
<tlrr ver="3.0.0">
  <msids>
    <msid>461011678298</msid>
  </msids>
  <start_time utc_off="+0300">20021003112700</start_time>
  <tlrr_event>
<ms_action type="MS_AVAIL"/>
</tlrr_event>
  <qop>
    <hor_acc>100</hor_acc>
  </qop>
  <geo_info>
    <CoordinateReferenceSystem>
      <Identifier>
        <code>4326</code>
        <codeSpace>EPSG</codeSpace>
        <edition>6.1</edition>
      </Identifier>
    </CoordinateReferenceSystem>
  </geo_info>
  <pushaddr>
    <url>http://location.application.com</url>
  </pushaddr>
  <loc_type type="CURRENT" />
  <prio type="HIGH" />
</tlrr>
```

4.3.6.2   Triggered Location Reporting Answer DTD

```
<!-- MLP_TLRA -->

<!ENTITY     % extension.param          "">

<!ELEMENT    tlra                       ((req_id | (result, add_info?))
                                        %extension.param;)>

<!ATTLIST    tlra
             ver CDATA                  #FIXED "3.0.0">
```

Example 1: TLRA if corresponding TLRR was successful

```
<tlra ver="3.0.0">
  <req_id>25293</req_id>
</tlra>
```

Example 2: TLRA if corresponding TLRR was in error

```
<tlra ver="3.0.0">
  <result resid="4">UNKNOWN SUBSCRIBER</result>
</tlra>
```

4.3.6.3   Triggered Location Report DTD

```
<!-- MLP_TLREP -->

<!ENTITY        % extension.param        "">

<!ELEMENT       tlrep                    (req_id, trl_pos+, time_remaining? %extension.param;)>
<!ATTLIST       tlrep
                ver CDATA                #FIXED "3.0.0">
```

Example

```
<tlrep ver="3.0.0">
  <req_id>25267</req_id>
  <trl_pos trl_trigger="PERIODIC">
    <msid>461011678298</msid>
    <pd>
      <time utc_off="+0300">20020813010423</time>
      <shape>
        <CircularArea srsName="www.epsg.org#4326">
          <coord>
            <X>35 35 24.139N</X>
            <Y>139 35 24.754E</Y>
          </coord>
          <radius>15</radius>
        </CircularArea>
      </shape>
    </pd>
  </trl_pos>
  <time_remaining>00010000</time_remaining>
</tlrep>
```

4.3.6.4   Triggered Location Reporting Stop Request DTD

```
<!-- MLP_TLRSR -->

<!ENTITY        % extension.param        "">

<!ELEMENT       tlrsr                    (req_id %extension.param;)>
<!ATTLIST       tlrsr
                ver CDATA                #FIXED "3.0.0">
```

Example

```
<tlrsr ver="3.0.0">
  <req_id>25293</req_id>
</tlrsr>
```

4.3.6.5   Triggered Location Reporting Stop Answer DTD

```
<!-- MLP_TLRSA -->

<!ENTITY      % extension.param        "">

<!ELEMENT     tlrsa                    ((req_id | (result, add_info?)) %extension.param;)>
<!ATTLIST     tlrsa
              ver CDATA                #FIXED "3.0.0">
```

Example

```
<tlrsa ver="3.0.0">
  <req_id>25293</req_id>

</tlrsa>
```

4.3.7   General Error Message Definition

When an LCS client attempts to invoke a service not defined in this specification, the location server will return a General Error Message. Sending a general error message (GEM) is no proper solution **by itself** because it can not always be expected that the client will understand this (MLP) response message, since - by sending an invalid request - the client showed that it may not be familiar with the proper set of MLP services. So additional error indications may be described in the different transport layer mappings.

```
<!-- MLP_GEM -->

<!ELEMENT   gem                              (result, add_info?)>
<!ATTLIST   gem
            ver CDATA                        #FIXED "3.0.0">

<!ENTITY   % mlp_loc.dtd                     SYSTEM "MLP_LOC_300.DTD">
%mlp_loc.dtd;
```

Example

```
<gem ver="3.0.0">
  <result resid="108">SERVICE NOT SUPPORTED</result>
  <add_info>
    The server does not support a service named 'skir'
  </add_info>
</gem>
```

5   Elements and attributes in DTD

5.1   add_info

Description:	
A text string containing additional information about a certain result.	
Type:	Element
Format:	Char string
Defined values:	-
Default value:	–
Example:	<add_info>EVENT</add_info>
Note: -	

5.2   alt

Description:	
The altitude of the MS in meters in respect of the ellipsoid which is used to be define the coordinates	
Type:	Element
Format:	Char String
Defined values:	[+|-] [0-9]+
Default value:	–
Example:	<alt>1200</alt>
Note: This element is present if altitude is possible to attain by the used positioning method.	

5.3   alt_acc

Description:	
Accuracy of altitude in meters	
Type:	Element
Format:	Char String
Defined values:	[0-9]+
Default value:	–
Example:	<alt_acc>200</alt_acc>
Note: -	

5.4   angle

Description:	
Specifies the angle (in angularUnit) of rotation of an ellipse measured clockwise from north	
Type:	Element
Format:	Char String
Defined values:	-
Default value:	–
Example:	`<angle>24.30</angle>`
Note: -	

5.5   angularUnit

Description:	
The angularUnit defines the unit for any angular value used in the shape description. For example the startAngle value in the CircularArcArea will be defined by this unit. If this unit is not included in a shape definition the angular unit defined in the CRS will be used.	
Type:	Element
Format:	Char String
Defined values:	Degrees
	Radians
Default value:	`Degrees`
Example:	`<angularUnit>Degrees</angularUnit>`
Note:.	

5.6   Box

Description:	
The Box element is used to encode extents	
Type:	Element
Format:	
Defined values:	-
Default value:	–

Example:	`<Box srsName="www.epsg.org#4004" gid="some_thing">`
	 `  <coord>`
	 `    <X>301628.312</X>`
	 `    <Y>451533.431</Y>`
	 `  </coord>`
	 `  <coord>`
	 `    <X>311628.312</X>`
	 `    <Y>461533.431</Y>`
	 `  </coord>`
	 `</Box>`

Note: -	

5.6.1   gid

Description:	
The gid is of XML attribute type ID and is used for references to elements within a single XML document. It allows XML technologies such as XPointer and xref to be used..	
Type:	attribute
Format:	Char String
Defined values:	
Default value:	
Example:	`<Box srsName="www.epsg.org#4004" gid="some_thing">`
Note:.This attribute is optional and is on all shape elements	

5.6.2   srsName

Description:	
srsName is a short hand method of defining the CoordinateReferenceSystem. It is a URI datatype that contains the codeSpace and code values, which are defined the same as in the CoordinateReferenceSystem..	
Type:	attribute
Format:	Char String
Defined values:	
Default value:	`www.epsg.org/#4326`
Example:	`<Box srsName="www.epsg.org/#4326">`
Note:. This attribute is optional and is on all shape elements. If the srsName is not included the WGS84 CRS is assumed.	

5.7   cc

Description:	
Specifies the country code.	
Type:	Element
Format:	Char String
Defined values:	1-3 digits e.g. 355 for Albania
Default value:	–
Example:	`<cc>355</cc>`
Note:	

5.8   cellid

Description:	
Identifies the Cell Identity	
Type:	Element
Format:	Char String
Defined values:	0-65535
Default value:	–
Example:	`<cellid>546</cellid>`
Note:	

5.9   CircularArcArea

Description:	
An arc is defined by a point of origin with one offset angle and one uncertainty angle plus one inner radius and one uncertainty radius.	
Type:	Element
Format:	Char String
Defined values:	-
Default value:	–

Example:	`<CircularArcArea srsName="www.epsg.org#4004" gid="some_thing">`
	`  <coord>`
	`    <X>301628.312</X>`
	`    <Y>451533.431</Y>`
	`  </coord>`
	`  <inRadius>280</inRadius>`
	`  <outRadius>360</outRadius>`
	`  <startAngle>5</startAngle>`
	`  <stopAngle>240</stopAngle>`
	`</CircularArcArea>`

Note:	

5.9.1   gid

See section 0.

5.9.2   srsName

see section 0.

5.10   CircularArea

Description:	
The set of points on the ellipsoid, which are at a distance from the point of origin less than or equal to "r".	
Type:	Element
Format:	Char String
Defined values:	-
Default value:	–
Example:	``` <CircularArea srsName="www.epsg.org#4004" gid="some_thing"> <coord> <X>301628.312</X> <Y>451533.431</Y> </coord> <radius>240</radius> </CircularArea> ```
Note:	

5.10.1   gid

See section 0.

5.10.2   srsName

See section 0.

5.11   code

Description:	
This is the unique identifier for the Coordinate ReferenceSystem as used by the authority cited in codeSpace	
Type:	Element
Format:	Char String
Defined values:	
Default value:	
Example:	`<code>4326</code>`
Note: .	

5.12   codeSpace

Description:	
The codeSpace is the authority which is responsible for the definition of the coordinate reference systems.	
Type:	Element
Format:	Char String
Defined values:	
Default value:	`www.epsg.org/...`
Example:	`<codeSpace>www.epsg.org</codeSpace>`
Note:.	

5.13   codeword

Description:	
Codeword is an access code defined per MS, used to protect location information of MS against unwanted location request. Only location requests with the correct codeword of a target MS are accepted.	
Type:	Element
Format:	Char String
Defined values:	-
Default value:	–
Example:	`<codeword>0918a7cb</codeword>`
Note: An error shall be returned if the number of codewords is not equal to the number of msid in an msid_range.	

5.14   distanceUnit

Description:	
The distanceUnit defines the linear unit for any distance used in the shape description. For example the radius value in the CircularArea will be defined by this unit. If this unit is not included in a shape definition the distance unit defined in the CRS will be used.	
Type:	Element
Format:	Char String
Defined values:	
Default value:	`meter`
Example:	`<distanceUnit>surveyfoot</distanceUnit>`
Note:.- values are defined by the CRS authority	

5.15 direction

Description:	
Specifies the direction, in degrees, that a positioned MS is moving in.	
Type:	Element
Format:	Char String
Defined values:	0-360
Default value:	–
Example:	`<direction>120</direction>`
Note: This element is present if direction is possible to attain by the used positioning method.	

5.16 edition

Description:	
The edition defines which version of the CRS database defined by the codeSpace authority is used..	
Type:	Element
Format:	Char String
Defined values:	
Default value:	
Example:	`<edition>6.0</edition>`
Note:.	

5.17   EllipticalArea

Description:	
A set of points on the ellipsoid, which fall within or on the boundary of an ellipse. This ellipse has a semi-major axis of length r1 oriented at angle A (0 to 180°) measured clockwise from north and a semi-minor axis of length r2.	
Type:	Element
Format:	Char String
Defined values:	-
Default value:	–

Example:	`<EllipticalArea srsName="www.epsg.org#4004" gid="some_thing">`
	  `<coord>`
	    `<X>301628.312</X>`
	    `<Y>451533.431</Y>`
	  `</coord>`
	  `<angle>240</angle>`
	  `<semiMajor>150</semiMajor>`
	  `<semiMinor>275</semiMinor>`
	  `<angularUnit>degrees</angularUnit>`
	`</EllipticilArea>`

Note:	

5.17.1   gid

See section 0.

5.17.2   srsName

see section 0.

5.18   eme_event

Description:	
Specifies the events that initiated the positioning of the MS at an emergency call.	
Type:	Element
Format:	-
Defined values:	-
Default value:	-
Example:	`<eme_event eme_trigger="EME_ORG">`
Note: -	

5.18.1 eme_trigger

Description:		
Specifies the trigger that initiated the positioning of the MS at an emergency call.		
Type:	Attribute	
Format:	Char string	
Defined values:	EME_ORG	An emergency service user originated an emergency call
	EME_REL	An emergency service user released an emergency call
Default value:	–	
Example:	`<eme_event eme_trigger="EME_ORG">`	
Note: -		

5.19 esrd

Description:		
This element specifies Emergency Services Routing Digits (ESRD).		
Type:	Element	
Format:	Char string	
Defined values:	–	
Default value:	–	
Example:	`<esrd>761287612582</esrd>`	
Note: -		

5.19.1 type

Description:		
Defines the origin of the ESRD		
Type:	Attribute	
Format:	Char string	
Defined values:	NA	Indicates that the ERSD is defined as the North American ESRD (NA-ERSD). NA-ESRD is a telephone number in the North American Numbering Plan that can be used to identify a North American emergency services provider and it's associated Location Services client. The NA-ESRD also identifies the base station, cell site or sector from which a North American emergency call originates
Default value:	NA	
Example:	`<esrd type="NA">12345678</ersd>`	
Note: Currently only NA is specified. It is expected that other origins will be specified in the future		

5.20   esrk

Description:	
This element specifies the Services Routing Key (ESRK).	
Type:	Element
Format:	Char string
Defined values:	–
Default value:	–
Example:	`<esrk>928273633343</esrk>`
Note: -	

5.20.1   type

Description:		
Defines the origin of the ESRK		
Type:	Attribute	
Format:	Char string	
Defined values:	NA	Indicates that the ERSK is defined as the North American ESRK (NA-ERSK). NA-ESRK is a telephone number in the North American Numbering Plan that is assigned to an emergency services call for the duration of the call. The NA-ESRK is used to identify (e.g. route to) both the emergency services provider and the switch currently serving the emergency caller. During the lifetime of an emergency services call, the NA-ESRK also identifies the calling subscriber.
Default value:	NA	
Example:	`<esrk type="NA">12345678</ersk>`	
Note: Currently only NA is specified. It is expected that other origins will be specified in the future		

5.21   GeometryCollection

Description:	
A collection of shapes.	
Type:	Element
Format:	Char String
Defined values:	-
Default value:	–

Example:	`<GeometryCollection srsName="www.epsg.org#4004" gid="some_thing">`
	  `<shape>`
	    `...`
	  `</shape>`
	`</GeometryCollection>`

Note:	

5.21.1   gid

See section 0.

5.21.2   srsName

See section 0.

5.22   hor_acc

Description:	
Requested horizontal accuracy in meters	
Type:	Element
Format:	Char String
Defined values:	[0-9]+
Default value:	–
Example:	`<hor_acc>200</hor_acc>`
Note: -	

5.23   id

Description:	
A string defining the name of a registered user performing a location request. In an answer the string represents the name of a location server.	
Type:	Element
Format:	Char string
Defined values:	–
Default value:	–
Example:	`<id>TheTruckCompany</id>`
Note: - This element is implementation specific.	

5.24   inRadius

Description:	
The inner radius is the geodesic distance (in distannceUnit) between center of the circle (that the arc is a part of) and arc closest to the center	
Type:	Element
Format:	Char String
Defined values:	[0-9]+
Default value:	–
Example:	`<inRadius>100</inRadius>`
Note: If the inner radius is 0 (zero) the area described represents a circle sector.	

5.25   interval

Description:			
Specifies the interval between two responses in case of a TLRR that indicates timer controlled, periodic responses.			
Type:	Element		
Format:	Char string The interval is expressed as `ddhhmmss` where:		
	String	**Description**	
	`dd`	Number of days between responses	
	`hh`	Number of hours between responses	
	`mm`	Number of minutes between responses	
	`ss`	Number of seconds between responses	
Defined values:	–		
Default value:	–		
Example:	`<interval>00010000</interval>`		
Note: -			

5.26 lac

Description:	
Identifies the Location Area Code	
Type:	Element
Format:	Char String
Defined values:	1-65535
Default value:	–
Example:	`<lac>234</lac>`
Note: - Location Area Code (LAC) which is a fixed length code (of 2 octets) identifying a location area within a GSM PLMN. This part of the location area identification can be coded using a full hexadecimal representation Except for the following reserved hexadecimal values: 0000, and FFFE	

5.27 lev_conf

Description:	
This parameter indicates the probability in percent that the MS is located in the position area that is returned.	
Type:	Element
Format:	Char String
Defined values:	0-100
Default value:	–
Example:	`<lev_conf>80</lev_conf>`
Note: -	

5.28   LinearRing

Description:	
A linear ring is a closed, simple piece-wise linear path which is defined by a list of coordinates that are assumed to be connected by straight line segments.	
Type:	Element
Format:	Char String
Defined values:	-
Default value:	–

Example:	```<LinearRing srsName="www.epsg.org#4004" gid="some_thing">```
	  `<coord>`
	    `<X>301628.312</X>`
	    `<Y>451533.431</Y>`
	  `</coord>`
	  `<coord>`
	    `<X>401628.312</X>`
	    `<Y>481533.431</Y>`
	  `</coord>`
	  `<coord>`
	    `<X>332628.312</X>`
	    `<Y>461533.431</Y>`
	  `</coord>`
	  `<coord>`
	    `<X>301628.312</X>`
	    `<Y>451533.431</Y>`
	  `</coord>`
	`</LinearRing>`

Note:	

5.28.1   gid

See section 0.

5.28.2   srsName

See section 0.

5.29   LineString

Description:	
A line string is a piece-wise linear path which is defined by a list of coordinates that are assumed to be connected by straight line segments.	
Type:	Element
Format:	Char String
Defined values:	-
Default value:	–
Example:	```<LineString srsName="www.epsg.org#4004" gid="some_thing">```

```
<LineString srsName="www.epsg.org#4004" gid="some_thing">
   <coord>
     <X>301628.312</X>
     <Y>451533.431</Y>
   </coord>
   <coord>
     <X>401628.312</X>
     <Y>481533.431</Y>
   </coord>
   <coord>
     <X>332628.312</X>
     <Y>461533.431</Y>
   </coord>
</LineString>
```

Note:	

5.29.1   gid

See section 0.

5.29.2   srsName

See section 0.

5.30   ll_acc

Description:	
Longitude and latitude accuracy in seconds.	
Type:	Element
Format:	Char String
Defined values:	-
Default value:	–
Example:	`<ll_acc>7.5</ll_acc>`
Note: -	

5.31   lmsi

Description:	
A local identity allocated by the VLR to a given subscriber for internal management of data in the VLR as defined in 29.002	
Type:	Element
Format:	Char String
Defined values:	-
Default value:	–
Example:	`<lmsi>2344512344565</lmsi>`
Note: - The LMSI consists of 4 octets	

5.32   loc_type

Description:	
Defines the type of location requested.	
Type:	Element
Format:	Void
Defined values:	-
Default value:	–
Example:	`<loc_type type="INITIAL" />`
Note: -	

5.32.1 type

Description:		
Defines the type of location requested		
Type:	Attribute	
Format:	Char string	
Defined values:	CURRENT	After a location attempt has successfully delivered a location estimate, the location estimate is known as the current location at that point in time.
	LAST	The current location estimate is generally stored in the network until replaced by a later location estimate and is known as the last known location. The last known location may be distinct from the initial location., i.e. more recent.
	CURRENT_OR_LAST	If a location attempt has successfully delivered, the current location is returned. Otherwise the last known location stored in the network is returned.
	INITIAL	In an originating emergency call, the location estimate at the commencement of the call set-up is known as the initial location.
Default value:	CURRENT	
Example:	`<loc_type type="INITIAL" />`	
Note: -		

5.33 max_loc_age

Description:	
This states the maximum allowable age in seconds of a location sent as a response to a location request. This location information may have been cached somewhere in the system from a previous location update.	
Type:	Element
Format:	Char string
Defined values:	Maximum number of seconds (must be >= 0)
Default value:	Implementation specific.
Example:	`<max_loc_age>3600</max_loc_age>`
Note: -	

5.34 mcc

Description:	
Specifies the mobile country code (MCC).	
Type:	Element
Format:	Char String
Defined values:	3 digits, e.g. 234 for the UK
Default value:	–
Example:	`<mcc>234</mcc>`
Note:	

5.35   mnc

Description:	
Specifies the mobile network code.	
Type:	Element
Format:	Char string
Defined values:	Up to 3 digits e.g. 15 for Vodafone
Default value:	-
Example:	`<mnc>215</mnc>`
Note: -	

5.36   ms_action

Description:	
Specifies the trigger that initiated the positioning of the MS.	
Type:	Element
Format:	Void
Defined values:	–
Default value:	–
Example:	`<ms_action type="MS_AVAIL" />`
Note: -	

5.36.1   type

Description:		
Specifies the trigger that initiated the positioning of the MS.		
Type:	Attribute	
Format:	Char string	
Defined values:	`MS_AVAIL`	The positioning is triggered by the MS available notification when the MS regains radio connection with the network if the connection was previously lost. For more information refer to 3GPP TS 23.271 rel. 4.
Default value:	–	
Example:	`<ms_action type="MS_AVAIL" />`	
Note: -		

5.37   msid

Description:	
This element represents an identifier of a mobile subscriber	
Type:	Element
Format:	Char string
Defined values:	–
Default value:	–
Example:	`<msid>460703057640</msid>`
Note: - When appropriate the MSID type format should confirm to the full standardised international representation of the MSID type, without any additional unspecified characters or spaces.	
As an example the GSM/3GPP identifiers should conform to the 3GPP CN TS 23.003, 'Numbering, Addressing and Identification' specification.	

5.37.1   type

Description:		
Type of identifier for the mobile subscriber		
Type:	Attribute	
Format:	Char string	
Defined values:	MSISDN	Mobile Station International ISDN Number
	IMSI	International Mobile Subscriber Identity
	IMEI	International Mobile station Equipment Identity
	MIN	Mobile Identification Number
	MDN	Mobile Directory Number
	EME_MSID	Emergency MSID
	ASID	Anonymous Subscriber Identity
	IPV4	Mobile station IP address (Version 4)
	OPE_ID	Operator specific Identity
	IPV6	Mobile station IP address (Version 6)
	SESSID	Session identifier relating to the user, which may be anonymous
Default value:	MSISDN	
Example:	`<msid type="IMSI">`	
Note: -		

5.37.2   enc

Description:		
Type of encoding for MSID identifier for the mobile subscriber		
Type:	Attribute	
Format:	Char string	
Defined values:	ASC	Normal textual format
	CRP	Encrypted format: In some countries the Network Operator (where is placed the Location Server) isn't allowed to send to a LCS client the private information of an MS like MSISDN.
		The Network Operator can send out to LCS client the Encrypted MSID, since only the Network Operator is the only entity able to decode this information, the LCS client will be never able to break the privacy of the MS.
Default value:	ASC	
Example:	`<msid type="IMSI" enc="ASC">`	
Note: -		

5.38   MultiLineString

Description:	
A collection of line strings.	
Type:	Element
Format:	Char String
Defined values:	-
Default value:	–

Example:	`<MultiLineString srsName="www.epsg.org#4004" gid="some_thing">`
	  `<LineString>`
	    `...`
	  `</LineString>`
	`</MultiLineString>`

Note:	

5.38.1   gid

See section 0.

5.38.2   srsName

see section 0.

5.39   MultiPoint

Description:	
A collection of points.	
Type:	Element
Format:	Char String
Defined values:	-
Default value:	–

Example:	`<MultiPoint srsName="www.epsg.org#4004" gid="some_thing">`
	`   <Point>`
	`      ...`
	`   </Point>`
	`</MultiPoint>`

Note:	

5.39.1   gid

See section 0.

5.39.2   srsName

See section 0.

5.40   MultiPolygons

Description:	
A collection of polygons.	
Type:	Element
Format:	Char String
Defined values:	-
Default value:	–

Example:	`<MultiPolygon srsName="www.epsg.org#4004" gid="some_thing">`
	`   <Polygon>`
	`      ...`
	`   </Polygon>`
	`</MultiPolygon>`

Note:	

5.40.1   gid

See section 0.

5.40.2   srsName

see section 0.

5.41   ndc

Description:	
Specifies the network destination code.	
Type:	Element
Format:	Char string
Defined values:	Up to 4 digits e.g. 7785 for Vodafone
Default value:	–
Example:	`<ndc>215</ndc>`
Note: -	

5.42   nmr

Description:	
Network specific measurement result for the target MS.	
Type:	Element
Format:	Char string
Defined values:	For examples see relevant standards documents.
Default value:	–
Example:	
Note: Measurement Results are encoded as 34 hexadecimal characters representing, 17 binary octets, in accordance with the Measurement Result information element described within GSM 04.18.	

5.43   radius

Description:	
The uncertainty radius is the radius (in distanceUnit) of the uncertainty; this is the geodesic distance between the arc and the position point.	
Type:	Element
Format:	Char String
Defined values:	[0-9]+
Default value:	-
Example:	`<radius>850</radius>`
Note: -	

5.44 startAngle

Description:	
The start angle is the angle (in angularUnit) between North and the first defined radius.	
Type:	Element
Format:	Char string
Defined values:	-
Default value:	–
Example:	`<off_angle>60</off_angle>`
Note: -	

5.45 stopAngle

Description:	
The stop angle is the angle (in angularUnit) between the first and second defined radius.	
Type:	Element
Format:	Char string
Defined values:	-
Default value:	–
Example:	`<incl_angle>180</incl_angle>`
Note: -	

5.46 Point

Description:	
A geographic 2D coordinate	
Type:	Element
Format:	Char String
Defined values:	-
Default value:	–

Example:	`<Point srsName="www.epsg.org#4004" gid="some_thing">`
	`  <coord>`
	`    <X>301628.312</X>`
	`    <Y>451533.431</Y>`
	`  </coord>`
	`</Point>`

Note:	

5.46.1   gid

See section 0.

5.46.2   srsName

See section 0.

5.47   Polygon

Description:	
A connected surface. Any pair of points in the polygon can be connected to one another by a path. The boundary of the Polygon is a set of LinearRings. We distinguish the outer (exterior) boundary and the inner (interior) boundaries; the LinearRings of the interior boundary cannot cross one another and cannot be contained within one another.	
Type:	Element
Format:	Char String
Defined values:	-
Default value:	–

Example:	`<Polygon srsName="www.epsg.org#4004" gid="some_thing">`
	`   <outerBoundaryIs>`
	`      ...`
	`   </outerBoundaryIs >`
	`</Polygon>`

Note:	

5.47.1   gid

See section 0.

5.47.2   srsName

See section 0.

5.48 prio

Description:	
Defines the priority of a location request	
Type:	Element
Format:	Void
Defined values:	-
Default value:	-
Example:	`<prio />`
Note: -	

5.48.1 type

Description:		
Defines the priority of a location request		
Type:	Attribute	
Format:	Char string	
Defined values:	NORMAL	The request is handled with normal priority
	HIGH	The request is handled with high priority
Default value:	NORMAL	
Example:	`<prio type="HIGH" />`	
Note: -		

5.49 pwd

Description:	
The password for the registered user performing a location request.	
In an answer the string represents the password for a location server.	
Type:	Element
Format:	Char string
Defined values:	
Default value:	–
Example:	`<pwd>the5pwd</pwd>`
Note: -	

5.50   outRadius

Description:	
The radius of a circle furthest away from the position in a CircularArcArea. The value is in the distanceUnit	
Type:	Element
Format:	Char String
Defined values:	[0-9]+
Default value:	–
Example:	`<outRadius>120</outRadius>`
Note: -	

5.51   req_id

Description:	
Unique identification of a request	
Type:	Element
Format:	Char string
Defined values:	–
Default value:	–
Example:	`<req_id>435.23.01</req_id>`
Note: -	

5.52   resp_req

Description:	
This attribute represents response time requirement.	
Type:	Element
Format:	Void
Defined values:	–
Default value:	-
Example:	`<resp_req type="NO_DELAY" />`
Note: -	

5.52.1   type

Description:		
This attribute represents response time requirement		
Type:	Attribute	
Format:	Char String	
Defined values:	NO_DELAY	No delay: The server should immediately return any location estimate that it currently has.
	LOW_DELAY	Low delay: Fulfilment of the response time requirement takes precedence over fulfilment of the accuracy requirement.
	DELAY_TOL	Delay tolerant: Fulfilment of the accuracy requirement takes precedence over fulfilment of the response time requirement.
Default value:	DELAY_TOL	
Example:	<resp_req />	
Note: - The interpretation of these parameters is defined in 3GPP specifications 22.071 and 29.002. The use of this element together with resp_timer is for further study.		

5.53   resp_timer

Description:	
Defines a timer for the response time within which the current location should be obtained and returned to the LCS Client.	
Type:	Element
Format:	Char String
Defined values:	Maximum number of seconds (must be >= 0)
Default value:	The default value is defined in the location server and will be implementation specific
Example:	<resp_timer>45</resp_timer>
Note: - The use of this element together with resp_reg is for further study	

5.54   result

Description:	
A text string indicating the result of the request or an individual positioning	
Type:	Element
Format:	Char string
Defined values:	See chapter 6.1
Default value:	–
Example:	<result resid=0>OK</result>
Note: -	

5.54.1   resid

Description:	
This attribute represents a numeric representation of a result message	
Type:	Attribute
Format:	Char String
Defined values:	[0-9]+ See chapter 6.1
Default value:	–
Example:	`<result resid=0>OK</result>`
Note: -	

5.55   semiMajor

Description:	
Specifies the length (in distanceUnit) of the semi-major axis of an ellipse.	
Type:	Element
Format:	Char String
Defined values:	[0-9]+
Default value:	–
Example:	`<semiMajor>560</semiMajor>`
Note: -	

5.56   semiMinor

Description:	
Specifies the length (in distanceUnit) of the semi-minor axis of an ellipse.	
Type:	Element
Format:	Char String
Defined values:	[0-9]+
Default value:	–
Example:	`<semiMinor>560</semiMinor>`
Note: -	

5.57 serviceid

Description:	
Specifies an id that is used by an entity to identify the service or application that is accessing the network.	
Type:	Element
Format:	Char String
Defined values:	-
Default value:	–
Example:	`<serviceid>0005</serviceid>`
Note:	

5.58 requestmode

Description:	
Defines the type of the service that has been requested by the ASP.	
Type:	Element
Format:	Void
Defined values:	-
Default value:	–
Example:	`<requestmode />`
Note:	

5.58.1 type

Description:		
Defines the type of the service that has been requested by the ASP		
Type:	Attribute	
Format:	Char string	
Defined values:	PASSIVE	The service is one that is not directly initiated by the user.
	ACTIVE	The service is one that the user is initiating personally.
	SESSION	The Service is one that has an established session with the user
Default value:	PASSIVE	
Example:	`<requestmode type="ACTIVE" />`	
Note: The default value is set to PASSIVE, as this is likely to be the one that is most restrictively defined by the user.		

5.59   session

Description:	
This element should be presented in location request when the LCS Client is making has an active session with the User Equipment, this will be either the number called by the UE or the APN on which the UE established the session.	
Type:	Element
Format:	Char String
Defined values:	-
Default value:	–
Example:	`<session>447073100177</session>`
Note: This information may be required for privacy validation of the location request by the VMSC, SGSN or MSC server	

5.59.1   type

Description:		
Defines the type of the session that is established between the User Equipment and LCS Client		
Type:	Attribute	
Format:	Char string	
Defined values:	APN	Access Point Name.
	DIAL	The number dialed by the user to access the LCS client.
Default value:	–	
Example:	`<session type="DIAL" />`	
Note:		

5.60   sessionid

Description:	
Specifies an id that can be used by an entity to support privacy mechanisms, a sessionid may replace the need to use an ID and PWD to use the location services.	
In a request when a client and sessionid are present together the session id may indicate the number dialed by the end user to access the service or the APN through which the original session was established that initiated the service. In an answer it indicates the sessionid that the entity can use on subsequent requests.	
In this casethe sessionid could be a generated alphanumeric string and can be time-limited.	
Type:	Element
Format:	Char String
Defined values:	-
Default value:	–
Example:	`<sessionid>34eg6.876.76h4</sessionid>`
Note:	

5.61   speed

Description:	
The speed of the MS in m/s.	
Type:	Element
Format:	Char String
Defined values:	[0-9]+
Default value:	–
Example:	`<speed>23</speed>`
Note: This element is present if speed is possible to attain by the used positioning method.	

5.62   start_time

Description:		
This element defines the absolute start time in a range of times.		
Type:	Element	
Format:	Char String	
	The time is expressed as yyyyMMddhhmmss where:	
	String	Description
	yyyy	Year
	MM	Month
	dd	Day
	hh	Hours
	mm	Minutes
	ss	Seconds
Defined values:	–	
Default value:	–	
Example:	`<start_time>20010630142810</start_time>`	
Note: -		

5.62.1   utc_off

Description:	
Specifies the UTC offset in hours and minutes. Positive values indicate time zones east of Greenwich.	
Type:	Attribute
Format:	Char string
Defined values:	[+|-] ?0000-1400
Default value:	–
Example:	`<start_time utc_off="+0200">20020813010423</start_time>`
Note: utc_off is specified as 'HHMM', where 'HH' can range between 0-14 and 'MM' between '0-59'. All other values shall result in error 105, 'Format error'.	

5.63   stop_time

Description:		
This element defines the absolute stop time in a range of times.		
Type:	Element	
Format:	Char String The time is expressed as yyyyMMddhhmmss where:	
	String	**Description**
	yyyy	Year
	MM	Month
	dd	Day
	hh	Hours
	mm	Minutes
	ss	Seconds
Defined values:	–	
Default value:	–	
Example:	`<stop_time>20020630142810</stop_time>`	
Note: -		

5.63.1   utc_off

See section 5.62.1

5.64   subclient

Description:	
Identifies the ASPs, resellers and portals in the chain of service providers between the network and the end-user	
Type:	Element
Format:	-
Defined values:	-
Default value:	-

Example:	`<subclient last_client="NO">`
	  `<id>TheASP</id>`
	   `<serviceid>0006</serviceid>`
	`</subclient>`

Note: -	

5.64.1   last_client

Description:		
Identifies whether the SUBCLIENT is the last one in the chain or not		
Type:	Attribute	
Format:	Char String	
Defined values:	YES	This is the last client – the one that the end-user is actually communicating with
	NO	This is not the last client
Default value:	NO	
Example:	`<subclient last_client="YES">`	
Note: -		

5.65   ta

Description:	
This Radio Access Network element that can arguably be used to offer enhanced positioning. (Timing Advance)	
Type:	Element
Format:	Char string
Defined values:	0-63
Default value:	0
Example:	`<ta>3</ta>`
Note: Further Information regarding this element can be found in the relevant GSM Specifications	

5.66   time

Description:	
In a location answer this element indicates the time when the positioning was performed.	
Type:	Element
Format:	Char String The time is expressed as yyyyMMddhhmmss where:

String	Description
yyyy	Year
MM	Month
dd	Day
hh	Hours
mm	Minutes
ss	Seconds

Defined values:	–
Default value:	–
Example:	`<time>20010630142810</time>`
Note: -	

5.66.1   utc_off

See section 5.62.1

5.67   time_remaining

Description:	
Defines the time remaining until the location server terminates the current triggered location service. The time for which the service is valid is either specified by the client using start time and stop time, or is a network operator specific default value where no s stop time is defined or where the stop time exceeds the allowed value by the location server involved.	
Type:	Element
Format:	Char String The time is expressed as ddhhmmss where:

String	Description
dd	Day
hh	Hours
mm	Minutes
ss	Seconds

Defined values:	–
Default value:	The default value is defined in the location server
Example:	`<time_remaining>00010000</time_remaining>`
Note: -	

5.68   trl_pos

Description:	
Specifies the position of the MS at a triggered location report.	
Type:	Element
Format:	-
Defined values:	-
Default value:	–

Example:	`<tlr_pos trl_trigger="PERIODIC">`
	    `<msid>4711</msid>`
	    `<poserr>`
	        `<result resid=1>SYSTEM FAILURE</result>`
	        `<time utc_off="0100">20011127104532</time>`
	    `</poserr>`
	`</trl_pos>`

Note: -	

5.68.1   trl_trigger

Description:		
Specifies the trigger that initiated the positioning of the MS at a triggered location report.		
Type:	Attribute	
Format:	Char string	
Defined values:	`PERIODIC`	The positioning is triggered when the periodical timer expired
	`MS_AVAIL`	The positioning is triggered by the MS presence notification
Default value:	–	
Example:	`<tlr_pos trl_trigger="PERIODIC">`	
Note: -		

5.69   url

Description:	
Specifies the location to which a response to a TLRR or an asynchronous SLIR should be sent to	
Type:	Element
Format:	Char string
Defined values:	–
Default value:	–
Example:	`<url>http://location.client.com/Response/</url>`
Note: - URL is part of pushaddr element which may also contain id and pwd. These elements are used by the LCS Client to inform the Location Server what credentials to use when 'pushing' a location report to the LCS Client in case of an asynchronous service.	

5.70   vlrno

Description:	
Uniquely specifies a VLR within a network.	
Type:	Element
Format:	Char String
Defined values:	In GSM this is the Global Title address. The Global Title is in the same format as an E.164 number.
Default value:	–
Example:	<vlrno>1541154871</vlrno>
Note:	

5.71   vmscno

Description:	
Uniquely specifies a VMSC within a network.	
Type:	Element
Format:	Char String
Defined values:	In GSM this is the Global Title address. The Global Title is in the same format as an E.164 number.
Default value:	–
Example:	<vmscno>1541154871</vmscno>
Note:	

5.72   X

Description:	
The first ordinate in a coordinate system	
Type:	Element
Format:	Char string
Defined values:	–
Default value:	–
Example:	<X>33498.23</X>
Note: -	

5.73 Y

Description:	
Second ordinate in a coordinate.system. This is optional if it is a linear coordinate system.	
Type:	Element
Format:	Char string
Defined values:	–
Default value:	–
Example:	<Y>33498.23</Y>
Note: -	

5.74 Z

Description:	
third ordinate in a coordinate.system. This is optional if it is a 2D coordinate system.	
Type:	Element
Format:	Char string
Defined values:	–
Default value:	–
Example:	<Z>33498.23</Z>
Note: -	

5.75   Service attributes

5.75.1   res_type

Description:		
Defines a response type at the Standard Location Immediate Service. This attribute applies to the Standard Immediate Location Request message.		
Type:	Attribute	
Format:	Char string	
Defined values:	SYNC	An LCS Client requests to receive the location response in one response
	ASYNC	An LCS Client request to receive the location responses one by one using some connections initiated by the location Server
Default value:	SYNC	
Example:	<slir ver="3.0.0" res_type="SYNC">	
Note: -		

5.75.2   ver

Description:	
Defines the version of the location protocol. This attribute is valid for ALL messages	
Type:	Element
Format:	Char string
Defined values:	[0-9].[0-9].[0-9]
Default value:	–
Example:	<slia ver="3.0.0">
Note: -	

6 Result codes

6.1 Result codes

This table defines the result codes that indicate the result of the request or individual positioning. The error codes are divided in ranges:

0	-	99	Location server specific errors
100	-	199	Request specific errors
200	-	299	Network specific errors
300	-	499	Reserved for future use
500	-	599	Vendor specific errors

Note: For privacy reasons it might be needed to not report certain specific errors. In this case it is up to the implementation or configuration of the location server which errors will be reported.

Resid	Slogan	Description
0	OK	No error occurred while processing the request.
1	SYSTEM FAILURE	The request can not be handled because of a general problem in the location server or the underlying network.
2	UNSPECIFIED ERROR	An unspecified error used in case none of the other errors applies. This can also be used in case privacy issues prevent certain errors from being presented
3	UNAUTHORIZED APPLICATION	The requesting location-based application is not allowed to access the location server or a wrong password has been supplied.
4	UNKNOWN SUBSCRIBER	Unknown subscriber. The user is unknown, i.e. no such subscription exists.
5	ABSENT SUBSCRIBER	Absent subscriber. The user is currently not reachable.
6	POSITION METHOD FAILURE	Position method failure. The location service failed to obtain the user's position.
101	CONGESTION IN LOCATION SERVER	The request can not be handled due to congestion in the location server.
102	CONGESTION IN MOBILE NETWORK	The request can not be handled due to congestion in the mobile network.
103	UNSUPPORTED VERSION	The Location server does not support the indicated protocol version.
104	TOO MANY POSITION ITEMS	Too many position items have been specified in the request.
105	FORMAT ERROR	A protocol element in the request has invalid format. The invalid element is indicated in ADD_INFO.
106	SYNTAX ERROR	The position request has invalid syntax. Details may be indicated in ADD_INFO.
107	PROTOCOL ELEMENT NOT SUPPORTED	A protocol element specified in the position request is not supported by the Location Server. The element is indicated in ADD_INFO.
108	SERVICE NOT SUPPORTED	The requested service is not supported in the Location Server. The service is indicated in ADD_INFO.
109	PROTOCOL ELEMENT ATTRIBUTE NOT SUPPORTED	A protocol element attribute is not supported in the Location Server. The attribute is indicated in ADD_INFO.
110	INVALID PROTOCOL ELEMENT VALUE	A protocol element in the request has an invalid value. The element is indicated in ADD_INFO.
111	INVALID PROTOCOL ELEMENT ATTRIBUTE	A protocol element attribute in the request has a wrong value. The

	VALUE	element is indicated in ADD_INFO.
112	PROTOCOL ELEMENT VALUE NOT SUPPORTED	A specific value of a protocol element is not supported in the Location Server. The element and value are indicated in ADD_INFO.
113	PROTOCOL ELEMENT ATTRIBUTE VALUE NOT SUPPORTED	A specific value of a protocol element attribute is not supported in the Location Server. The attribute and value are indicated in ADD_INFO.
201	QOP NOT ATTAINABLE	The requested QoP cannot be provided.
202	POSITIONING NOT ALLOWED	The subscriber does not allow the application to position him/her for whatever reason (privacy settings in location server, LCS privacy class).
204	DISALLOWED BY LOCAL REGULATIONS	The location request is disallowed by local regulatory requirements.
207	MISCONFIGURATION OF LOCATION SERVER	The location server is not completely configured to be able to calculate a position.
500 – 599		Vendor specific errors

7   References

References are either specific (identified by date of publication, edition
number, version number, etc.) or non-specific:

- For a specific reference, subsequent revisions do not apply.

- For a non-specific reference, the latest version applies.

7.1   References (Normative)

[1]   Hypertext Transfer Protocol –HTTP/1.1
           RFC 2616, June 1999
           Available at http://www.ietf.org

[2]   The TLS Protocol Version 1.0
           RFC 2246, January 1999
           Available at http://www.ietf.org

[3]   Extensible Markup Language (XML) 1.0
           W3C Recommendation: REC-xml-20001006
           Available at http://www.w3c.org

[4]   Internet Assigned Numbers Authority (IANA)
           http://www.iana.org/

[5]   US-ASCII. Coded Character Set - 7-Bit American Standard Code for
           Information Interchange. Standard ANSI X3.4-1986, ANSI, 1986.

7.2   References (Informative)

[6]   GSM 02.71: "Digital cellular telecommunications system (Phase 2+);
           Location Services (LCS); Service description; Stage 1".

[7]   GSM 03.71: "Digital cellular telecommunications system (Phase 2+);
           Location Services (LCS); Functional description; Stage 2".

[8]   GSM 09.02: "Digital cellular telecommunications system (Phase 2+);
           Mobile Application Part (MAP) specification".

[9]   3GPP TS 22.071: "Location Services (LCS); Service description,
           Stage 1".

[10]   3GPP TS 23.171: "Functional stage 2 description of location services
           in UMTS"

[11]   3GPP TS 23.271: "Functional stage 2 description of LCS"

[12]   3GPP TS 23.032: " Universal Geographical Area Description (GAD)"

[13]   3GPP TS 29.002: "Digital cellular telecommunications system
        (Phase 2+); Mobile Application Part (MAP) specification".

[14]   3GPP TS 29.198-6 "Open Service Access (OSA) Application
        Programming Interface (API); Part 6: Mobility"

[15]   Parlay API 2.1 Mobility Interfaces v1.1.1.
        Available on the Parlay web-site at http://www.parlay.org

[16]   ITU-T E.164: "The international public telecommunication numbering
        plan

[17]   TR-45 J-STD-036 "Enhanced Wireless 9-1-1 Phase 2 Document"

[18]   IS-41D: " Cellular Radiotelecommunications Intersystem Operations",
        June 1997

[19]   OpenGIS© Consortium Abstract Specification Topic 2: 01-063R2 at
        the public OGC document repository
        http://www.opengis.org/techno/abstract/02-102.pdf.

[20]   OpenGIS© Consortium Recommendation Paper 01-014r5:
        Recommended Definition Data for Coordinate Reference Systems
        and Coordinate Transformations available at
        http://www.opengis.org/techno/discussions/01-014r5.pdf

[21]   OpenGIS© Consortium Impementation Specification: Geography
        Markup Language V 2.0 available at http://www.opengis.net/gml/01-
        029/GML2.html

[22]   OpenGIS© Consortium Abstract Specification Topic 1 Feature
        Geometry : 010101 at the public document repository
        http://www.opengis.org/techno/abstract/01-101.pdf.

8   Appendix A (informative): Adaptation to 3GPP LCS

8.1   Version mapping between 3GPP TS23.271 and this specification

The following table shows the version number of this specification (LIF TS101) fully conforming to a certain version of 3GPP TS23.271, i.e. the version of this specification for the correct reference in a certain version of the 3GPP specification.

3GPP TS23.271 version number	Conforming version number of LIF TS101
Release 5	Version 3

Note: In case there are versions not appearing in this table, it should be interpreted that such update did not affect the other specification. That is, the version number not appearing in the table should apply to the conformance mapping for the closest smaller version number in the table.

8.2   The terminology mapping table with 3GPP LCS Specifications

The following is a list of the terms in MLP used differently from the ones defined for 3GPP:

Term		Notes
MLP	3GPP	
Location Server	LCS Server	
MS (Mobile Station)	UE	
MSID (Mobile Station Identifier)	Identification of the target UE	
MPC (Mobile Positioning Centre)		There is no term applicable to 3GPP.

8.3   The corresponding terms used for the location procedures in 3GPP LCS Definition

The following is a list of terms defined in MLP corresponding to the 3GPP LCS definition in TS23.271 for the location procedures.

Location procedures defined in 3GPP(23.271)		Services defined in MLP
Circuit Switched Mobile Terminating Location Request CS-MT-LR	LCS Service Request	Standard Location Immediate Request
	LCS Service Response	Standard Location Immediate Answer
CS-MT-LR without HLR Query - applicable to North America Emergency Calls only	LCS Service Request	Emergency Location Immediate Request
	LCS Service Response	Emergency Location Immediate Answer
Packet Switched Mobile Terminating Location Request PS-MT-LR	LCS Service Request	Standard Location Immediate Request
	LCS Service Response	Standard Location Immediate Answer
Network Induced Location Request NI-LR	Location Information	Emergency Location Report
Packet Switched Network Induced Location Request PS-NI-LR	Location Information	Emergency Location Report
Mobile Terminating Deferred Location Request	LCS Service Request	Triggered Location Reporting Request
	LCS Service Response (Provide Subscriber Location ack)	Triggered Location Reporting Answer
	LCS Service Response (Subscriber Location Report)	Triggered Location Report
Combined Periodical/Deferred Mobile Terminating Location Request	LCS Service Request	Triggered Location Reporting Request
	LCS Service Response (Provide Subscriber Location ack)	Triggered Location Reporting Answer
	LCS Service Response (Subscriber Location Report)	Triggered Location Report
Cancellation of a Deferred Location Request	LCS Cancel Service Request	Triggered Location Reporting Stop Request
	LCS Cancel Service Response	Triggered Location Reporting Stop Answer
Mobile Originating Location Request, Circuit Switched CS-MO-LR	Location Information	Standard Location Report
Mobile Originating Location Request, Packet Switched PS-MO-LR	Location Information	Standard Location Report

8.4   Error Mapping (informative)

The following list provides a mapping between the errors defined for LCS in MAP (see [13]) and MLP (see section 6)

MAP error	MLP resid
Unknown subscriber	4
Unidentified Subscriber	4
Absent Subscriber	5
System failure	1
Facility Not Supported	6
Unexpected Data Value	1
Data missing	1
Unauthorised LCS Client with detailed reason	3
Position method failure with detailed reason.	6
Illegal Subscriber	2
Illegal Equipment	2
Unauthorized requesting network	2

9 Appendix B - HTTP Mapping

This section describes how to use MLP over the HTTP transport mechanism using "HTTP/1.1".

HTTP is a request/response protocol involving a server and a client. In the context of MLP, the client is referred to as the LCS Client and the server is the Location Server (GMLC/MPC). For more information about HTTP, refer to http://www.w3.org and ref [1].

The Location Server should provide two socket ports for operation, one for encryption with SSL/TLS and one without. The reason for having one insecure port is that encryption can consume resources, and if the client is in a secure domain there might not be a need for encryption. Applications residing in an insecure domain, i.e. on the Internet, may use the secure port to ensure the security and privacy of the location information.

For further information about SSL/TLS see ref [2].

Four port numbers have been selected and proposed as standard ports for location servers implementing MLP. These ports are registered with IANA (Internet Assigned Numbers Authority, see ref [4]). The four port numbers are:

- lif-mlp   9210/tcp   LIF Mobile Locn Protocol
- lif-mlp   9210/udp   LIF Mobile Locn Protocol
- lif-mlp-s   9211/tcp   LIF Mobile Locn Secure
- lif-mlp-s   9211/udp   LIF Mobile Locn Secure

A Location Server can choose to introduce any other socket based or HTTP transparent technology for secure transfers. Any such technology should be provided over a different port than the four mentioned above.

9.1 Location Services using HTTP

An LCS Client requests a Location Service by issuing an HTTP POST request towards the Location Server. For more information about HTTP POST, see ref. [1]. The request line syntax is shown below.

Request-line: POST *SP* host *SP* HTTP/1.1 *CRLF*

The request must include the entity-header Content-length field as part of the request. The message body of the request should include the XML formatted request and should have the length specified by the LCS Client in the Content-length field.

If the request is a deferred request (triggered or periodic) the result is delivered to the client through an HTTP POST operation issued by the Location Server. This implies that the client must be able to receive HTTP POST requests and be able to give a valid response.

All Location Services are invoked by sending a request using HTTP POST to a certain URI. An example of an URI is shown below.

http://host:port/LocationQueryService/

The response to the invocation of a Location Service is returned using an HTTP response.

If the LCS client requests standard location of asynchronous mode, triggered or periodic reporting of location, the Location Server will return the answer by performing an HTTP POST operation towards the client. The client must specify the URI that the answer should be posted to. This is done in the service request or by having it in the LCS client profile that can be stored in the Location Server.

The answer will be included in the message body and the Content-length entity will be set to the length of the answer.

When an LCS client attempts to invoke a service request that is not defined in this specification, the Location Server shall return a General Error Message (GEM) in a HTTP '404' error reponse:

Status-Line: HTTP/1.1 *SP* 404 *SP* Not Found *CRLF*

9.2   Request and Response Encapsulation

A request and a response consist of a header part and a body part so to be able to make a location request with a single XML document the header and the body are encapsulated in the same service initiation DTD. The context header holds the authentication and authorization data pertinent to a particular location request. The body part is described in the sections 4.3.2 - 4.3.6.

9.2.1   Service Initiation DTD

```
<!-- MLP_SVC_INIT -->

<!ENTITY    % extension.message      "">

<!ELEMENT   svc_init                 (hdr, (slir | eme_lir | tlrr | tlrsr
                                     %extension.message;)))>
<!ATTLIST   svc_init
            ver CDATA                #FIXED "3.0.0">

<!ENTITY    % mlp_ctxt.dtd           SYSTEM "MLP_CTXT_300.DTD">
%mlp_ctxt.dtd;
<!ENTITY    % mlp_id.dtd             SYSTEM "MLP_ID_300.DTD">
%mlp_id.dtd;
<!ENTITY    % mlp_func.dtd           SYSTEM "MLP_FUNC_300.DTD">
%mlp_func.dtd;
<!ENTITY    % mlp_qop.dtd            SYSTEM "MLP_QOP_300.DTD">
%mlp_qop.dtd;
<!ENTITY    % mlp_loc.dtd            SYSTEM "MLP_LOC_300.DTD">
%mlp_loc.dtd;
<!ENTITY    % mlp_shape.dtd          SYSTEM "MLP_SHAPE_300.DTD">
%mlp_shape.dtd;
<!ENTITY    % mlp_gsm_net_param.dtd  SYSTEM "MLP_GSM_NET_300.DTD">
%mlp_gsm_net_param.dtd;


<!ENTITY    % mlp_hdr.dtd            SYSTEM "MLP_HDR_300.DTD">
%mlp_hdr.dtd;
<!ENTITY    % mlp_slir.dtd           SYSTEM "MLP_SLIR_300.DTD">
%mlp_slir.dtd;
<!ENTITY    % mlp_eme_lir.dtd        SYSTEM "MLP_EME_LIR_300.DTD">
%mlp_eme_lir.dtd;
<!ENTITY    % mlp_tlrr.dtd           SYSTEM "MLP_TLRR_300.DTD">
%mlp_tlrr.dtd;
<!ENTITY    % mlp_tlrsr.dtd          SYSTEM "MLP_TLRSR_300.DTD">
%mlp_tlrsr.dtd;
```

Example

```
<?xml version="1.0" ?>
<!DOCTYPE svc_init SYSTEM "MLP_SVC_INIT_300.DTD">
<svc_init ver="3.0.0">
  <hdr ver="3.0.0">
    ...
  </hdr>
  <slir ver="3.0.0">>
    ...
  </slir
</svc_init>
```

9.2.2 Service Result DTD

```
<!-- MLP_SVC_RESULT -->

<!ENTITY    % extension.message        "">
<!ELEMENT   svc_result                 (hdr?, (slia | slirep | slrep | eme_lia | emerep |
                                        tlra | tlrep | tlrsa %extension.message;)))>

<!ATTLIST   svc_result
            ver CDATA                  #FIXED "3.0.0">

<!ENTITY    % mlp_ctxt.dtd             SYSTEM "MLP_CTXT_300.DTD">
%mlp_ctxt.dtd;
<!ENTITY    % mlp_id.dtd               SYSTEM "MLP_ID_300.DTD">
%mlp_id.dtd;
<!ENTITY    % mlp_func.dtd             SYSTEM "MLP_FUNC_300.DTD">
%mlp_func.dtd;
<!ENTITY    % mlp_qop.dtd              SYSTEM "MLP_QOP_300.DTD">
%mlp_qop.dtd;
<!ENTITY    % mlp_loc.dtd              SYSTEM "MLP_LOC_300.DTD">
%mlp_loc.dtd;
<!ENTITY    % mlp_shape.dtd            SYSTEM "MLP_SHAPE_300.DTD">
%mlp_shape.dtd;
<!ENTITY    % mlp_gsm_net_param.dtd    SYSTEM "MLP_GSM_NET_300.DTD">
%mlp_gsm_net_param.dtd;


<!ENTITY    % mlp_hdr.dtd              SYSTEM "MLP_HDR_300.DTD">
%mlp_hdr.dtd;
<!ENTITY    % mlp_slia.dtd             SYSTEM "MLP_SLIA_300.DTD">
%mlp_slia.dtd;
<!ENTITY    % mlp_slirep.dtd           SYSTEM "MLP_SLIREP_300.DTD">
%mlp_slirep.dtd;
<!ENTITY    % mlp_slrep.dtd            SYSTEM "MLP_SLREP_300.DTD">
%mlp_slrep.dtd;
<!ENTITY    % mlp_eme_lia.dtd          SYSTEM "MLP_EME_LIA_300.DTD">
%mlp_eme_lia.dtd;
<!ENTITY    % mlp_emerep.dtd           SYSTEM "MLP_EMEREP_300.DTD">
%mlp_emerep.dtd;
<!ENTITY    % mlp_tlra.dtd             SYSTEM "MLP_TLRA_300.DTD">
%mlp_tlra.dtd;
<!ENTITY    % mlp_tlrep.dtd            SYSTEM "MLP_TLREP_300.DTD">
%mlp_tlrep.dtd;
<!ENTITY    % mlp_tlrsa.dtd            SYSTEM "MLP_TLRSA_300.DTD">
%mlp_tlrsa.dtd;
```

Example

```
<?xml version="1.0" ?>
<!DOCTYPE svc_init SYSTEM "MLP_SVC_RESULT_300.DTD">
<svc_result ver="3.0.0">
  <slia ver="3.0.0">
    ...
  </slia>
</svc_result>
```

9.2.3   **Message Sequence Diagram**

The following HTTP sequence is used for all the defined service
requests/responses in MLP.

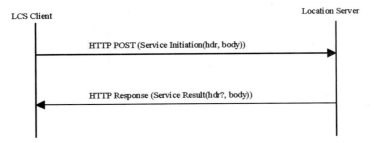

The following HTTP sequence diagram is used for all defined reports in MLP.

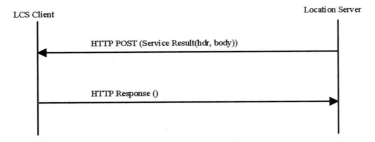

The following HTTP sequence diagram is used in the case of a General Error
Message.

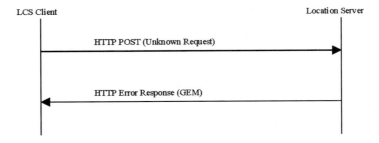

10   Appendix C: Geographic Information

10.1   Coordinate Reference systems (Informative)

The study of determining the relative positions on or close to the surface of the earth is a complex science, referred to as geodesy. A complete definition of Coordinate Reference systems is not within the scope of this standard. This section includes a brief overview of the subject. For more details see the OpenGIS© Consortium Abstract Specification Topic 2 [19.

10.1.1   The Geoid, ellipsoids and datums

The Geoid is a physically realizable surface defined by the set of points with equal gravity potential approximately at the Mean Sea Level. While this surface is measurable it is not easy to define mathematically. In order to use known mathematics, the Geoid is approximated by an ellipsoid (spheroid).

There are many ellipsoids, each defined to best approximate some part of the Geoid. These ellipsoids are defined by an ellipse that is rotated about the major axis. There are many methods for defining an ellipse, the most common used in Geodesy the length of the semi-major axis and the flattening. This defines a mathematical ellipsoid for calculations. it does not provide enough information to locate the ellipsoid with respect to the Geoid or other ellipsoids. To locate the ellipsoid in space a datum is defined. Some of the common ellipsoids are WGS84, Bessel1841, Clark 1866.

A datum is the ellipsoid with it's position in space. The position is defined by the origin and orientation of the ellipsoid with respect to the Geoid. Different datums locate latitude, longitude at different positions in space. For example ellipsoids Samboja, CH1903 and Stockholm are each based on Bessel1841, the National Geodetic Network and World Geodetic System 1984 are based on WGS84.

10.1.2   Coordinate systems

A coordinate system is the link between the datum and the coordinate values. It defines all of the information about the axes system that defines the values. The names of the axes, their units (formats), the order of ordinates ((Easting, Northing) versus (Northing, Easting)) and the angle between the axes are defined by the coordinate system.

10.1.2.1   Cartesian coordinate systems

A Cartesian coordinate system is defined by values of $(x,y,(z))$. x is the distance from the x-axis, y is the distance from the y-axis, z the distance from the z-axis. The axis are orthogonal to each other. The unit used for x, y, z are a distance unit, such as meter. These coordinate systems are used for flat 'planar' descriptions of points. In general they are used over small areas where a projection method has been used to minimize distortions of the geography in the area.

10.1.2.2   Ellipsoid coordinates

More global geographic calculations need to take the surface of the earth into account. So we need a second coordinate system that describes each position relative to other points and lines on the earth's surface.

Each point can then be described as set of values (longitude, latitude) or (longitude, latitude, altitude) giving a point on the ellipsoid or relative to the ellipsoid we choose to describe the earth. The longitude tells us how far east we have to move on the equator from the null-meridian, the latitude tells us how far north to move from the equator and the altitude tells us how far above the ellipsoid to go to finally reach the location. Negative values direct us to go in the opposite direction.

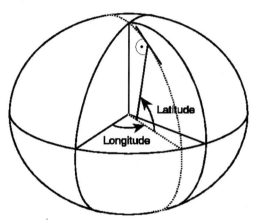

10.1.3   Coordinate Reference Systems

The two coordinate reference systems relevant to this protocol are Geographic 2D Coordinate Reference Systems and Projected Coordinate Reference Systems.

Geographic 2D Coordinate Reference Systems describe locations on the ellipsoid. They are used for large national or continental geodetic networks. In particular GPS uses the Geographic 2D Coordinate Reference System WGS84. This uses the World Geodetic System 1984 based on the WGS84 ellipsoid. The coordinate axes have units of decimal degrees (or DMSH) with ordinate order (Northing, Easting). This Coordinate Reference System is the default for all basic MLP service requests and responses. A GMLC is only required to support WGS84. The GMLC geographies that are defined with altitude are modeled in this protocol as geographies in a Geographic 2D CRS with a separate altitude element, not as a Geographic 3D CRS. The geographies are planar and carrying a constant z value is not desirable.

There are several ways to convert ellipsoid coordinates to 2 dimensional cartesian coordinates. These are called projection methods. Each method is designed to minimize some type of distortion in the mapping for the ellipsoid to the 2D Cartesian coordinate system.

Projected Coordinate Reference Systems are used for map display, to allow Cartesian mathematics and for Advanced Location Services.

10.2   Coordinate Reference System Transformations (Informative)

A transformation is used to define a point in one CRS into the appropriate values in a second CRS. When the datums are the same, the transformation can frequently be defined by equations. A transformation from one datum to another is usually done with a least squares approximation. Transformation equations are available in from several places, transformation services are also available.

10.3   Methodology for defining CRSs and transformations in this protocol (Normative)

The MLP protocol defines the CRS by citing an authority and the unique reference identifier for the CRS defined by this authority. This leaves the definition of many CRS used over the world to be defined by a group of geodesy experts. This methodology is used by the OpenGIS© Consortium and the ISO TC 211 working group for well-known CRS. The encoding used is from the OpenGIS© Consortium Recommendation Paper 01-014r5: Recommended Definition Data for Coordinate Reference Systems and Coordinate Transformations [20].

The MLP protocol may use the {EPSG} authority as an example. Support of other authority is for further study. This database is defined by a Microsoft Access database which can be found at www.epsg.org. An xml version of this database will be available at http://www.opengis.net/gml/srs/epsg.xml in the future.

The default WGS84 CRS is defined to be 4326 by the EPSG authority. Other examples are 326xx define the UTM xx N zones.

Coordinate Reference System transformation are done by an advance Location Service request. The implementation of this service is determined by the provider.

10.4   Supported coordinate systems and datum

All MLP implementations must support at least the WGS84 Coordinate Reference System.

10.5   **Shapes representing a geographical position**

There are a number of shapes used to represent a geographic area that
describes where a mobile subscriber is located. There are additional shapes
that are required for advanced MLP services. The standards bodies for
geographic data for advanced MLP services such as routing, geocoding,
coordinate conversion, and map display are the Location Interoperability
Forum, the OpenGIS© Consortium and the ISO TC211 working group. The
current public XML specification defining geography from these groups is
GML V211 [21]. These two groups work together and from are working towards a
GML V3 with additional geometry and topology types. The geometry required
for the MLP is the GMLV211 with additional polygon types with boundaries
that contain circles, ellipses or circular arcs. GML V3 will define the linear
curves segments to allow the these polygons to be defined. These
boundaries will be defined as special cases of polygons, using the given
interpolation methods. The following geographies are defined in this protocol.
The relevant OGC Abstract Specification is Topic 1 [22].

10.5.1   **Ellipsoid point**

This a point on the ellipsoid and is modeled as a point in a Geographic 2D
Coordinate Reference Systems.

10.5.2   **Ellipsoid point with uncertainty circle**

An ellipsoid point with uncertainty circle is characterized by the coordinates of
an ellipsoid point (the origin) and a radius, "r". It describes the set of points on
the ellipsoid, which are at a distance from the point of origin less than or equal
to "r". This shape can be used to indicate points on the Earth surface, or near
the Earth surface. This shape is a special case of a polygon with no interior
boundaries.

The typical use of this shape is to indicate a point when its position is known
only with a limited accuracy.

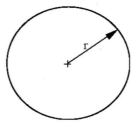

10.5.3   **Ellipsoid point with uncertainty ellipse**

The shape of an "ellipsoid point with uncertainty ellipse" is characterized by
the following:

•   The coordinates of an ellipsoid point (the origin)

- The distances r1 and r2

- The angle of orientation A

It describes formally the set of points on the ellipsoid, which fall within or on the boundary of an ellipse. This ellipse has a semi-major axis of length r1 oriented at angle A (0 to 180°) measured clockwise from north and a semi-minor axis of length r2. The distances being the geodesic distance over the ellipsoid, i.e., the minimum length of a path staying on the ellipsoid and joining the two points, as shown in figure below.

As for the ellipsoid point, this can be used to indicate points on the Earth's surface, or near the Earth's surface, of same latitude and longitude. This shape is a special case of a polygon with no interior boundaries.

The typical use of this shape is to indicate a point when its position is known only with a limited accuracy, but the geometrical contributions to uncertainty can be quantified.

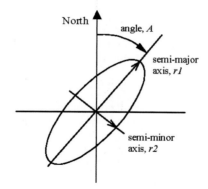

10.5.4   Ellipsoid point with uncertainty arc

The shape of an "ellipsoid point with uncertainty arc" is characterized by the following:

- The coordinates of an ellipsoid point (the origin)

- The inner radius(r) and uncertainty radius(r),

- The offset angle (θ) and included angle (β)

An arc is defined by a point of origin with one offset angle and one uncertainty angle plus one inner radius and one uncertainty radius. In this case the striped area describes the actual arc area. The smaller arc defines the inner radius(r) and the difference between inner and the outer arc defines the uncertainty radius(r). This shape is a special case of a polygon with no interior boundaries.

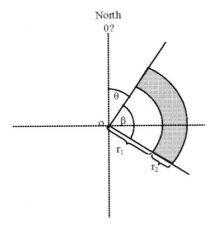

10.5.5   Polygon

A **Polygon** is a connected surface. Any pair of points in the polygon can be connected to one another by a path. The boundary of the Polygon is a set of LinearRings. We distinguish the outer (exterior) boundary and the inner (interior) boundaries; the LinearRings of the interior boundary cannot cross one another and cannot be contained within one another. There must be at most one exterior boundary and zero or more interior boundary elements. The ordering of LinearRings and whether they form clockwise or anti-clockwise paths is not important. The minimum number of points allowed in a LinearRing is 3.

A **LinearRing** is a closed, simple piece-wise linear path which is defined by a list of coordinates that are assumed to be connected by straight line segments. The last coordinate must be coincident with the first coordinate and at least four coordinates are required (the three to define a ring plus the fourth duplicated one). This geometry is only used in the construction of a Polygon.

For basic MLP services polygons are the number of interior bondaries MUST be 0. Also to conform to 3GPP TS 23032 the maximum number of points allowed in an exterior boundary is 15. The points shall be connected in the order that they are given.

The described area is situated to the right of the exterior boundaries and left of the interior boundaries with the downward direction being toward the Earth's center and the forward direction being from a point to the next.

NOTE: This definition does not permit connecting lines greater than roughly 20 000 km. If such a need arises, the polygon can be described by adding an intermediate point.

Computation of geodesic lines is not simple. Approximations leading to a maximum distance between the computed line and the geodesic line of less than 3 meters are acceptable.

10.5.6   LineString

A **LineString** is a piece-wise linear path defined by a list of coordinates that are assumed to be connected by straight line segments. A closed path is indicated by having coincident first and last coordinates. At least two coordinates are required.

10.5.7   Box

The **Box** element is used to encode extents. Each <Box> element encloses a sequence of two <coord> elements containing exactly two coordinate tuples; the first of these is constructed from the minimum values measured along all axes, and the second is constructed from the maximum values measured along all axes

10.5.8   Geometries Collections

These are geometry objects that contain 2 or more primative geometry objects. These collections can either be homogenous, a set of points, or heterogeneous, a point, circularArea and a LineString.

Geometry collections are not valid for the basic MLP services.

Platform for Privacy Preferences (P3P)

THE PLATFORM FOR PRIVACY PREFERENCES 1.0 (P3P1.0)
SPECIFICATION

W3C WORKING DRAFT 28 SEPTEMBER 2001

This Version:

http://www.w3.org/TR/2001/WD-P3P-20010928

Latest Public Version:

http://www.w3.org/TR/P3P/

Previous Version:

http://www.w3.org/TR/2001/WD-P3P-20010924

Editor:

Massimo Marchiori, W3C/MIT/UNIVE (massimo@w3.org)

Authors:

Lorrie Cranor, AT&T

Marc Langheinrich, ETH Zurich

Massimo Marchiori, W3C/MIT/UNIVE

Martin Presler-Marshall, IBM

Joseph Reagle, W3C/MIT

Abstract

This is the specification of the Platform for Privacy Preferences (P3P). This document, along with its normative references, includes all the specification necessary for the implementation of interoperable P3P applications.

Status of This Document

This section describes the status of this document at the time of its publication. Other documents may supersede this document. The latest status of this document series is maintained at the W3C.

This is the 28 September 2001 Last Call Working Draft of the Platform for Privacy Preferences 1.0 (P3P1.0) Specification, for review by W3C members and other interested parties. This Draft has been produced by the P3P Specification Working Group [*member only*] as part of the P3P Activity. The 24 September Last Call Draft was republished to include a missing change already approved by the Working Group (an embedded DATASCHEMA is now child of POLICIES rather than of POLICY).

The current draft is a revision of the 15 December 2000 Candidate Recommendation Draft. Due to substantive changes based on feedback from implementers, the Working Group has agreed [*members only*] to return this draft to Last Call to invite comments and input from W3C Members and the community at large. A change log with a summary of the modifications occurred from the 15 December 2000 Candidate Recommendation is included at the end of this document for convenience. Because these changes are based on solid implementation experience, the P3P Specification Group maintains all the Candidate Recommendation milestones and expects to request Proposed Recommendation after all those milestones and comments from this Last Call are properly addressed.

The last call review period ends 15 October 2001. Please send review comments before the review period ends to www-p3p-public-comments@w3.org (publicly archived).

The milestones are:

1. at least one P3P user agent implementation integrated into an HTTP user agent capable of fetching HTML files that includes all of the functionality required and recommended by this specification
2. a second P3P user agent implementation of each specified function (these functions may be demonstrated across several partial P3P implementations or they may be demonstrated in a second full P3P implementation)
3. at least one special-purpose tool for generating P3P policies and policy reference files
4. at least one tool for converting full P3P policies to compact policies
5. at least 10 P3P-enabled production web sites
6. at least one web site that illustrates each of the example scenarios in Section 2.5 of the P3P1.0 specification as well as at least one web site that uses mini-policies (these may be either production web sites or demonstration sites)

Furthermore, before requesting that this specification be advanced to Proposed Recommendation status, the Working Group will:

1. Prepare a W3C Note describing RDF data models representing P3P policies and policy reference files.
2. Submit an Internet Draft to the IETF describing the P3P header and request that an RFC be issued documenting this header.

3. Prepare a set of test policies and policy reference files that user agent implementers can use to demonstrate that their implementations behave correctly. This should include examples of policies that contain syntax errors.

4. Specify the appropriate behavior for user agents upon encountering a policy with invalid syntax.

The working group also encourages implementers to explore the possibility of implementations in web proxies and mobile devices, as well as implementations that can import user preferences using the [APPEL] language.

A list of current public W3C Working Drafts can be found at *http://www.w3.org/TR*.

TABLE OF CONTENTS

1 Introduction
  1.1 The P3P1.0 Specification
    1.1.1 Goals and Capabilities of P3P1.0
    1.1.2 Example of P3P in Use
    1.1.3 P3P Policies
    1.1.4 P3P User Agents
    1.1.5 Implementing P3P1.0 on Servers
    1.1.6 Future Versions of P3P
  1.2 About this Specification
  1.3 Terminology
2 Referencing Policies
  2.1 Overview and Purpose of Policy References
  2.2 Locating Policy Reference Files
    2.2.1 Well-Known Location
    2.2.2 HTTP Headers
    2.2.3 The HTML link Tag
    2.2.4 HTTP ports and other protocols
  2.3 Policy Reference File Syntax and Semantics
    2.3.1 Example Policy Reference File
    2.3.2 Policy Reference File Definition
      2.3.2.1 Policy reference file processing
        2.3.2.1.1 Significance of order
        2.3.2.1.2 Wildcards in policy reference files
      2.3.2.2 The META and POLICY-REFERENCES elements
      2.3.2.3 Policy reference file lifetimes and the EXPIRY element
        2.3.2.3.1 Motivation and mechanism
        2.3.2.3.2 The EXPIRY element
        2.3.2.3.3 Use of HTTP headers
        2.3.2.3.4 Error handling for policy reference file lifetimes
      2.3.2.4 The POLICY-REF element
      2.3.2.5 The INCLUDE and EXCLUDE elements

2.3.2.6 The HINT element
2.3.2.7 The COOKIE-INCLUDE and COOKIE-EXCLUDE elements
2.3.2.8 The METHOD element
2.3.3 Applying a Policy to a URI
2.3.4 Forms and Related Mechanisms
2.4 Additional Requirements
2.4.1 Non-ambiguity
2.4.2 Multiple Languages
2.4.3 The Safe Zone
2.4.4 Non-discrimination of Policies
2.4.5 Security of Policy Transport
2.4.6 Policy Updates
2.4.7 Absence of Policy Reference File
2.4.8 Asynchronous Evaluation
2.5 Example Scenarios
3 Policy Syntax and Semantics
3.1 Example policies
3.1.1 English language policies
3.1.2 XML encoding of policies
3.2 Policies
3.2.1 The POLICIES element
3.2.2 The POLICY element
3.2.3 The TEST element
3.2.4 The ENTITY element
3.2.5 The ACCESS element
3.2.6 The DISPUTES element
3.2.7 The REMEDIES element
3.3 Statements
3.3.1 The STATEMENT element
3.3.2 The CONSEQUENCE element
3.3.3 The NON-IDENTIFIABLE element
3.3.4 The PURPOSE element
3.3.5 The RECIPIENT element
3.3.6 The RETENTION element
3.3.7 The DATA-GROUP and DATA elements
3.4 Categories and the CATEGORIES element
3.5 Extension Mechanism: the EXTENSION element
3.6 Import and Export of User Preferences
4 Compact Policies
4.1 Referencing Compact Policies
4.2 Compact Policies Vocabulary
4.2.1 Compact ACCESS
4.2.2 Compact DISPUTES

4.2.3 Compact REMEDIES

4.2.4 Compact NON-IDENTIFIABLE

4.2.5 Compact PURPOSE

4.2.6 Compact RECIPIENT

4.2.7 Compact RETENTION

4.2.8 Compact CATEGORIES

4.2.9 Compact TEST

4.3 Compact Policy Scope

4.4 Compact Policy Lifetime

4.5 Transforming a P3P Policy to a Compact Policy

4.6 Transforming a Compact Policy to a P3P Policy

5 Data schemas

5.1 Natural Language Support for Data Schemas

5.2 Data Structures

5.3 The DATA-DEF and DATA-STRUCT elements

5.3.1 Categories in P3P Data Schemas

5.3.2 P3P Data Schema Example

5.3.3 Use of data element names

5.4 Persistence of data schemas

5.5 Basic Data Structures

5.5.1 Dates

5.5.2 Names

5.5.3 Logins

5.5.4 Certificates

5.5.5 Telephones

5.5.6 Contact Information

5.5.6.1 Postal

5.5.6.2 Telecommunication

5.5.6.3 Online

5.5.7 Access Logs and Internet Addresses

5.5.7.1 URI

5.5.7.2 ipaddr

5.5.7.3 Access Log Information

5.5.7.4 Other HTTP Protocol Information

5.6 The base data schema

5.6.1 User Data

5.6.2 Third Party Data

5.6.3 Business Data

5.6.3 Dynamic Data

5.7 Categories and Data Elements/Structures

5.7.1 Fixed-Category Data Elements/Structures

5.7.2 Variable-Category Data Elements/Structures

5.8 Using Data Elements

6 Appendices

Appendix 1: References (Normative)
Appendix 2: References (Non-normative)
Appendix 3: The P3P base data schema Definition (Normative)
Appendix 4: XML Schema Definition (Normative)
Appendix 5: XML DTD Definition (Non-normative)
Appendix 6: ABNF Notation (Non-normative)
Appendix 7: P3P Guiding Principles (Non-normative)
Appendix 8: Working Group Contributors (Non-normative)

1 INTRODUCTION

The Platform for Privacy Preferences Project (P3P) enables Web sites to express their privacy practices in a standard format that can be retrieved automatically and interpreted easily by user agents. P3P user agents will allow users to be informed of site practices (in both machine- and human-readable formats) and to automate decision-making based on these practices when appropriate. Thus users need not read the privacy policies at every site they visit.

Although P3P provides a technical mechanism for ensuring that users can be informed about privacy policies before they release personal information, it does not provide a technical mechanism for making sure sites act according to their policies. Products implementing this specification MAY provide some assistance in that regard, but that is up to specific implementations and outside the scope of this specification. However, P3P is complementary to laws and self-regulatory programs that can provide enforcement mechanisms. In addition, P3P does not include mechanisms for transferring data or for securing personal data in transit or storage. P3P may be built into tools designed to facilitate data transfer. These tools should include appropriate security safeguards.

1.1 The P3P1.0 Specification

The P3P1.0 specification defines the syntax and semantics of P3P privacy policies, and the mechanisms for associating policies with Web resources. P3P policies consist of statements made using the P3P vocabulary for expressing privacy practices. P3P policies also reference elements of the P3P base data schema—a standard set of data elements that all P3P user agents should be aware of. The P3P specification includes a mechanism for defining new data elements and data sets, and a simple mechanism that allows for extensions to the P3P vocabulary.

1.1.1 Goals and Capabilities of P3P1.0

P3P version 1.0 is a protocol designed to inform Web users of the data-collection practices of Web sites. It provides a way for a Web site to encode its data-collection and data-use practices in a machine-readable XML format known as a *P3P policy*. The P3P specification defines:

- A standard schema for data a Web site may wish to collect, known as the "P3P base data schema"
- A standard set of uses, recipients, data categories, and other privacy disclosures

- An XML format for expressing a privacy policy
- A means of associating privacy policies with Web pages or sites, and cookies
- A mechanism for transporting P3P policies over HTTP

The goal of P3P version 1.0 is twofold. First, it allows Web sites to present their data-collection practices in a standardized, machine-readable, easy-to-locate manner. Second, it enables Web users to understand what data will be collected by sites they visit, how that data will be used, and what data/uses they may "opt-out" of or "opt-in" to.

1.1.2 Example of P3P in Use

As an introduction to P3P, let us consider one common scenario that makes use of P3P. Claudia has decided to check out a store called CatalogExample, located at *http://www.catalog.example.com/*. Let us assume that CatalogExample has placed P3P policies on all their pages, and that Claudia is using a Web browser with P3P built in.

Claudia types the address for CatalogExample into her Web browser. Her browser is able to automatically fetch the P3P policy for that page. The policy states that the only data the site collects on its home page is the data found in standard HTTP access logs. Now Claudia's Web browser checks this policy against the preferences Claudia has given it. Is this policy acceptable to her, or should she be notified? Let's assume that Claudia has told her browser that this is acceptable. In this case, the home page is displayed normally, with no pop-up messages appearing. Perhaps her browser displays a small icon somewhere along the edge of its window to tell her that a privacy policy was given by the site, and that it matched her preferences.

Next, Claudia clicks on a link to the site's online catalog. The catalog section of the site has some more complex software behind it. This software uses cookies to implement a "shopping cart" feature. Since more information is being gathered in this section of the Web site, the Web server provides a separate P3P policy to cover this section of the site. Again, let's assume that this policy matches Claudia's preferences, so she gets no pop-up messages. Claudia continues and selects a few items she wishes to purchase. Then she proceeds to the checkout page.

The checkout page of CatalogExample requires some additional information: Claudia's name, address, credit card number, and telephone number. Another P3P policy is available that describes the data that is collected here and states that her data will be used only for completing the current transaction, her order.

Claudia's browser examines this P3P policy. Imagine that Claudia has told her browser that she wants to be warned whenever a site asks for her telephone number. In this case, the browser will pop up a message saying that this Web site is asking for her telephone number, and explaining the contents of the P3P statement. Claudia can then decide if this is acceptable to her. If it is acceptable, she can continue with her order; otherwise she can cancel the transaction.

Alternatively, Claudia could have told her browser that she wanted to be warned only if a site is asking for her telephone number and was going to give it to third parties and/or use it for uses other than completing the current transaction. In that case, she would have received no prompts from her browser at all, and she could proceed with completing her order.

Note that this scenario describes one hypothetical implementation of P3P. Other types of user interfaces are also possible.

1.1.3 P3P Policies

P3P policies use an XML encoding of the P3P vocabulary to provide contact information for the legal entity making the representation of privacy practices in a policy, enumerate the types of data or data elements collected, and explain how the data will be used. In addition, policies identify the data recipients, and make a variety of other disclosures including information about dispute resolution, and the address of a site's human-readable privacy policy. P3P policies must cover all relevant data elements and practices (but note that legal issues regarding law enforcement demands for information are not addressed by this specification; it is possible that a site that otherwise abides by its policy of not redistributing data to others may be required to do so by force of law). P3P declarations are positive, meaning that sites state what they do, rather than what they do not do. The P3P vocabulary is designed to be descriptive of a site's practices rather than simply an indicator of compliance with a particular law or code of conduct. However, user agents may be developed that can test whether a site's practices are compliant with a law or code.

P3P policies represent the practices of the site. Intermediaries such as telecommunication providers, Internet service providers, proxies, and others may be privy to the exchange of data between a site and a user, but their practices may not be governed by the site's policies.

1.1.4 P3P User Agents

P3P1.0 user agents can be built into Web browsers, browser plug-ins, or proxy servers. They can also be implemented as Java applets or JavaScript; or built into electronic wallets, automatic form-fillers, or other user data management tools. P3P user agents look for references to a P3P policy at a well-known location, in P3P headers in HTTP responses, and in P3P `link` tags embedded in HTML content. These references indicate the location of a relevant P3P policy. User agents can fetch the policy from the indicated location, parse it, and display symbols, play sounds, or generate user prompts that reflect a site's P3P privacy practices. They can also compare P3P policies with privacy preferences set by the user and take appropriate actions. P3P can perform a sort of "gate keeper" function for data transfer mechanisms such as electronic wallets and automatic form fillers. A P3P user agent integrated into one of these mechanisms would retrieve P3P policies, compare them with users' preferences, and authorize the release of data only if a) the policy is consistent with the user's preferences and b) the requested data transfer is consistent with the policy. If one of these conditions is not met, the user might be informed of the discrepancy and given an opportunity to authorize the data release themselves.

1.1.5 Implementing P3P1.0 on Servers

Web sites can implement P3P1.0 on their servers by translating their human-readable privacy policies into P3P syntax and then publishing the resulting files along with a policy reference file that indicates the parts of the site to which the policy applies. Automated tools can assist site operators in performing this translation. P3P1.0 can be implemented on existing HTTP/1.1-compliant Web servers without requiring additional or upgraded software. Servers may publish their policy reference files at a well-known location, or they may reference their P3P policy reference files in HTML content using a `link` tag. Alternatively, compatible servers may be configured to insert a P3P extension header into all HTTP responses that indicates the location of a site's P3P policy reference file.

Web sites have some flexibility in how they use P3P: they can opt for one P3P policy for their entire site or they can designate different policies for different parts of their sites. A P3P policy MUST cover all data generated or exchanged as part of a site's HTTP interactions with visitors. In addition, some sites may wish to write policies that cover all data an entity collects, regardless of how the data is collected.

1.1.6 Future Versions of P3P

Significant sections were removed from earlier drafts of the P3P1.0 specification in order to facilitate rapid implementation and deployment of a P3P first step. A future version of the P3P specification might incorporate those features after P3P1.0 is deployed. Such specification would likely include improvements based on feedback from implementation and deployment experience as well as four major components that were part of the original P3P vision but not included in P3P1.0:

- a mechanism to allow sites to offer a choice of P3P policies to visitors
- a mechanism to allow visitors (through their user agents) to explicitly agree to a P3P policy
- mechanisms to allow for non-repudiation of agreements between visitors and Web sites
- a mechanism to allow user agents to transfer user data to services

1.2 About This Specification

This document, along with its normative references, includes all the specification necessary for the implementation of interoperable P3P applications.

The [ABNF] notation used in this specification is specified in RFC2234 and summarized in Appendix 6. However, note that in the case of XML syntax, such ABNF syntax is only a grammar representative of the XML syntax (for example, all the syntactic flexibilities of XML are also implicitly included; e.g., whitespace rules, quoting using either single quote (`'`) or double quote (`"`), character escaping, comments, case sensitivity, order of attributes). All the XML syntax defined in this specification MUST conform to the XML Schema for P3P (see Appendix 4), which is the normative definition. For non-XML syntax (like, for example, in HTTP headers), the ABNF notation is the normative one.

In the sections that follow a number of XML elements are introduced. Each element is given in angle brackets (`"<element>"`), followed by a list of valid attributes. All listed attributes are optional, except when tagged as *mandatory*. Note that many XML elements are shown in the BNF with separate beginning and ending tags to allow optional elements inside them. If no elements are included, then, following standard XML rules, a self-closing element may be used instead.

The following key words are used throughout the document and have to be read as interoperability requirements. This specification uses words as defined in RFC2119 [KEY] for defining the significance of each particular requirement. These words are:

MUST or MUST NOT

This word or the adjective "required" means that the item is an absolute requirement of the specification.

SHOULD or SHOULD NOT

This word or the adjective "recommended" means that there may exist valid reasons in particular circumstances to ignore this item, but the full implications should be understood and the case carefully weighed before choosing a different course.

MAY

This word or the adjective "optional" means that this item is truly optional. One vendor may choose to include the item because a particular marketplace requires it or because it enhances the product, for example; another vendor may omit the same item.

1.3 Terminology

Character

Strings consist of a sequence of zero or more characters, where a character is defined as in the XML Recommendation [XML]. A single character in P3P thus corresponds to a single Unicode abstract character with a single corresponding Unicode scalar value (see [UNICODE]).

Data Element

An individual data entity, such as last name or telephone number. For interoperability, P3P1.0 specifies a base set of data elements.

Data Category

A significant attribute of a data element or data set that may be used by a trust engine to determine what type of element is under discussion, such as physical contact information. P3P1.0 specifies a set of data categories.

Data Set

A known grouping of data elements, such as `"user.home-info.postal"`. The P3P1.0 base data schema specifies a number of data sets.

Data Schema

A collection of data elements and sets defined using the P3P1.0 DATASCHEMA element. P3P1.0 defines a standard data schema called the *P3P base data schema*.

Data Structure

A hierarchical description of a set of data elements. A data set can be described according to its data structure. P3P1.0 defines a set of basic data structures that are used to describe the data sets in the P3P base data schema.

Equable Practice

A practice that is very similar to another in that the purpose and recipients are the same or more constrained than the original, and the other disclosures are not substantially different. For example, two sites with otherwise similar practices that follow different—but similar—sets of industry guidelines.

Identified Data

Data that reasonably can be used by the data collector to identify an individual.

Policy

A collection of one or more privacy statements together with information asserting the identity, URI, assurances, and dispute resolution procedures of the service covered by the policy.

Practice

The set of disclosures regarding data usage, including purpose, recipients, and other disclosures.

Preference

A rule, or set of rules, that determines what action(s) a user agent will take. A preference might be expressed as a formally defined computable statement (e.g., the [APPEL] preference exchange language).

Purpose

The reason(s) for data collection and use.

Repository

A mechanism for storing user information under the control of the user agent.

Safe Zone

Part of a Web site where the service provider performs only minimal data collection, and any data that is collected is used only in ways that would not reasonably identify an individual.

Service

A program that issues policies and (possibly) data requests. By this definition, a service may be a server (site), a local application, a piece of locally active code, such as an ActiveX control or Java applet, or even another user agent.

Service Provider (Data Controller, Legal Entity)

The person or legal entity which offers information, products or services from a Web site, collects information, and is responsible for the representations made in a practice statement.

Statement

A P3P statement is a set of privacy practice disclosures relevant to a collection of data elements.

URI

A Uniform Resource Identifier used to locate Web resources. For definitive information on URI syntax and semantics, see [URI]. URIs that appear within XML or HTML have to be treated as specified in [CHARMODEL], section Character Encoding in URI References. This does not apply to URIs appearing in HTTP header fields; the URIs there should always be fully escaped.

User

An individual (or group of individuals acting as a single entity) on whose behalf a service is accessed and for which personal data exists.

User Agent

A program whose purpose is to mediate interactions with services on behalf of the user under the user's preferences. A user may have more than one user agent, and agents need not reside on the user's desktop, but *any agent must be controlled by and act on behalf of only the user.* The trust relationship between a user and his or her agent may be governed by constraints outside of P3P. For instance, an agent may be trusted as a part of the user's operating system or Web client, or as a part of the terms and conditions of an ISP or privacy proxy.

2 REFERENCING POLICIES

2.1 Overview and Purpose of Policy References

Locating a P3P policy is one of the first steps in the operation of the P3P protocol. Services use policy references to state what policy applies to a specific URI or set of URIs. User agents use policy references to locate the privacy policy that applies to a page, so that they can process that policy for the benefit of their user.

Policy references are used extensively as a performance optimization. P3P policies are typically several kilobytes of data, while a URI that references a privacy policy is typically less than 100 bytes. In addition to the bandwidth savings, policy references also reduce the need for computation: policies can be uniquely associated with URIs, so that a user agent need only parse and process a policy once rather than process it with every document to which the policy applies. Furthermore, by placing the information about relevant policies in a centralized location, Web site administration is simplified.

A policy reference file is used to associate P3P policies with certain regions of URI-space. The policy reference file is an XML (see [XML]) file that can specify the policy for a single Web document, portions of a Web site, or for an entire site. The policy reference file may refer to one or more P3P policies; this allows for a single reference file to cover an entire site, even if different P3P policies apply to different portions of the site. The policy reference file is used to make any or all of the following statements:

- The URI where a P3P policy is found
- The URIs or regions of URI-space covered by this policy
- The URIs or regions of URI-space not covered by this policy
- The regions of URI-space for embedded content on other servers that are covered by this policy
- The cookies that are or are not covered by this policy
- The access methods for which this policy is applicable
- The period of time for which these claims are considered to be valid

All of these statements are made in the body of the policy reference file.

2.2 Locating Policy Reference Files

This section describes the mechanisms used to indicate the location of a policy reference file. Detailed syntax is also given for the supported mechanisms.

The location of the policy reference file can be indicated using one of three mechanisms. The policy reference file may be located in a predefined "well-known" location, or a document may indicate a policy reference file through an HTML `link` tag, or through an HTTP header.

Note that if user agents support retrieving HTML content over HTTP, they MUST handle all three mechanisms listed above interchangeably. See also the requirements for non-ambiguity.

Note that policies are applied at the level of HTTP entities. An entity, retrieved by fetching a URI, has a P3P policy associated with it. A "page" from the user's perspective may be composed of multiple HTTP entities; each entity may have its own P3P policy associated with it. As a practical note, however, placing many different P3P policies on different entities on a single page may make rendering the page and informing the user of the relevant policies difficult for user agents. Additionally, services are recommended to attempt to craft their policy reference files such that a single policy reference file covers any given "page"; this will speed up the user's browsing experience.

For a user agent to process the policy that applies to a given entity, it must locate the policy reference file for that entity, fetch the policy reference file, parse the policy reference file, fetch any required P3P policies, and then parse the P3P policy or policies.

This document does not specify how P3P policies may be associated with documents retrieved by means other than HTTP. However, it does not preclude future development of mechanisms for associating P3P policies with documents retrieved over other protocols. Furthermore, additional methods of associating P3P policies with documents retrieved using HTTP may be developed in the future.

2.2.1 Well-Known Location

Web sites using P3P SHOULD place a policy reference file in a "well-known" location. To do this, a policy reference file would be placed in the site's /w3c directory, under the name p3p.xml. Thus a user agent could request this policy reference file by using a GET request for the resource /w3c/p3p.xml.

Note that sites are not required to use this mechanism; however, by using this mechanism, sites can ensure that their P3P policy will be accessible to user agents before any other resources are requested from the site. This will reduce the need for user agents to access the site using safe zone practices. Additionally, if a site chooses to use this mechanism, the policy reference file located in the well-known location is not required to cover the entire site. For example, sites where not all of the content is under the control of a single organization MAY choose not to use this mechanism, or MAY choose to post a policy reference file which covers only a limited portion of the site.

Use of the well-known location for a policy reference file does not preclude use of other mechanisms for specifying a policy reference file. Portions of the site MAY use any of the other supported mechanisms to specify a policy reference file, so long as the non-ambiguity requirements are met.

For example, imagine a shopping-mall Web site run by the MallExample company. On their Web site (mall.example.com), companies offering goods or services at the mall would get a company-specific subtree of the site, perhaps in the path /companies/*company-name*. The MallExample company may choose to put a policy reference file in the well-known location which covers all of their site except the /companies subtree. Then if the ShoeStoreExample company has some content in /companies/shoestoreexample, they could use one of the other mechanisms to indicate the location of a policy reference file covering their portion of the mall.example.com site.

One case where using the well-known location for policy reference files is expected to be particularly useful is in the case of a site which has divided its content across several hosts. For example, consider a site which uses a different logical host for all of its Web-based applications than for its static HTML content. The other mechanisms allowed for specifying the location of a policy reference file require that some URI on the host being accessed must be fetched to locate the policy reference file. However, the well-known location mechanism has no such requirement. Consider the example of an HTML form located on www.example.com. Imagine that the action URI on that form points to server cgi.example.com. The policy reference file that covers the form is unable to make any statements about the action URI that processes the form. However, the site administrator publishes a policy reference file at http://cgi.example.com/w3c/p3p.xml that covers the action URI, thus enabling a user agent to easily locate the P3P policy that applies to the action URI before submitting the form contents.

2.2.2 HTTP Headers

Any document retrieved by HTTP MAY point to a policy reference file through the use of a new response header, the P3P header ([P3P-HEADER]). If a site is using P3P headers, it SHOULD include this on responses for all appropriate request methods, including HEAD and OPTIONS requests.

The P3P header gives one or more comma-separated directives. The syntax follows:

```
[1]   p3p-header            =   `P3P: ` p3p-header-field *(`,` p3p-
                                header-field)

[2]   p3p-header-field      =   policy-ref-field | compact-policy-field
                                | extension-field

[3]   policy-ref-field      =   `policyref="` URI `"`

[4]   extension-field       =   token
                                [`=` (token | quoted-string) ]
```

Here, URI is defined as per RFC 2396 [URI], token and quoted-string are defined by [HTTP1.1].

In keeping with the rules for other HTTP headers, the name of the P3P header may be written with any casing. The contents should be specified using the casing precisely as specified in this document.

The policyref directive gives a URI which specifies the location of a policy reference file which may reference the P3P policy covering the document that pointed to the reference file, and possibly others as well. When the policyref attribute is a relative URI, that URI is interpreted relative to the request URI. Note that fetching the URI given in the policyref directive MAY result in a 300-class HTTP return code (redirection); user agents MUST interpret those redirects with normal HTTP semantics. Services should note, of course, that use of redirects will increase the time required for user agents to find and interpret their policies. The policyref URI MUST NOT be used for any other purpose beyond locating and referencing P3P policies.

The compact-policy-field is used to specify "compact policies." This is described in Section 4.

User agents which find unrecognized directives (in the extension-fields) MUST ignore the unrecognized directives. This is to allow easier deployment of future versions of P3P.

Example 2.1:

---

1. Client makes a GET request.

```
GET /index.html HTTP/1.1
Host: catalog.example.com
Accept: */*
Accept-Language: de, en
User-Agent: WonderBrowser/5.2 (RT-11)
```

2. Server returns content and the P3P header pointing to the policy of the page.

```
HTTP/1.1 200 OK
P3P: policyref="http://catalog.example.com/P3P/
    PolicyReferences.xml"
Content-Type: text/html
Content-Length: 7413
Server: CC-Galaxy/1.3.18
```

2.2.3 The HTML `link` Tag

Servers MAY serve HTML content with embedded `link` tags that indicate the location of the relevant P3P policy reference file. This use of P3P does not require any change in the server behavior.

The `link` tag encodes the policy reference information that could be expressed using the P3P header. The link tag takes the following form:

```
[5]   p3p-link-tag        =    `<link rel="P3Pv1" href="`  URI  `">`
```

Here, URI is defined as per RFC 2396 [URI].

When the `href` attribute is a relative URI, that URI is interpreted relative to the request URI.

In order to illustrate with an example the use of the `link` tag, we consider the policy reference expressed in Example 2.1 using HTTP headers. That example can be equivalently expressed using the `link` tag with the following piece of HTML:

```
<link rel="P3Pv1"
    href="http://catalog.example.com/P3P/PolicyReferences.xml">
```

Finally, note that since the `p3p-link-tag` is embedded in an HTML document, its character encoding will be the same as that of the HTML document. In contrast to P3P policy and policy reference documents (see section 2.3 and section 3 below), the `p3p-link-tag` need not be encoded using [UTF-8]. Note also that the `link` tag is not case sensitive

2.2.4 HTTP Ports and Other Protocols

The mechanisms described here MAY be used for HTTP transactions over any underlying protocol. This includes plain-text HTTP over TCP/IP connections as well as encrypted HTTP over SSL connections, as well as HTTP over any other communications protocol network designers wish to implement.

URLs MAY contain TCP/IP port numbers, as specified in RFC 2396 [URI]. For the purposes of P3P, the different ports on a single host MUST be considered to be separate "sites." Thus, for example, the policy reference file at the well-known location for www.example.com on port 80 (*http://www.example.com/w3c/p3p.xml*) would not give any information about the policies which apply to www.example.com when accessed over SSL (as the SSL communication would take place on a different port, 443 by default).

This document does not specify how P3P policies may be associated with documents retrieved by means other than HTTP. However, it does not preclude future development of mechanisms for associating P3P policies with documents retrieved over other protocols. Furthermore, additional methods of associating P3P policies with documents retrieved using HTTP may be developed in the future.

2.3 Policy Reference File Syntax and Semantics

This section explains the contents of policy reference files in detail.

2.3.1 Example Policy Reference File

Consider the case of a Web site wishing to make the following statements:

1. P3P policy /P3P/Policies.xml#first applies to the entire site, except the subtrees /catalog, /cgi-bin, and /servlet.
2. P3P policy /P3P/Policies.xml#second applies to all documents in the /catalog directory (and its subdirectories).
3. P3P policy /P3P/Policies.xml#third applies to all documents in the /cgi-bin and /servlet directories (and their subdirectories), except for /servlet/unknown.
4. No statement is made about what P3P policy applies to /servlet/unknown.
5. These statements are valid for 2 days.

These statements could be represented by the following piece of XML:

Example 2.2:

```
<META xmlns="http://www.w3.org/2001/09/P3Pv1">
 <POLICY-REFERENCES>
  <EXPIRY max-age="172800"/>

    <POLICY-REF about="/P3P/Policies.xml#first">
      <INCLUDE>/*</INCLUDE>
      <EXCLUDE>/catalog/*</EXCLUDE>
      <EXCLUDE>/cgi-bin/*</EXCLUDE>
      <EXCLUDE>/servlet/*</EXCLUDE>
    </POLICY-REF>

    <POLICY-REF about="/P3P/Policies.xml#second">
      <INCLUDE>/catalog/*</INCLUDE>
    </POLICY-REF>

    <POLICY-REF about="/P3P/Policies.xml#third">
      <INCLUDE>/cgi-bin/*</INCLUDE>
      <INCLUDE>/servlet/*</INCLUDE>
      <EXCLUDE>/servlet/unknown</EXCLUDE>
    </POLICY-REF>

 </POLICY-REFERENCES>
</META>
```

Note this example also includes via EXPIRY a relative expiry time in the document (cf. Section 2.3.2.3.2).

2.3.2 Policy Reference File Definition

This section defines the syntax and semantics of P3P policy reference files. All policies MUST be encoded using [UTF-8]. P3P servers MUST encode their policy references using this syntax. P3P user agents MUST be able to parse this syntax.

One significant point to make about the syntax of policy reference files is that the syntax defined here does not have an extension mechanism. The syntax for P3P policies has a powerful extension mechanism, but that mechanism is not supported for policy reference files.

2.3.2.1 Policy Reference File Processing

2.3.2.1.1 Significance of Order
A policy reference file may contain multiple POLICY-REF elements. If it does contain more than one element, they MUST be processed by user agents in the order given in the file. When a user agent is attempting to determine what policy applies to a given URI, it MUST use the first POLICY-REF element in the policy reference file which applies to that URI.

Note that each POLICY-REF may contain multiple INCLUDE, EXCLUDE, METHOD, COOKIE-INCLUDE, and COOKIE-EXCLUDE elements and that all of these elements within a given POLICY-REF MUST be considered together to determine whether the POLICY-REF applies to a given URI. Thus, it is not sufficient to find an INCLUDE element that matches a given URI, as EXCLUDE or METHOD elements may serve as modifiers that cause the POLICY-REF not to match.

2.3.2.1.2 Wildcards in Policy Reference Files
Policy reference files make statements about what policy applies to a given URI. Policy reference files support a simple wildcard character to allow making statements about regions of URI-space. The character asterisk ("*") is used to represent a sequence of 0 or more of any character. No other special characters (such as those found in regular expressions) are supported. Note that since the asterisk is also a legal character in URIs ([URI]), some special conventions have to be followed when encoding such "extended URIs" in a policy reference file:

- URIs represented in policy-ref files MUST be properly escaped, as in [URI].
- P3P user agents MUST escape any characters which should be escaped, as according to [URI], before attempting to match a URI for a policy.
- P3P user agents MUST un-escape any escaped sequences which resolve to URI-legal characters, according to [URI], before attempting to match a URI for a policy, EXCEPT
- Literal '*'s in URIs MUST be escaped by P3P user agents before attempting to match a URI for a policy.
- P3P user agents MUST ignore any URI pattern that does not conform to [URI]

The wildcard character MAY be used in the INCLUDE and EXCLUDE elements, in the COOKIE-INCLUDE and COOKIE-EXCLUDE elements, and in the HINT element.

2.3.2.2 The META and POLICY-REFERENCES Elements

The META element contains a complete policy reference file. Optionally, one POLICIES element can follow. Additionally, other XML markup MAY follow the POLICY-REFERENCES (or POLICIES, if present) element, although that markup MUST be ignored by any P3P1.0 user agent.

<POLICY-REFERENCES>

This element MAY contain one or more POLICY-REF (policy reference) elements. It MAY also contain one EXPIRY element (indicating their expiration time), and one or more HINT element.

```
[6]    prf           =    `<META xmlns="http://www.w3.org/2001/09/
                          P3Pv1">`
                          policyrefs
                          [policies]
                          PCDATA
                          "</META>"

[7]    policyrefs    =    "<POLICY-REFERENCES>"
                          [expiry]
                          *policyref
                          *hint
                          "</POLICY-REFERENCES>"
```

Here PCDATA is defined in [XML].

2.3.2.3 Policy Reference File Lifetimes and the EXPIRY Element

2.3.2.3.1 Motivation and Mechanism   It is desirable for servers to inform user agents about how long they can use the claims made in a policy reference file. By enabling clients to cache the contents of a policy reference file, it reduces the time required to process the privacy policy associated with a Web page. This also reduces load on the network. In addition, clients that don't have a valid policy reference file for a URI will need to use "safe zone" practices for their requests. If clients have policy reference files that they know are still valid, then they can make more informed decisions on how to proceed.

In order to achieve these benefits, policy reference files SHOULD contain an EXPIRY element, which indicates the lifetime of the policy reference file. If the policy reference file does not contain an EXPIRY element, then it is given a 24-hour lifetime.

The lifetime of a policy reference file tells user agents how long they can rely on the claims made in the policy reference file. By setting the lifetime of a policy reference file, the publishing site agrees that the policies mentioned in the policy reference file are appropriate for the lifetime of the policy reference file. For example, if a policy reference file has a lifetime of 3 days, then a user agent need not reload that file for 3 days, and can assume that the references made in that policy reference file are good for 3 days. All of the policy references made in a single policy reference file will receive the same lifetime. The only way to specify different lifetimes for different policy references is to use separate policy reference files.

The same mechanism used to indicate the lifetime of a policy reference file is also used to indicate the lifetime of a P3P policy. Thus P3P POLICIES elements SHOULD have an EXPIRY element associated with them as well. This lifetime applies to all P3P policies contained within that POLICIES element. If there is no EXPIRY element associated with a P3P policy, then it is given a 24-hour lifetime.

When picking a lifetime for policies and policy reference files, sites need to pick a lifetime which balances two competing concerns. One concern is that the lifetime ought to be long enough to allow user agents to receive significant benefits from caching. The other concern is that the site would like to be able to change their policy for new data collection without waiting for an extremely long lifetime to expire. It is expected that lifetimes in the range of 1–7 days would be a reasonable balance between these two competing desires. Sites also need to remember the policy update requirements when updating their policies.

When a policy reference file has expired, the information in the policy reference file MUST NOT be used by a user agent until that user agent has successfully revalidated the policy reference file, or has fetched a new copy of the policy reference file.

Note that while user agents are not obligated to revalidate policy reference files or policy files that have not expired, they MAY choose to revalidate those files before their expiry period has passed, in order to reduce the need for using "safe zone" practices. A valid P3P user agent implementation doesn't need to contain a cache for policies and policy reference files, though the implementation will have a better performance if it does.

2.3.2.3.2 The EXPIRY Element
The EXPIRY element can be used in a policy reference file and/or in a POLICIES element to state how long the policy reference file (or policies) remains valid. The expiry is given as either an absolute expiry time, or a relative expiry time. An absolute expiry time is a time, given in GMT, until which the policy reference file (or policies) is valid. A relative expiry time gives a number of seconds for which the policy reference file (or policies) is valid. This expiry time is relative to the time the policy reference file (or policies) was requested or last revalidated by the client. This computation MUST be done using the time of the original request or revalidation, and the current time, with both times generated from the client's clock. Revalidation is defined in section 13.3 of [HTTP1.1].

The minimum amount of time for any relative expiry time is 24 hours, or 86400 seconds. Any relative expiration time shorter than 86400 seconds MUST be treated as being equal to 86400 seconds in a client implementation. If a client encounters an absolute expiration time that is in the past, it MUST act as if NO policy reference file (or policy) is available. See section 2.4.7 "Absence of Policy Reference File" for the required procedure in such cases.

```
[8]      expiry        =      "<EXPIRY" (absdate|reldate) "/>"

[9]      absdate       =      `date="` HTTP-date `"`

[10]     reldate       =      `max-age="` delta-seconds `"`
```

Here, HTTP-date is defined in section 3.3.1 of [HTTP1.1], and delta-seconds is defined in section 3.3.2 of [HTTP1.1].

2.3.2.3.3 Requesting Policies and Policy Reference Files   In a real-world network, there may be caches which will cache the contents of policies and policy reference files. This is good for increasing the overall network performance, but may have deleterious effects on the operation of P3P if not used correctly. There are two specific concerns:

1. When a user agent receives a policy reference file (or policy), if it was served from a network cache, the user agent needs to know how long the policy reference file or policy resided in the network cache. This time MUST be subtracted from the lifetime of the policy or policy reference file which uses relative expiry.

2. When a user agent needs to revalidate a policy reference file (or policy), it needs to make sure that the revalidation fetches a current version of the policy reference file (or policy). For example, consider the case where a user agent holds a policy reference file with a 1 day relative expiry. If the user agent refetches it from a network cache, and the file has been residing in the network cache for 3 days, then the resulting file is useless.

HTTP 1.1 [HTTP1.1] contains powerful cache-control mechanisms to allow clients to place requirements on the operations of network caches; these mechanisms can resolve the problems mentioned above. The specific method will be discussed below.

HTTP 1.0, however, does not provide those more sophisticated cache control mechanisms. An HTTP 1.0 network cache will, in all likelihood, compute a cache lifetime for the policy reference file (or policies) based on the file's last-modified date; the resulting cache lifetime could be significantly longer than the lifetime specified by the EXPIRY element. The network cache could then serve the policy reference file (or policies) to clients beyond the lifetime in the EXPIRY; the result would be that user-agents would receive a useless policy reference file (or policies).

The second problem with HTTP 1.0 network caches is that a user agent has no way to know how long the reference file may have been stored by the network cache. If the policy reference file (or policies) relies on relative expiry, it would then be impossible for the user agent to determine if the reference file's lifetime has already expired, or when it will expire.

Thus, if a user agent is requesting a policy reference file or a policy, and does not know for certain that there are no HTTP 1.0 caches in the path to the origin server, then the request must force an end-to-end revalidation. This can be done with the Pragma: no-cache HTTP request-header. Note that neither HTTP nor P3P define a way to determine if there is a HTTP 1.0-compliant cache in any given network path, so unless the user agent has this information derived from an outside source, it MUST force the end-to-end revalidation.

If the user agent has some way to know that all caches in the network path to the origin server are compliant with HTTP 1.1 (or that there are no caches in the network path to the origin server), then the client MUST do the following:

1. Use cache-control request-headers to ensure that the received response is not older than its lifetime. This is done with the max-age cache-control setting, with a maximum age significantly less than the lifetime of the policy reference file (or policies). For example, a user agent could send Cache-Control: max-age=43200, thus ensuring that the response is no more than 12 hours old.
2. Subtract the age of the response from the lifetime of the policy reference file (or policies), if it uses a relative expiry time. The age of the response is given by the Age: HTTP response-header.

Note that it is impossible for a client to accurately predict the amount of latency that may affect an HTTP request. Thus, if the policy reference file covering a request is going to expire soon, clients MAY wish to consider warning their users and/or revalidating the policy reference file before continuing with the request.

2.3.2.3.4 Error Handling for Policy Reference File and Policy Lifetimes   The following situations have their semantics specifically defined:

1. An absolute expiry date in the past renders the policy reference file (or policies) useless, as does an invalid or malformed expiry date, whether relative or absolute. In this case, user agents MUST act as if NO policy reference file (or policies) is available. See section 2.4.7 "Absence of Policy Reference File" for the required procedure in such cases.
2. A relative expiration time shorter than 86400 seconds (1 day) is considered to be equal to 86400 seconds.
3. When a policy reference file contains more than one EXPIRY element, the first one takes precedence for determining the lifetime of the policy reference file.

2.3.2.4 The POLICY-REF Element

A policy reference file may refer to multiple P3P policies, specifying information about each. The POLICY-REF element describes attributes of a single P3P policy. Elements within the POLICY-REF element give the location of the policy and specify the areas of URI-space (and cookies) that each policy covers.

POLICY-REF
Contains information about a single P3P policy.

- about (mandatory attribute)
  URI reference ([URI]), where the fragment identifier part denotes the *name* of the policy (given in its name attribute), and the URI part denotes the URI where the policy resides. If this is a relative URI reference, it is interpreted relative to the URI of the policy reference file.

```
[11]    policy-ref      =    `<POLICY-REF about="` URI-reference `">`
                             *include
                             *exclude
                             *cookie-include
                             *cookie-exclude
                             *method-element
                             `</POLICY-REF>`
```

Here, URI is defined as per RFC 2396 [URI].

2.3.2.5 The INCLUDE and EXCLUDE Elements

Each INCLUDE or EXCLUDE element specifies one local URI or set of local URIs. A set of URIs is specified if the wildcard character '*' is used in the URI-pattern. These elements are used to specify the portion of the Web site that is covered by the policy referenced by the enclosing POLICY-REF element.

When INCLUDE (and optionally, EXCLUDE) elements are present in a POLICY-REF element, it means that the policy specified in the about attribute of the POLICY-REF element applies to all the URIs at the requested host corresponding to the local-URI(s) matched by any of the INCLUDEs, but not matched by an EXCLUDE element.

A policy referenced in a policy reference file can be applied only to URIs on the DNS (Domain Name System) host that reference it. The INCLUDE and EXCLUDE elements MUST specify URI patterns relative to the root of the DNS host to which they are applied. This requirement does NOT apply to the location of the P3P policy file (the about attribute on the POLICY-REF element).

If a METHOD element (section 2.3.2.8) specifies one or more methods for an enclosing policy reference, it follows that all methods *not* mentioned are consequently *not* covered by this policy. In the case that this is the only policy reference for a given URI prefix, user agents MUST assume that NO policy is in effect for all methods NOT mentioned in the policy reference file. It is legal but pointless to supply a METHOD element without any INCLUDE or COOKIE-INCLUDE elements.

It is legal, but pointless, to supply an EXCLUDE element without any INCLUDE elements; in that case, the EXCLUDE element MUST be ignored by user agents.

Note that the set of URIs specified with INCLUDE and EXCLUDE does not include cookies that might be triggered when requesting one of such URIs: in order to associate policies with cookies, the COOKIE-INCLUDE and COOKIE-EXCLUDE elements are needed.

```
[12]    include     =    "<INCLUDE>" relativeURI "</INCLUDE>"

[13]    exclude     =    "<EXCLUDE>" relativeURI "</EXCLUDE>"
```

Here, relativeURI is defined as per RFC 2396 [URI], with the addition that the '*' character is to be treated as a wildcard, as defined in section 2.3.2.1.2.

2.3.2.6 The `HINT` Element

Policy reference hints are a performance optimization that can be used under certain conditions. A DNS host may declare a policy reference for itself using the well-known location, the P3P response header, or the HTML `link` tag. The host MAY further provide a hint to additional policy references, such as those declared by other hosts. For example, an HTML page might hint at policy references for its hyperlinks, embedded content, and form submission URIs. User agents MAY use the hint mechanism to discover policy references before requesting the affected URIs when the policy references are not available from the well-known location.

Any policy reference file MAY contain zero or more policy reference hints. Each hint is contained in a `HINT` element, and consists of single host or domain of hosts to which the hinted policy reference can be applied. When using a hint applicable to multiple hosts, the policy reference is expected in the same relative location on each host, but the content may vary according to the host. Therefore, a user agent that finds a policy reference on a particular host via the hint mechanism MUST NOT apply it to another host.

The `domain` attribute is used to domain-match (possibly using the '*' wildcard) the host(s) to which the hinted policy reference file can be applied. The `path` attribute specifies the location of the hinted policy reference files relative to the applicable host rather than the policy reference file containing the hint.

Here is an example of `HINT` elements that hint at the location of policy reference files on the host example.org and on any host in the domain shop.example.com:

Example 2.3:

```
<HINT domain="example.org" path="/mypolicy/p2.xml"/>
<HINT domain="*.shop.example.com" path="/w3c/prf.xml"/>
```

If a hinted policy reference file is not found, expired, or otherwise invalid, the user agent MUST ignore the hint. Before using a hinted policy reference, the user agent MUST check the well-known location and give precedence to any policy references directly declared by the host, with the well-known location taking the highest precedence. If a hinted policy reference is not directly declared by the host as expected, the user agent MAY ignore it.

```
[14]    hint    =    `<HINT domain="` HN `" path="` token `/>`
```

Here, `HN` and `token` are defined as per RFC 2965 [STATE], with the addition that in `HN` the '*' character is to be treated as a wildcard, as defined in section 2.3.2.1.2.

2.3.2.7 The `COOKIE-INCLUDE` and `COOKIE-EXCLUDE` Elements

The `COOKIE-INCLUDE` and `COOKIE-EXCLUDE` elements are used to associate policies to cookies.

A cookie policy MUST cover any data (within the scope of P3P) that is stored in that cookie or linked via that cookie. It MUST also reference all purposes associated with data stored in that cookie or enabled by that cookie. In addition, any data/purpose stored or linked via a cookie MUST also be put in the cookie policy. In addition, if that linked data is collected by HTTP, then the policy that covers that GET/POST/whatever request must cover that data collection. For example, when CatalogExample asks customers to fill out a form with their name, billing, and shipping information, the P3P policy that covers the form submittal will disclose that CatalogExample collects this data and explain how it is used. If CatalogExample sets a cookie so that it can recognize its customers and observe their behavior on its web site, it would have a separate policy for this cookie. However, if this cookie is also linked to the user's name, billing, and shipping information—perhaps so CatalogExample can generate custom catalog pages based on where the customer lives—then that data must also be disclosed in the cookie policy.

For the purpose of this specification, state management mechanisms use either SET-COOKIE or SET-COOKIE2 headers, and cookie-namespace is defined as the value of the NAME, VALUE, Domain and Path attributes, specified in [COOKIES] and [STATE].

Each COOKIE-INCLUDE or COOKIE-EXCLUDE element can be used to match (similarly to INCLUDE and EXCLUDE) the NAME, VALUE, Domain and Path components of a cookie, expressing the cookies which are covered by the policy specified by the about attribute when the cookies are set from the documents on the Web site where the policy reference file resides:

COOKIE-INCLUDE (resp. COOKIE-EXCLUDE)

Include (resp. exclude) cookies that match the name, value, domain and path attributes.

- name: match the NAME portion of the cookie
- value: match the VALUE portion of the cookie
- domain: match the Domain portion of the cookie
- path: match the Path portion of the cookie

All four attributes are optional. If an attribute is absent, the COOKIE-INCLUDE (resp. COOKIE-EXCLUDE) will match cookies that have that attribute set to any value.

When COOKIE-INCLUDE (and optionally, COOKIE-EXCLUDE) elements are present in a POLICY-REF element, the policy specified in the about attribute of the POLICY-REF element applies to every cookie that is matched by any COOKIE-INCLUDE's, and not matched by a COOKIE-EXCLUDE element.

A site MUST NOT declare policies for cookies unless the cookies are set by its own site. User agents MUST accordingly interpret COOKIE-INCLUDE and COOKIE-EXCLUDE elements in a policy reference file to determine the policy that applies to cookies. Note that COOKIE-INCLUDE and COOKIE-EXCLUDE are the only mechanisms for associating policies with cookies in policy reference files (see Section 4).

The policy that applies to a cookie applies until the policy expires, even if the associated policy reference file expires prior to policy expiry (but after the cookie was set). If the policy associated with a cookie has expired, then the user agent SHOULD reevaluate the cookie policy before sending the cookie. In addition, user agents MUST use only non-expired policies and policy reference files when evaluating new set-cookie events.

Example 2.4 states that `/P3P/Policies.xml#first` applies to all cookies.

Example 2.4:

```
<META xmlns="http://www.w3.org/2001/09/P3Pv1">
 <POLICY-REFERENCES>
    <POLICY-REF about="/P3P/Policies.xml#first">
       <COOKIE-INCLUDE name="*" value="*" domain="*" path="*"/>
    </POLICY-REF>
 </POLICY-REFERENCES>
</META>
```

Example 2.5 states that `/P3P/Policies.xml#first` applies to all cookies, except cookies with the cookie name value of `"obnoxious-cookie"`, a domain value of `".example.com"`, and a path value of `"/"`, and that `/P3P/Policies.xml#second` applies to all cookies with the cookie name of `"obnoxious-cookie"`, a domain value of `".example.com"`, and a path value of `"/"`.

Example 2.5:

```
<META xmlns="http://www.w3.org/2001/09/P3Pv1">
 <POLICY-REFERENCES>
    <POLICY-REF about="/P3P/Policies.xml#first">
       <COOKIE-INCLUDE name="*" value="*" domain="*" path="*"/>
       <COOKIE-EXCLUDE name="obnoxious-cookie" value="*"
          domain=".example.com" path="/"/>
    </POLICY-REF>
    <POLICY-REF about="/P3P/Policies.xml#second">
       <COOKIE-INCLUDE name="obnoxious-cookie" value="*"
          domain=".example.com" path="/"/>
    </POLICY-REF>
 </POLICY-REFERENCES>
</META>
```

```
[15]   cookie-include   =   "<COOKIE-INCLUDE"
                [` name="` token `"`]    ; matches the
            cookie's NAME
                [` value="` token `"`]   ; matches the
            cookie's VALUE
                [` domain="` token `"`]  ; matches the
            cookie's Domain
                [` path="` token `"`]    ; matches the
            cookie's Path
                "/>"
```

```
[16]   cookie-exclude   =    "<COOKIE-EXCLUDE"
                              [` name="` token `"`]    ; matches the
                       cookie's NAME
                              [` value="` token `"`]   ; matches the
                       cookie's VALUE
                              [` domain="` token `"`] ; matches the
                       cookie's Domain
                              [` path="` token `"`]    ; matches the
                       cookie's Path
                              "/>"
```

Here, token, NAME, VALUE, Domain and Path are defined as per RFC 2965 [STATE], with the addition that the '*' character is to be treated as a wildcard, as defined in section 2.3.2.1.2.

Note that [STATE] states default values for the domain and path attributes of cookies: these should be used in the comparison if those attributes are not found in a specific cookie. Also, conforming to [STATE], if an explicitly specified Domain value does not start with a full stop ("."), the user agent MUST prepend a full stop for it; and, note that every Path begins with the "/" symbol.

2.3.2.8 The METHOD Element

By default, a policy reference applies to the stated URIs regardless of the method used to access the resource. However, a Web site may wish to define different P3P policies depending on the method to be applied to a resource. For example, a site may wish to collect more data from users when they are performing PUT or DELETE methods than when performing GET methods.

The METHOD element in a policy reference file is used to state that the enclosing policy reference only applies when the specified methods are used to access the referenced resources. The METHOD element may be repeated to indicate multiple applicable methods. If the METHOD element is not present in a POLICY-REF element, then that POLICY-REF element covers the resources indicated regardless of the method used to access them.

So, to state that /P3P/Policies.xml#first applies to all documents in the subtree /docs/ for GET and HEAD methods, while /P3P/Policies.xml#second applies for PUT and DELETE methods, the following policy reference would be written:

Example 2.6:

```
<META xmlns="http://www.w3.org/2001/09/P3Pv1">
 <POLICY-REFERENCES>
    <POLICY-REF about="/P3P/Policies.xml#first">
      <INCLUDE>/docs/*</INCLUDE>
      <METHOD>GET</METHOD>
      <METHOD>HEAD</METHOD>
    </POLICY-REF>
    <POLICY-REF about="/P3P/Policies.xml#second">
```

```
        <INCLUDE>/docs/*</INCLUDE>
        <METHOD>PUT</METHOD>
        <METHOD>DELETE</METHOD>
     </POLICY-REF>
  </POLICY-REFERENCES>
</META>
```

Note that HTTP requires the same behavior for GET and HEAD requests, thus it is inappropriate to specify different P3P policies for these methods. The syntax for the METHOD element is:

```
[17]    method-element          =    `<METHOD>` Method `</METHOD>`
```

Here, Method is defined in the section 5.1.1 of [HTTP1.1].

Finally, note that the METHOD element is designed to be used in conjunction with INCLUDE or COOKIE-INCLUDE elements. A METHOD element by itself will never apply a POLICY-REF to a URI.

2.3.3 Applying a Policy to a URI

A policy reference file specifies the policy which applies to a given URI. The meaning of this is that the indicated policy describes all effects of performing any of the methods listed in the policy reference file against the given URI.

There is a general rule which describes what it means for a P3P policy to cover a URI: *the referenced policy MUST cover actions that the user's client software is expected to perform as a result of requesting that URI*. Obviously, the policy must describe all data collection performed by site as a result of processing the request for the URI. Thus, if a given URI is covered for terms of GET requests, then the policy given by the policy reference file MUST describe all data collection performed by the site when that URI is fetched. Likewise, if a URI is covered for POST requests, then any data collection that occurs as a result of posting a form or other content to that URI MUST be described by the policy.

The concept of "actions that the client software is expected to perform" includes the setting of client-side cookies or other state-management mechanisms invoked by the response. If executable code is returned when a URI is requested, then the P3P policy covering that URI MUST cover certain actions which will occur when that code is executed. The covered actions are any actions which could take place without the user explicitly invoking them. If explicit user action causes data to be collected, then the P3P policy covering the URI for that action would disclose that data collection.

Some specific examples:

1. Fetching a URI returns an HTML page which contains a form, and the form contents are sent to a second URI when the user clicks a "Submit" button. The P3P policy covering the second URI MUST disclose all data collected by the form. The

P3P policy covering the first URI (the URI the form was loaded from) MAY or MAY NOT disclose any of the data that will be collected on the form.

2. An HTML page includes JavaScript code which tracks how long the page is displayed and whether the user moved the mouse over a certain object on the page; when the page is unloaded, the JavaScript code sends that information to the server where the HTML page originated. The activity of the JavaScript code MUST be covered by the P3P policy of the HTML page. The reasoning is that this activity takes place without the user's knowledge or consent, and it occurs automatically as a result of loading the page.

3. A response is an installable image for an electronic mail program. In order to use the email program, the user must run an installation program, start the email program, and use its facilities. The P3P policy covering URI from where the email program was downloaded is not required to make a statement about the data which could be collected by using the email program. Installing and running the email program is clearly outside the Web browsing experience, so it is not covered by this specification. A separate protocol could be designed to allow downloaded applications to present a P3P policy, but this is outside the scope of this specification.

4. An HTML page containing a form includes a reference to an executable which provides a custom client-side control. The data in the control is submitted to a site when the form is submitted. In this case, the URI for the HTML page and the URI for the custom control is not required to make a statement about the data the custom control represents. However, the URI to which the form contents are posted MUST cover the data from the custom control, just as it would cover any other data collected by processing the form. This behavior is similar to the way HTML forms are handled when they use only standard HTML controls: the control itself collects no data, and the data is collected when the form is posted. Note that this example assumes that the form is only posted when the user actively presses a "submit" or similar button. If the form were posted automatically (for example, by some JavaScript code in the page), then this example would be similar to example #2, and the data collected by the form MUST be described in the P3P policy which covers the HTML form.

5. Requests to a URI are redirected to a third party. If the first party embeds previously collected personal data in the query string or other part of the redirect URI, the privacy policy for the first party's URI MUST describe the types of data transmitted and include the third party as a recipient.

2.3.4 Forms and Related Mechanisms

Forms deserve special consideration, as they often link to CGI scripts or other server-side applications in their action URIs. It is often the case that those action URIs are covered by a different policy than the form itself.

If a user agent is unable to find a matching include-rule for a given *action URI* in the policy reference file that was referenced from the page, it SHOULD assume that *no* policy is in effect. Under these circumstances, user agents SHOULD check the well-known location on the host of the action URI to attempt to find a policy reference file that covers the action URI. If this does not provide a P3P policy to cover the action URI, then a user agent MAY try to retrieve the policy reference file by using the HINT mechanism on the action URI, and/or by issuing a HEAD request to the action URI before actually submitting any data in order to find the policy in effect. Services SHOULD ensure that server-side applications can properly respond to such HEAD requests and return the corresponding policy reference link in the headers. In case the underlying application does not understand the HEAD request and *no* policy has been predeclared for the action URI in question, user agents MUST assume that no policy is in effect and SHOULD inform the user about this or take the corresponding actions according to the user's preferences.

Note that services might want to make use of the <METHOD> element in order to declare policies for server-side applications that only cover a subset of supported methods, e.g., POST or GET. Under such circumstances, it is acceptable that the application in question only supports the methods given in the policy reference file (i.e., HEAD requests need not be supported). User agents SHOULD NOT attempt to issue a HEAD request to an action URI if the relevant methods specified in the form's method attribute have been properly predeclared in the page's policy reference file.

In some cases, *different* data is collected at the *same* action URI depending on some selection in the form. For example, a search service might offer to both search for people (by name and/or email) and (arbitrary) images. Using a set of radio buttons on the form, a single server-side application located at one and the same action URI handles both cases and collects the required information necessary for the search. If a service wants to predeclare the data collection practices of the server-side application it MAY declare *all* of the data collection practices in a *single* policy file (using a <INCLUDE> declaration matching the action URI). In this case, user agents MUST assume that all data elements are collected under every circumstance. This solution offers the convenience of a single policy but might not properly reflect the fact that only parts of the listed data elements are collected at a time. Services SHOULD make sure that a simple HEAD request to the action URI (i.e., without any arguments, especially without the value of the selected radio button) will return a policy that covers all cases.

Note that if a form is handled through use of the GET method, then the action URI reflects the choice of form elements selected by the user. In some cases, it will be possible to make use of the wildcard syntax allowed in policy reference files to specify different policies for different uses of the same form action-handler URI. Therefore, user agents MUST include the query-string portion of URIs when making comparisons with INCLUDE and EXCLUDE elements in policy reference files.

2.4 Additional Requirements

2.4.1 Non-ambiguity

User agents need to be able to determine unambiguously what policy applies to a given URI. Therefore, sites SHOULD avoid declaring more than one non-expired policy for a given URI. In some rare cases sites MAY declare more than one non-expired policy for a given URI, for example, during a transition period when the site is changing its policy. In those cases, the site will probably not be able to determine reliably which policy any given user has seen, and thus it MUST honor all policies (this is also the case for compact policies, cf. Section 4.1 and Section 4.6). Sites MUST be cautious in their practices when they declare multiple policies for a given URI, and ensure that they can actually honor all policies simultaneously.

If a policy reference file at the well-known location declares a non-expired policy for a given URI, this policy applies, regardless of any conflicting policy reference files referenced through HTTP headers or HTML link tags.

If an HTTP response includes references to more than one policy reference file, P3P user agents MUST ignore all references after the first one.

If an HTML file includes HTML `link` tag references to more than one policy reference file, P3P user agents MUST ignore all references after the first one.

If a user agent discovers more than one non-expired P3P policy for a given URI (for example because a page has both a P3P header and a `link` tag that reference different policy reference files, or because P3P headers for two pages on the site reference different policy reference files that declare different policies for the same URI), the user agent MAY assume any (or all) of these policies apply as the site MUST honor all of them.

2.4.2 Multiple Languages

Multiple language versions (translations) of the same policy can be offered by the server using the HTTP `"Content-Language"` header to properly indicate that a particular language has been used for the policy. This is useful so that human-readable fields such as entity and consequence can be presented in multiple languages. The same mechanism can also be used to offer multiple language versions for data schemas.

Whenever `Content-Language` is used to distinguish policies at the same URI that are offered in multiple languages, the policies MUST have the same meaning in each language. Two policies (or two data schemas) are taken to be identical if

- All formal (not natural language) protocol elements are semantically identical (i.e., attribute order does not matter, the presence or absence of a default value does not matter, but attribute values matter)
- All natural language protocol elements correspond one-to-one, and for each correspondence, one is a careful translation of the other.

Due to the use of the `Accept-Language` mechanism, implementers should take note that user agents may see different language versions of a policy or policy reference file despite sending the same `Accept-Language` request header if a new language version of a policy or data schema has been added.

2.4.3 The "Safe Zone"

P3P defines a special set of "safe zone" practices, which SHOULD be used by all P3P-enabled user agents and services for the communications which take place as part of fetching a P3P policy or policy reference file. In particular, requests to the well-known location for policy reference files SHOULD be covered by these "safe zone" practices. Communications covered by the safe zone practices SHOULD have only minimal data collection, and any data that is collected is used only in non-identifiable ways.

To support this safe zone, P3P user agents SHOULD suppress the transmission of data unnecessary for the purpose of finding a site's policy until the policy has been fetched. Therefore safe-zone practices for user agents include the following requirements:

- User agents SHOULD NOT send the HTTP `Referer` header in the safe zone
- User agents SHOULD NOT accept cookies from safe-zone requests
- User agents MAY also wish to refrain from sending user agent information or cookies accepted in a previous session on safe zone requests
- User agent implementers need to be aware that there is a privacy trade-off with using the `Accept-Language` HTTP header in the safe zone. Sending the correct `Accept-Language` header will allow fetching a P3P policy in the user's preferred natural language (if available), but does expose a certain amount of information about the identity of the user. User agents MAY wish to allow users to decide when these headers should be sent.

Safe-zone practices for servers include the following requirements:

- Servers SHOULD NOT require the receipt of an HTTP `Referer` header, cookies, user agent information, or other information unnecessary for responding to requests in the safe zone
- If the communications is taking place over a secure connection (such as SSL), then the server SHOULD NOT require an identity certificate from the user agent for safe zone requests
- In addition, servers SHOULD NOT use in an identifiable way any information collected while serving a safe zone request

Note that the safe zone requirements do not say that sites cannot keep identifiable information—only that they SHOULD NOT use in an identifiable way any information collected while serving a policy file. Tracking down the source of a denial of service attack, for example, would be a legitimate reason to use this information and ignore this recommendation.

2.4.4 Non-Discrimination of Policies

Servers SHOULD make every effort to help user agents find P3P policies. In particular, servers SHOULD place a policy reference file at the well-known location whenever possible. When the `P3P` HTTP header is used as an alternative, servers SHOULD:

- Reference a policy in response to any request: P3P-compliant sites SHOULD include a link to a policy reference file for a Webresource whenever possible.
- Support HTTP `HEAD` requests: P3P-compliant servers SHOULD support `HEAD` requests for any documents that can be retrieved with `GET` requests. Whenever technically feasible, servers should give a valid response to a `HEAD` request for documents that are normally accessed by other HTTP methods as well (such as `POST`).

2.4.5 Security of Policy Transport

P3P policies and references to P3P policies SHOULD NOT, in themselves, contain any sensitive information. This means that there are no additional security requirements for transporting a reference to a P3P policy beyond the requirements of the document it is associated with; so, if an HTML document would normally be served over a non-encrypted session, then the P3P protocol would not require nor recommend that the document be served over an encrypted session when a reference to a P3P policy is included with that document.

2.4.6 Policy Updates

Note that when a Web site changes its P3P policy, the old policy applies to data collected when it was in effect. It is the responsibility of the site to keep records of past P3P policies and policy reference files along with the dates when they were in effect, and to apply these policies appropriately.

If a site wishes to apply a new P3P policy to previously collected data, it MUST provide appropriate notice and opportunities for users to accept the new policy that are consistent with applicable laws, industry guidelines, or other privacy-related agreements the site has made.

2.4.7 Absence of Policy Reference File

If no policy reference file is available for a given site, user agents MUST assume (an empty) policy reference file exists at the well-known location with a 24 hour expiry, and therefore if the user returns to the site after 24 hours, the user agent MUST attempt to fetch a policy reference file from the well-known location again. User agents MAY check the well-known location more frequently, or upon a certain event such as the user clicking a browser refresh button. Sites MAY place a policy reference file at the well-known location that indicates that no policy is available, but set the expiry such that user agents know they need not check every 24 hours.

2.4.8 Asynchronous Evaluation

User agents MAY asynchronously fetch and evaluate P3P policies. That is, P3P policies need not necessarily be fetched and evaluated prior to other HTTP transactions. This behavior may be dependent on the user's preferences and the type of request being made. Until a policy is evaluated, the user agent SHOULD treat the site as if it has no privacy policy. Once the policy has been evaluated, the user agent SHOULD apply the user's preferences. To promote deterministic behavior, the user agent SHOULD defer application of a policy until a consistent point in time. For example, a web browser might apply a user's preferences just after the user agent completes a navigation, or when confirming a form submission.

2.5 Example Scenarios

As an aid to sites deploying P3P, several example scenarios are presented, along with descriptions of how P3P is used on those sites.

Scenario 1: Web site basic.example.com uses a variety of images, all of which it hosts. It also includes some forms, which are all submitted directly to the site. This site can declare a single P3P policy for the entire site (or if different privacy policies apply to different parts of the site, it can declare multiple P3P policies). As long as all of the images and form action URIs are in directories covered by the site's P3P policy, user agents will automatically recognize the images and forms as covered by the site's policy.

Scenario 2: Web site busy.example.com uses a content distribution network called cdn.example.com to host its images so as to reduce the load on its servers. Thus, all of the images on the site have URIs at cdn.example.com. CDN acts as an agent to Busy in this situation, and collects no data other than log data. This log data is used only for Web site and system administration in support of providing the services that Busy contracted for. Busy's privacy policy applies to the images hosted by CDN, so Busy uses the HINT element in its policy reference file to point to a suitable policy reference file at CDN, indicating that such images are covered by example.com P3P policy.

Scenario 3: Web site busy.example.com also has a contract with an advertising company called clickads.example.com to provide banner ads on its site. The contract allows Clickads to set cookies so as to make sure each user doesn't see a given ad more than three times. Clickads collects statistics on how many users view each ad and reports them to the companies whose products are being advertised. But these reports do not reveal information about any individual users. As was the case in Scenario 2, Busy's privacy policies applies to these ads hosted by Clickads, so Busy uses the HINT element in its policy reference file to point to a suitable policy reference file at Clickads, indicating that Busy P3P policy applies to such embedded content served by clickads.example.com. The companies whose products are being advertised need not be mentioned in the Busy privacy policy because the only data they are receiving is aggregate data.

Scenario 4: Web site busy.example.com also has a contract with funchat.example.com to host a chat room for its users. When users enter the chat room they are actually leaving the Busy site. However, the chat room has the Busy logo and is actually covered by the Busy privacy policy. In this instance Funchat is acting as an agent for Busy, but—unlike the previous examples—their content is not embedded in the Busy site. Busy can use the HINT element in its policy reference file to point to a suitable Funchat policy reference file, that indicates that Funchat chat room is covered by Busy privacy policy, therefore facilitating a smoother transition to the chat room.

Scenario 5: Web site bigsearch.example.com has a form that allows users to type in a search query and have it performed on their choice of search engines located on other sites. When a user clicks the "submit" button, the search query is actually submitted directly to these search engines—the action URI is not on bigsearch.example.com but rather on the search engine selected by the user. Bigsearch can declare the privacy policies for these search engines by using the HINT element to point to their corresponding policy reference files. So when a user clicks the "submit" button, their user agent can check its privacy policy before posting any data. In order to make this search choice mechanism work, Bigsearch might actually have a form with an action URI on its own site, which redirects to the appropriate search engine. In this case, the user agent should check the search engine privacy policy upon receiving the redirect response.

Scenario 6: Web site bigsearch.example.com also has a form that allows users to type in a search query and have it simultaneously performed on ten different search engines. Bigsearch submits the queries, gets back the results from each search engine, removes the duplicates, and presents the results to the user. In this case, the user interacts only with Bigsearch. Thus, the only P3P policy involved is the one that covers the Bigsearch Web site. However, Bigsearch must disclose that it shares the users' search queries with third parties (the search Web sites), unless Bigsearch has a contract with these search engines and they act as agents to Bigsearch.

Scenario 7: Web site bigsearch.example.com also has banner advertisements provided by a company called adnetwork.example.com. Adnetwork uses cookies to develop profiles of users across many different Web sites so that it can provide them with ads better suited to their interests. Because the data about the sites that users are visiting is being used for purposes other than just serving ads on the Bigsearch Web site, Adnetwork cannot be considered an agent in this context. Adnetwork must create its own P3P policy and use its own policy reference file to indicate what content it applies to. In addition, Bigsearch may optionally use the `HINT` element in its policy reference file to indicate that the Adnetwork P3P policy reference file applies to these advertisements. Bigsearch should only do this if Adnetwork has told it what P3P policy applies to these advertisements and has agreed to notify Bigsearch if the policy reference needs to be changed.

Scenario 8: Web site busy.example.com uses cookies throughout its web site. It discloses a cookie policy, separate from its regular P3P policy to cover these cookies. It uses the `COOKIE-INCLUDE` element in its policy reference file to declare the appropriate policy for these cookies. As a performance optimization, it also makes available a compact policy by sending a P3P header that includes this compact policy whenever it sets a cookie.

Scenario 9: Web site config.example.com provides a service in which they optimize various kinds of web content based on each user's computer and Internet configuration. Users go to the Config web site and answer questions about their computer, monitor, and Internet connection. Config encodes the responses and stores them in a cookie. Later, when the user is visiting Busy—a Web site that has contracted with Config—whenever the browser requests content that can be optimized (certain images, audio files, etc.), Busy will redirect the user to Config, which will read the user's cookie, and deliver the appropriate content. In this case, Config should declare a privacy policy that describes the kinds of data collected and stored in its cookies, and how that data is used. It should use a `COOKIE-INCLUDE` element in its policy reference file to declare the policy for the cookies. It will probably reference Busy's P3P policy for the actual images or audio files delivered, as it is acting much like CDN acts in scenario 2. Busy will probably also use `HINT` elements in its policy reference file to reference the policy for the Config-delivered content.

3 POLICY SYNTAX AND SEMANTICS

P3P policies are encoded in XML. They may also be represented using the RDF data model ([RDF]); however, an RDF representation is not included in this specification. (Such a representation is planned to be made available as a W3C Note prior to submitting P3P as a Proposed Recommendation, together with a suitable RDF encoding of the policy reference file).

Section 3.1 begins with an example of an English language privacy policy and a corresponding P3P policy. P3P policies include general assertions that apply to the entire policy as well as specific assertions—called *statements*—that apply only to the handling of particular types of data referred to by *data references*. Section 3.2 describes the POLICY element and policy-level assertions. Section 3.3 describes statements and data references.

3.1 Example Policies

3.1.1 English Language Policies

The following are two examples of English-language privacy policy to be encoded as a P3P policy. Both policies are for one example company, CatalogExample, which has different policies for those browsing their site and those actually purchasing products. Example 3.1. is provided in both English and as a more formal description using P3P element and attribute names.

Example 3.1: CatalogExample's Privacy Policy for Browsers
_____

At CatalogExample, we care about your privacy. When you come to our site to look for an item, we will only use this information to improve our site and will not store it with information we could use to identify you.

CatalogExample, Inc. is a licensee of the PrivacySealExample Program. The PrivacySealExample Program ensures your privacy by holding Web site licensees to high privacy standards and confirming with independent auditors that these information practices are being followed.

Questions regarding this statement should be directed to:

```
CatalogExample
4000 Lincoln Ave.
Birmingham, MI 48009 USA
email: catalog@example.com
Telephone 248-EXAMPLE (248-392-6753)
```

If we have not responded to your inquiry or your inquiry has not been satisfactorily addressed, you can contact PrivacySealExample at *http://www.privacyseal.example.org.* CatalogExample will correct all errors or wrongful actions arising in connection with the privacy policy.

What We Collect and Why:

When you browse through our site we collect:

• the basic information about your computer and connection to make sure that we can get you the proper information and for security purposes.

• aggregate information on what pages consumers access or visit to improve our site.

Data retention:

We purge every two weeks the browsing information that we collect.

Here is Example 3.1 in a more formal description, using the P3P element and attribute names [with the section of the spec that was used cited in brackets for easy reference]:

- Disclosure URI:
  http://www.catalog.example.com/PrivacyPracticeBrowsing.html
  [3.2.2 Policy]

- Entity: CatalogExample
  4000 Lincoln Ave.
  Birmingham, MI 48009
  USA
  catalog@example.com
  +1 (248) 392-6753
  [3.2.4 Entity]

- Access to Identifiable Information: None
  [3.2.5 Access]

- Disputes:
  resolution type: independent
  service: *http://www.privacyseal.example.org*
  description: PrivacySealExample
  [3.2.6 Disputes]

- Remedies: we'll correct any harm done wrong
  [3.2.7 Remedies]

- We collect:
  dynamic.clickstream
  dynamic.http
  [4.5 Base data schema]

- For purpose: Web site and system administration, research and development
  [3.3.4 Purpose]

- Recipients: Only ourselves and our agents
  [3.3.5 Recipients]

- Retention: As long as appropriate for the stated purposes
  [3.3.6 Retention]

(Note also that the site's human-readable privacy policy MUST mention that data is purged every two weeks, or provide a link to this information.)

Example 3.2: CatalogExample's Privacy Policy for Shoppers

At CatalogExample, we care about your privacy. We will never share your credit card number or any other financial information with any third party. With your permission only, we will share information with carefully selected marketing partners that meet either the preferences that you've specifically provided or your past purchasing habits. The more we know about your likes and dislikes, the better we can tailor offerings to your needs.

CatalogExample is a licensee of the PrivacySealExample Program. The PrivacySealExample Program ensures your privacy by holding Web site licensees to high privacy standards and confirming with independent auditors that these information practices are being followed.

Questions regarding this statement should be directed to:

```
CatalogExample
4000 Lincoln Ave.
Birmingham, MI 48009 USA
email: catalog@example.com
Telephone +1 248-EXAMPLE (+1 248-392-6753)
```

If we have not responded to your inquiry or your inquiry has not been satisfactorily addressed, you can contact PrivacySealExample: *http://privacyseal.example.org/privacyseal*. CatalogExample will correct all errors or wrongful actions arising in connection with the privacy policy.

When you browse through our site we collect:

- the basic information about your computer and connection to make sure that we can get you the proper information and for security purposes; and

- aggregate information on what pages consumers access or visit to improve our site

If you choose to purchase an item we will ask you for more information including:

- your name and address so that we can have your purchase delivered to you and so we can contact you in the future;

- your email address and telephone number so we can contact you;

- a login and password to use to update your information at any time in the future; and

- financial information to complete your purchase (you may choose to store this for future use)

- optionally, you can enter other demographic information so that we can tailor services to you in the future.

Also on this page we will give you the option to choose if you would like to receive email, telephone calls or written service from CatalogExample or from our carefully selected marketing partners who maintain similar privacy practices. If you would like to receive these solicitations simply check the appropriate boxes. You can choose to stop participating at any time simply by changing your preferences.

Changing and Updating personal information

Consumers can change all of their personal account information by going to the preferences section of CatalogExample at *http://catalog.example.com/preferences.html*. You can change your address, telephone number, email address, password as well as your privacy settings.

Cookies

CatalogExample uses cookies only to see if you have been an CatalogExample customer in the past and, if so, customize services based on your past browsing habits and purchases. We do not store any personal data in the cookie nor do we share or sell any of the information with other parties or affiliates.

Data retention

We will keep the information about you and your purchases for as long as you remain our customer. If you do not place an order from us for one year we will remove your information from our databases.

3.1.2 XML Encoding of Policies

The following pieces of [XML] capture the information as expressed in the above two examples. P3P policies are statements that are properly expressed as well-formed XML. The policy syntax will be explained in more detail in the sections that follow.

XML Encoding of Example 3.1:

```
<POLICY name="forBrowsers"
    discuri="http://www.catalog.example.com/
        PrivacyPracticeBrowsing.html">
 <ENTITY>
  <DATA-GROUP>
   <DATA ref="#business.name">CatalogExample</DATA>
   <DATA ref="#business.contact-info.postal.street">4000 Lincoln
       Ave.</DATA>
   <DATA ref="#business.contact-info.postal.city">Birmingham</DATA>
   <DATA ref="#business.contact-info.postal.stateprov">MI</DATA>
   <DATA ref="#business.contact-info.postal.postalcode">48009</DATA>
   <DATA ref="#business.contact-info.postal.country">USA</DATA>
   <DATA ref="#business.contact-info.online.email">
       catalog@example.com</DATA>
   <DATA ref="#business.contact-info.telecom.telephone.intcode">1
       </DATA>
   <DATA ref="#business.contact-info.telecom.telephone.loccode">248
       </DATA>
   <DATA ref="#business.contact-info.telecom.telephone.number">
       3926753</DATA>
  </DATA-GROUP>
 </ENTITY>
 <ACCESS><nonident/></ACCESS>
 <DISPUTES-GROUP>
  <DISPUTES resolution-type="independent"
    service="http://www.PrivacySeal.example.org"
    short-description="PrivacySeal.example.org">
   <IMG src="http://www.PrivacySeal.example.org/Logo.gif"
       alt="PrivacySeal's logo"/>
   <REMEDIES><correct/></REMEDIES>
  </DISPUTES>
 </DISPUTES-GROUP>
 <STATEMENT>
  <PURPOSE><admin/><develop/></PURPOSE>
  <RECIPIENT><ours/></RECIPIENT>
  <RETENTION><stated-purpose/></RETENTION> <!-- Note also that the
  site's human-readable  privacy policy MUST mention that data is
  purged every two weeks, or provide a link to this information. -->
  <DATA-GROUP>
   <DATA ref="#dynamic.clickstream"/>
   <DATA ref="#dynamic.http"/>
  </DATA-GROUP>
 </STATEMENT>
</POLICY>
```

XML Encoding of Example 3.2:

```
<POLICY name="forShoppers"
    discuri="http://www.catalog.example.com/Privacy/
        PrivacyPracticeShopping.html"
    opturi="http://catalog.example.com/preferences.html">
 <ENTITY>
  <DATA-GROUP>
   <DATA ref="#business.name">CatalogExample</DATA>
   <DATA ref="#business.contact-info.postal.street">4000 Lincoln
       Ave.</DATA>
   <DATA ref="#business.contact-info.postal.city">Birmingham</DATA>
   <DATA ref="#business.contact-info.postal.stateprov">MI</DATA>
   <DATA ref="#business.contact-info.postal.postalcode">48009</DATA>
   <DATA ref="#business.contact-info.postal.country">USA</DATA>
   <DATA ref="#business.contact-info.online.email">
       catalog@example.com</DATA>
   <DATA ref="#business.contact-info.telecom.telephone.intcode">1
       </DATA>
   <DATA ref="#business.contact-info.telecom.telephone.loccode">248
       </DATA>
   <DATA ref="#business.contact-info.telecom.telephone.number">
       3926753</DATA>
  </DATA-GROUP>
 </ENTITY>
 <ACCESS><contact-and-other/></ACCESS>
 <DISPUTES-GROUP>
  <DISPUTES resolution-type="independent"
    service="http://www.PrivacySeal.example.org"
    short-description="PrivacySeal.example.org">
   <IMG src="http://www.PrivacySeal.example.org/Logo.gif" alt=
       "PrivacySeal's logo"/>
   <REMEDIES><correct/></REMEDIES>
  </DISPUTES>
 </DISPUTES-GROUP>
 <STATEMENT>
  <CONSEQUENCE>
    We record some information in order to serve your request
    and to secure and improve our Web site.
  </CONSEQUENCE>
  <PURPOSE><admin/><develop/></PURPOSE>
  <RECIPIENT><ours/></RECIPIENT>
  <RETENTION><stated-purpose/></RETENTION>
  <DATA-GROUP>
   <DATA ref="#dynamic.clickstream"/>
   <DATA ref="#dynamic.http.useragent"/>
  </DATA-GROUP>
 </STATEMENT>
 <STATEMENT>
  <CONSEQUENCE>
    We use this information when you make a purchase.
```

```
    </CONSEQUENCE>
    <PURPOSE><current/></PURPOSE>
    <RECIPIENT><ours/></RECIPIENT>
    <RETENTION><stated-purpose/></RETENTION>
    <DATA-GROUP>
     <DATA ref="#user.name"/>
     <DATA ref="#user.home-info.postal"/>
     <DATA ref="#user.home-info.telecom.telephone"/>
     <DATA ref="#user.business-info.postal"/>
     <DATA ref="#user.business-info.telecom.telephone"/>
     <DATA ref="#user.home-info.online.email"/>
     <DATA ref="#user.login.id"/>
     <DATA ref="#user.login.password"/>
     <DATA ref="#dynamic.miscdata">
      <CATEGORIES><purchase/></CATEGORIES>
     </DATA>
    </DATA-GROUP>
   </STATEMENT>
   <STATEMENT>
    <CONSEQUENCE>
      At your request, we will send you carefully selected marketing
      solicitations that we think you will be interested in.
    </CONSEQUENCE>
    <PURPOSE>
     <contact required="opt-in"/>
     <individual-decision required="opt-in"/>
     <tailoring required="opt-in"/>
    </PURPOSE>
    <RECIPIENT><ours/><same required="opt-in"/></RECIPIENT>
    <RETENTION><stated-purpose/></RETENTION>
    <DATA-GROUP>
     <DATA ref="#user.name" optional="yes"/>
     <DATA ref="#user.home-info.postal" optional="yes"/>
     <DATA ref="#user.home-info.telecom.telephone" optional="yes"/>
     <DATA ref="#user.business-info.postal" optional="yes"/>
     <DATA ref="#user.business-info.telecom.telephone" optional="yes"/>
     <DATA ref="#user.home-info.online.email" optional="yes"/>
    </DATA-GROUP>
   </STATEMENT>
   <STATEMENT>
    <CONSEQUENCE>
      We allow you to set a password so that you
      can access your own information.
    </CONSEQUENCE>
    <PURPOSE><individual-decision required="opt-in"/></PURPOSE>
    <RECIPIENT><ours/></RECIPIENT>
    <RETENTION><stated-purpose/></RETENTION>
    <DATA-GROUP>
     <DATA ref="#dynamic.miscdata">
      <CATEGORIES><uniqueid/></CATEGORIES>
     </DATA>
```

```
      </DATA-GROUP>
    </STATEMENT>
    <STATEMENT>
     <CONSEQUENCE>
        At your request, we will tailor our site and
        highlight products related to your interests.
     </CONSEQUENCE>
     <PURPOSE>
        <pseudo-decision required="opt-in"/>
        <tailoring required="opt-in"/>
     </PURPOSE>
     <RECIPIENT><ours/></RECIPIENT>
     <RETENTION><stated-purpose/></RETENTION>
     <DATA-GROUP>
      <DATA ref="#user.bdate.ymd.year" optional="yes"/>
      <DATA ref="#user.gender" optional="yes"/>
     </DATA-GROUP>
    </STATEMENT>
    <STATEMENT>
     <CONSEQUENCE>
        We tailor our site based on your past visits.
     </CONSEQUENCE>
     <PURPOSE><tailoring/><develop/></PURPOSE>
     <RECIPIENT><ours/></RECIPIENT>
     <RETENTION><stated-purpose/></RETENTION>
     <DATA-GROUP>
      <DATA ref="#dynamic.cookies">
       <CATEGORIES><state/></CATEGORIES>
      </DATA>
      <DATA ref="#dynamic.miscdata">
       <CATEGORIES><preference/></CATEGORIES>
      </DATA>
     </DATA-GROUP>
    </STATEMENT>
   </POLICY>
```

3.2 Policies
This section defines the syntax and semantics of P3P policies.

All policies MUST be encoded using [UTF-8]. P3P servers MUST encode their policies using this encoding. P3P user agents MUST be able to parse this syntax. User agents SHOULD NOT act upon or render policy data containing a syntax error. Furthermore, user agents MUST NOT act upon or render policy data containing a syntax error, unless the user has been informed in a meaningful way that there may be an error, or unless the user's preferences specify that errors may be ignored.

Policies have to be placed inside a POLICIES element.

3.2.1 The POLICIES Element
The POLICIES element is used to gather several P3P policies together in a single file. This is provided as a performance optimization: many policies can be collected with a sin-

gle request, improving network traffic and caching. Even, the POLICIES element can be placed in the well-known location, inside the META element: in this case, user agents need only fetch a single file, containing both the policy reference file and the policies.

The POLICIES element can optionally contain an EXPIRY element, indicating the expiration of the included policies, and an embedded data schema using the DATASCHEMA element (see Section 5).

Since policies are included in a POLICIES element, they MUST have a name attribute which is unique in the file. This allows policy references (in POLICY-REF elements) to link to that policy.

Example 3.3:

The file in http://www.example.com/Shop/policies.xml could have the following content:

```
<POLICIES xmlns="http://www.w3.org/2001/09/P3Pv1">
    <POLICY name="policy1" discuri="http://www.example.com/disc1"> ....
        </POLICY>
    <POLICY name="policy2" discuri="http://www.example.com/disc2"> ....
        </POLICY>
    <POLICY name="policy3" discuri="http://www.example.com/disc3"> ....
        </POLICY>
</POLICIES>
```

The files in http://www.example.com/Shop/CDs/* could then be associated to the second policy ("policy2") using the following policy reference file in http://www.example.com/w3c/p3p.xml:

```
<META xmlns="http://www.w3.org/2001/09/P3Pv1">
<POLICY-REFERENCES>
    <POLICY-REF about="/Shop/policies#policy2">
        <INCLUDE>/Shops/CDs/*</INCLUDE>
    </POLICY-REF>
  </POLICY-REFERENCES>
</META>
```

```
[18]    policies    =    `<POLICIES xmlns="http://www.w3.org/2001/09/
                         P3Pv1">`


                         [expiry]
                         [dataschema]
                         *policy
                         "</POLICIES>"
```

3.2.2 The POLICY Element

The POLICY element contains a complete P3P policy. Each P3P policy MUST contain exactly one POLICY element. The policy element MUST contain an ENTITY element that identifies the legal entity making the representation of the privacy practices contained in the policy. In addition, the policy element MUST contain an ACCESS element, and optionally STATEMENT elements, a DISPUTES-GROUP element, a P3P data schema, and one or more extensions.

<POLICY>

Includes one or more statements. Each statement includes a set of disclosures as applied to a set of data elements.

- name (*mandatory attribute*)
  name of the policy, used as a fragment identifier to be able to reference the policy.
- discuri (*mandatory attribute*)
  URI of the natural language privacy statement
- opturi
  URI of instructions that users can follow to request or decline to have their data used for a particular purpose (opt-in or opt-out). This attribute is *mandatory* for policies that contain a purpose with required attribute set to opt-in or opt-out.

```
[19]    policy         =    `<POLICY name=` quotedstring
                                ` discuri=` quoted-URI
                                [` opturi=` quoted-URI]
                                `>`
                            *extension
                            [test]
                            entity
                            access
                            [disputes-group]
                            *statement-block
                            *extension
                            `</POLICY>`

[20]    quoted-URI    =    `"` URI `"`
```

Here, URI is defined as per RFC 2396 [URI].

3.2.3 The TEST Element

The TEST element is used for testing purposes: the presence of TEST in a policy indicates that the policy is just an example, and as such, it MUST be ignored, and not be considered as a valid P3P policy.

```
[21]              test         =              "<TEST/>"
```

3.2.4 The ENTITY Element

The ENTITY element gives a precise description of the legal entity making the representation of the privacy practices.

```
<ENTITY>
```

Identifies the legal entity making the representation of the privacy practices contained in the policy.

The ENTITY element contains a description of the legal entity consisting of DATA elements referencing (all or some of) the fields of the business dataset: it MUST contain both the legal entity's name as well as contact information such as postal address, telephone number, email address, or other information that individuals may use to contact the entity about their privacy policy. Note that some laws and codes of conduct require entities to include a postal address or other specific information in their contact information.

```
[22]     entity                  =    "<ENTITY>"
                                       *extension
                                       entitydescription
                                       *extension
                                       "</ENTITY>"

[23]     entitydescription       =    "<DATA-GROUP>"
                                       `<DATA ref="#business.name"/>`
                                       PCDATA "</DATA>"
                                       *(`<DATA ref="#business.` string `"/
                                       >` PCDATA "</DATA>")
                                       "</DATA-GROUP>"
```

Here, string is defined as a sequence of characters (with " and & escaped) among the values that are allowed by the business dataset. PCDATA is defined as in [XML].

3.2.5 The ACCESS Element

The ACCESS element indicates whether the site provides access to various kinds of information.

```
<ACCESS>
```

The ability of the individual to view identified data and address questions or concerns to the service provider. Service providers MUST disclose one value for the access attribute. The method of access is not specified. Any disclosure (other than <all/>) is not meant to imply that access to all data is possible, but that some of the data may be accessible and that the user should communicate further with the service provider to determine what capabilities they have.

Note that service providers may also wish to provide capabilities to access information collected through means other than the Web at the discuri. However, the scope of P3P statements are limited to data collected through HTTP or other Web transport protocols.

Also, if access is provided through the Web, use of strong authentication and security mechanisms for such access is recommended; however, security issues are outside the scope of this document.

The ACCESS element must contain one of the following elements:

- `<nonident/>`
  Web site does not collect identified data.
- `<all/>`
  All Identified Data: access is given to all identified data.
- `<contact-and-other/>`
  Identified Contact Information and Other Identified Data: access is given to identified online and physical contact information as well as to certain other identified data.
- `<ident-contact/>`
  Identifiable Contact Information: access is given to identified online and physical contact information (e.g., users can access things such as a postal address).
- `<other-ident/>`
  Other Identified Data: access is given to certain other identified data (e.g., users can access things such as their online account charges).
- `<none/>`
  None: no access to identified data is given.

```
[24]   access                =  "<ACCESS>"
                                 access_disclosure
                                 *extension
                                 "</ACCESS>"

[25]   access_disclosure     =  "<nonident/>"              | ; Identified
                                                               Data is
                                                               Not Used
                                 "<all/>"                  | ; All
                                                               Identifiable
                                                               Information
                                 "<contact-and-other/>"    | ; Identified
                                                               Contact
                                                               Information
                                                               and Other
                                                               Identified
                                                               Data
                                 "<ident-contact/>"        | ; Identifiable
                                                               Contact
                                                               Information
                                 "<other-ident/>"          | ; Other
                                                               Identified
                                                               Data
                                 "<none/>"                   ; None
```

3.2.6 The DISPUTES Element

A policy SHOULD contain a DISPUTES-GROUP element, which contains one or more DISPUTES elements. These elements describe dispute resolution procedures that may be followed for disputes about a service's privacy practices. Each DISPUTES element can optionally contain a LONG-DESCRIPTION element, an IMG element, and a REMEDIES element. Service providers with multiple dispute resolution procedures should use a separate DISPUTES element for each. Since different dispute procedures have separate remedy processes, each DISPUTES element would need a separate LONG-DESCRIPTION, IMG tag and REMEDIES element, if they are being used.

```
<DISPUTES>
```
Describes dispute resolution procedures that may be followed for disputes about a service's privacy practices, or in case of protocol violation.
- resolution-type (*mandatory attribute*)
  Takes one of the following four values:
  1. Customer Service [service]
     Individual may complain to the Web site's customer service representative for resolution of disputes regarding the use of collected data. The description MUST include information about how to contact customer service.
  2. Independent Organization [independent]
     Individual may complain to an independent organization for resolution of disputes regarding the use of collected data. The description MUST include information about how to contact the third party organization.
  3. Court [court]
     Individual may file a legal complaint against the Web site.
  4. Applicable Law [law]
     Disputes arising in connection with the privacy statement will be resolved in accordance with the law referenced in the description.
- service (*mandatory attribute*)
  URI of the customer service Web page or independent organization, or URI for information about the relevant court or applicable law
- verification
  URI or certificate that can be used for verification purposes. It is anticipated that seal providers will provide a mechanism for verifying a site's claim that they have a seal.
- short-description
  A short human readable description of the name of the appropriate legal forum, applicable law, or third party organization; or contact information for customer service if not already provided at the service URI. No more than 255 characters.

The DISPUTES element can contain a LONG-DESCRIPTION element, where a human readable description is present: this should contain the name of the appropriate legal forum, applicable law, or third party organization; or contact information for customer service if not already provided at the service URI.

`<LONG-DESCRIPTION>`
This element contains a (possibly long) human readable description.

`<IMG>`
An image logo (for example, of the independent organization or relevant court)
- `src` (*mandatory attribute*)
  URI of the image logo
- `width`
  width in pixels of the image logo
- `height`
  height in pixels of the image logo
- `alt` (*mandatory attribute*)
  very short textual alternative for the image logo

```
[26]   disputes-group    =   "<DISPUTES-GROUP>"
                              1*dispute
                              *extension
                              "</DISPUTES-GROUP>"

[27]   dispute           =   "<DISPUTES"
                              " resolution-type=" '"'("service"|
                                  "independent"|"court"|"law")'"'
                              " service=" quoted-URI
                              [" verification=" quotedstring]
                              [" short-description=" quotedstring]
                              ">"
                              *extension
                              [longdescription]
                              [image]
                              [remedies]
                              *extension
                              "</DISPUTES>"

[28]   longdescription   =   <LONG-DESCRIPTION> PCDATA
                                  </LONG-DESCRIPTION>

[29]   image             =   "<IMG src=" quoted-URI
                              [" width=" `"` number `"`]
                              [" height=" `"` number `"`]
                              " alt=" quotedstring
                              "/>"

[30]   quotedstring      =   `"` string `"`
```

Here, string is defined as a sequence of characters (with " and & escaped), and PCDATA is defined as in [XML].

Note that there can be multiple assurance services, specified via multiple occurrences of DISPUTES within the DISPUTES-GROUP element. These fields are expected to be used in a number of ways, including representing that one's privacy practices are self assured, audited by a third party, or under the jurisdiction of a regulatory authority.

3.2.7 The REMEDIES Element

Each DISPUTES element SHOULD contain a REMEDIES element that specifies the possible remedies in case a policy breach occurs.

<REMEDIES>
Remedies in case a policy breach occurs.
The REMEDIES element must contain one or more of the following:
-
  Errors or wrongful actions arising in connection with the privacy policy will be remedied by the service.
-
  If the service provider violates its privacy policy it will pay the individual an amount specified in the human readable privacy policy or the amount of damages.
-
  Remedies for breaches of the policy statement will be determined based on the law referenced in the human readable description.

```
[31]        remedies         =        "<REMEDIES>"
                                       1*remedy
                                       *extension
                                       "</REMEDIES>"

[32]        remedy           =        "<correct/>"  |
                                       "<money/>"   |
                                       "<law/>"
```

3.3 Statements

Statements describe data practices that are applied to particular types of data.

3.3.1 The STATEMENT Element

The STATEMENT element is a container that groups together a PURPOSE element, a RECIPIENT element, a RETENTION element, a DATA-GROUP element, and optionally a CONSEQUENCE element and one or more extensions. All of the data referenced by the DATA-GROUP is handled according to the disclosures made in the other elements contained by the statement. Thus, sites may group elements that are handled the same way and create a statement for each group. Sites that would prefer to disclose separate purposes and other information for each kind of data they collect can do so by creating a separate statement for each data element.

`<STATEMENT>`
data practices as applied to data elements.

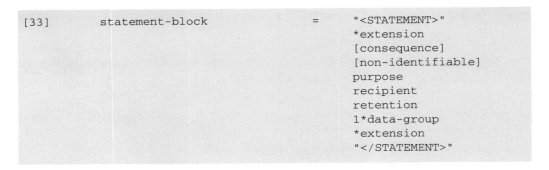

```
[33]        statement-block              =      "<STATEMENT>"
                                                *extension
                                                [consequence]
                                                [non-identifiable]
                                                purpose
                                                recipient
                                                retention
                                                1*data-group
                                                *extension
                                                "</STATEMENT>"
```

To simplify practice declaration, service providers may aggregate any of the disclo-sures (purposes, recipients, and retention) within a statement over data elements. Service providers MUST make such aggregations as an additive operation. For instance, a site that distributes your age to `ours` (ourselves and our agents), but distributes your postal code to `unrelated` (unrelated third parties), MAY say they distribute your name and postal code to `ours` and `unrelated`. Such a statement appears to distribute more data than actually hap-pens. It is up to the service provider to determine if their disclosure deserves specificity or brevity. Note that when aggregating disclosures across statements that include the NON-IDENTIFIABLE element, this element may be included in the aggregated statement only if it would otherwise appear in every statement if the statements were written separately.

Also, one must always disclose all options that apply. Consider a site with the sole purpose of collecting information for the purposes of `contact` (Contacting Visitors for Marketing of Services or Products). Even though this is considered to be for the `current` (Completion and Support of Current Activity) purpose, the site must state both `contact` and `current` purposes. Consider a site which distributes information to `ours` in order to redistribute it to `public`: the site must state both `ours` and `public` recipients.

Service providers often aggregate data they collect. Sometimes this aggregate data may be used for different purposes than the original data, shared more widely than the orig-inal data, or retained longer than the original data. For example many sites publish or dis-close to their advertisers statistics such as number of visitors to their Web site, percentage of visitors who fit into various demographic groups, etc. When aggregate statistics are used or shared such that it would not be possible to derive data for individual people or house-holds based on these statistics, no disclosures about these statistics are necessary in a P3P policy. However, services MUST disclose the fact that the original data is collected and declare any use that is made of the data before it is aggregated.

3.3.2 The CONSEQUENCE Element

STATEMENT elements may optionally contain a CONSEQUENCE element that can be shown to a human user to provide further explanation about a site's practices.

```
<CONSEQUENCE>
```
Consequences that can be shown to a human user to explain why the suggested practice may be valuable in a particular instance even if the user would not normally allow the practice.

```
[34]        consequence              =       "<CONSEQUENCE>"
                                             PCDATA
                                             "</CONSEQUENCE>"
```

3.3.3 The NON-IDENTIFIABLE Element

STATEMENT elements may optionally contain a NON-IDENTIFIABLE element, only when the requirements specified below are fulfilled.

```
<NON-IDENTIFIABLE/>
```
This is an element that can only be present in the statement, if there is no data or no identifiable data collected. Data is seen as non-identifiable in the sense of the present specification, if there is no reasonable way for the entity or a third party to attach the collected data to the identity of a natural person.

If the `<NON-IDENTIFIABLE/>` element is present, a human-readable explanation of how this is achieved MUST be included at the discuri.

```
[35]     non-identifiable              =       "<NON-IDENTIFIABLE/>"
```

3.3.4 The PURPOSE Element

Each STATEMENT element MUST contain a PURPOSE element that contains one or more purposes of data collection or uses of data. Sites MUST classify their data practices into one or more of the purposes specified below.

```
<PURPOSE>
```
Purposes for data processing relevant to the Web.

The PURPOSE element MUST contain one or more of the following:

- `<current/>`

  Completion and Support of Activity For Which Data Was Provided: Information may be used by the service provider to complete the activity for which it was provided, whether a one-time activity such as returning the results from a Web search, forwarding an email message, or placing an order; or a recurring activity such as providing a subscription service, or allowing access to an online address book or electronic wallet.

- `<admin/>`

  Web Site and System Administration: Information may be used for the technical support of the Web site and its computer system. This would include processing computer account information, information used in the course of securing and maintaining the site, and verification of Web site activity by the site or its agents.

- `<develop/>`
Research and Development: Information may be used to enhance, evaluate, or otherwise review the site, service, product, or market. This does not include personal information used to tailor or modify the content to the specific individual nor information used to evaluate, target, profile or contact the individual.

- `<tailoring/>`
One-time Tailoring: Information may be used to tailor or modify content or design of the site where the information is used only for a single visit to the site and not used for any kind of future customization. For example, an online store that suggests other items a visitor may wish to purchase based on the items he has already placed in his shopping basket.

- `<pseudo-analysis/>`
Pseudonymous Analysis: Information may be used to create or build a record of a particular individual or computer that is tied to a pseudonymous identifier, without tying identified data (such as name, address, phone number, or email address) to the record. This profile will be used to determine the habits, interests, or other characteristics of individuals *for purpose of research, analysis and reporting*, but it will not be used to attempt to identify specific individuals. For example, a marketer may wish to understand the interests of visitors to different portions of a Web site.

- `<pseudo-decision/>`
Pseudonymous Decision: Information may be used to create or build a record of a particular individual or computer that is tied to a pseudonymous identifier, without tying identified data (such as name, address, phone number, or email address) to the record. This profile will be used to determine the habits, interests, or other characteristics of individuals *to make a decision that directly affects that individual*, but it will not be used to attempt to identify specific individuals. For example, a marketer may tailor or modify content displayed to the browser based on pages viewed during previous visits.

- `<individual-analysis/>`
Individual Analysis: Information may be used to determine the habits, interests, or other characteristics of individuals and combine it with identified data *for the purpose of research, analysis and reporting*. For example, an online Web site for a physical store may wish to analyze how online shoppers make offline purchases.

- `<individual-decision/>`
Individual Decision: Information may be used to determine the habits, interests, or other characteristics of individuals and combine it with identified data *to make a decision that directly affects that individual*. For example, an online store suggests items a visitor may wish to purchase based on items he has purchased during previous visits to the Web site.

- `<contact/>`

  Contacting Visitors for Marketing of Services or Products: Information may be used to contact the individual, through a communications channel other than voice telephone, for the promotion of a product or service. This includes notifying visitors about updates to the Web site. This does not include a direct reply to a question or comment or customer service for a single transaction—in those cases, `<current/>` would be used. In addition, this does not include marketing via customized Web content or banner advertisements embedded in sites the user is visiting—these cases would be covered by the `<tailoring/>`, `<pseudo-analysis/>` and `<pseudo-decision/>`, or `<individual-analysis/>` and `<individual-decision/>` purposes.

- `<historical/>`

  Historical Preservation: Information may be archived or stored for the purpose of preserving social history as governed by an existing law or policy. This law or policy MUST be referenced in the `<DISPUTES>` element and MUST include a specific definition of the type of qualified researcher who can access the information, where this information will be stored and specifically how this collection advances the preservation of history.

- `<telemarketing/>`

  Contacting Visitors for Marketing of Services or Products Via Telephone: Information may be used to contact the individual via a voice telephone call for promotion of a product or service. This does not include a direct reply to a question or comment or customer service for a single transaction—in those cases, `<current/>` would be used.

- `<other-purpose>` *string* `</other-purpose>`

  Other Uses: Information may be used in other ways not captured by the above definitions. (A human readable explanation should be provided in these instances).

  Each type of purpose (with the exception of `current`) can have the following optional attribute:

  `required`

  Whether the purpose is a required practice for the site. The attribute can take the following values:

  - `always`: The purpose is always required; users cannot opt-in or opt-out of this use of their data. This is the default when no `required` attribute is present.
  - `opt-in`: Data may be used for this purpose only when the user affirmatively requests this use—for example, when a user asks to be added to a mailing list. An affirmative request requires users to take some action specifically to make the request. For example, when users fill out a survey, checking an additional box to request to be added to a mailing list would be considered an affirmative request. However, submitting a survey form that contains a pre-checked mailing list request box would not be considered an affirmative request. In addition, for any purpose that users may affirmatively request,

there must also be a way for them to change their minds later and decline—
this MUST be specified at the `opturi`.
- `opt-out`: Data may be used for this purpose unless the user requests that it
  not be used in this way. When this value is selected, the service MUST
  provide clear instructions to users on how to opt-out of this purpose at the
  `opturi`. Services SHOULD also provide these instructions or a pointer to
  these instructions at the point of data collection.

```
[36]    purpose       =    "<PURPOSE>"
                           1*purposevalue
                           *extension
                           "</PURPOSE>"

[37]    purposevalue  =    "<current/>"
                           | ; Completion and Support of Activity For
                               Which Data Was Provided
                           "<admin" [required]   "/>"
                           | ; Web Site and System Administration
                           "<develop" [required] "/>"
                           | ; Research and Development
                           "<tailoring" [required] "/>"
                           | ; One-time Tailoring
                           "<pseudo-analysis" [required] "/>"
                           | ; Pseudonymous Analysis
                           "<pseudo-decision" [required] "/>"
                           | ; Pseudonymous Decision
                           "<individual-analysis" [required] "/>"
                           | ; Individual Analysis
                           "<individual-decision" [required] "/>"
                           | ; Individual Decision
                           "<contact" [required] "/>"
                           | ; Contacting Visitors for Marketing of
                               Services or Products
                           "<historical" [required] "/>"
                           | ; Historical Preservation
                           "<telemarketing" [required] "/>"
                           | ; Telephone Marketing
                           "<other-purpose" [required] ">" PCDATA "
                               </other-purpose>"; Other Uses

[38]    required      =    " required=" `"` ("always"|"opt-in"|
                               "opt-out") `"`
```

Service providers MUST use the above elements to explain the purpose of data collection. Service providers MUST disclose *all that apply.* If a service provider does not disclose that a data element will be used for a given purpose, that is a representation that data will not be used for that purpose. Service providers that disclose that they use data for "`other`" purposes MUST provide human readable explanations of those purposes.

3.3.5 The RECIPIENT Element

Each STATEMENT element MUST contain a RECIPIENT element that contains one or more recipients of the collected data. Sites MUST classify their recipients into one or more of the six recipients specified.

    <RECIPIENT>

The legal entity, or domain, beyond the service provider and its agents where data may be distributed.

The RECIPIENT element MUST contain one or more of the following:

- <ours>

  Ourselves and/or our entities acting as our agents or entities for whom we are acting as an agent: An agent in this instance is defined as a third party that processes data only on behalf of the service provider for the completion of the stated purposes (e.g., the service provider and its printing bureau which prints address labels and does nothing further with the information).

- <delivery>

  Delivery services possibly following different practices: Legal entities *performing delivery services* that may use data for purposes other than completion of the stated purpose. This should also be used for delivery services whose data practices are unknown.

- <same>

  Legal entities following our practices: Legal entities who use the data on their own behalf under equable practices. (e.g., consider a service provider that grants the user access to collected personal information, and also provides it to a partner who uses it once but discards it. Since the recipient, who has otherwise similar practices, cannot grant the user access to information that it discarded, they are considered to have equable practices.)

- <other-recipient>

  Legal entities following different practices: Legal entities that are constrained by and accountable to the original service provider, but may use the data in a way not specified in the service provider's practices (e.g., the service provider collects data that is shared with a partner who may use it for other purposes. However, it is in the service provider's interest to ensure that the data is not used in a way that would be considered abusive to the users' and its own interests.).

- <unrelated>

  Unrelated third parties: Legal entities whose data usage practices are not known by the original service provider.

- <public>

  Public fora: Public fora such as bulletin boards, public directories, or commercial CD-ROM directories.

Each of the above tags can optionally contain:

- one or more `recipient-description` tags, containing a description of the recipient;
- with the exception of `<ours>`, a `required` attribute: this attribute is defined exactly as the analogous attribute in the `PURPOSE` tag, indicating whether opt-in/opt-out of sharing is available (and, its default value is `always`).

```
[39]    recipient         =   "<RECIPIENT>"
                                1*recipientvalue
                                *extension
                                "</RECIPIENT>"

[40]    recipientvalue    =   "<ours>" *recdescr
                                "</ours>
                                |  ; only ourselves and our agents
                                "<same" [required] ">" *recdescr
                                "</same>"
                                |  ; legal entities following our
                                     practices
                                "<other-recipient" [required] ">"
                                *recdescr
                                "</other-recipient>"
                                |  ; legal entities following different
                                     practices
                                "<delivery" [required] ">" *recdescr
                                "</delivery>"
                                |  ; delivery services following different
                                     practices
                                "<public" [required] ">" *recdescr
                                "</public>"
                                |  ; public fora
                                "<unrelated" [required] ">" *recdescr
                                "</unrelated>"                          ;
                                unrelated third parties

[41]    recdescr          =   "<recipient-description>"
                                PCDATA                                  ;
                                description of the recipient
                                "</recipient-description>"
```

Service providers MUST disclose *all the recipients that apply*. P3P makes no distinctions about how that data is released to the recipient; it simply requires that if data is released, then that sharing must be disclosed in the P3P policy. Examples of disclosing data which MUST be covered by a P3P statement include:

- Transmitting customer data as part of an order-fulfillment or billing process
- Leasing or selling mailing lists

- Placing personal information in URIs when redirecting requests to a third party
- Placing personal information in URIs which link to a third party

Note that in some cases the above set of recipients may not completely describe all the recipients of data. For example, the issue of transaction facilitators, such as shipping or payment processors, who are necessary for the completion and support of the activity but may follow different practices was problematic. Currently, only delivery services can be explicitly represented in a policy. Other such transaction facilitators should be represented in whichever category most accurately reflects their practices with respect to the original service provider.

A special element for delivery services is included, but not one for payment processors (such as banks or credit card companies) for the following reasons: Financial institutions will typically have separate agreements with their customers regarding the use of their financial data, while delivery recipients typically do not have an opportunity to review a delivery service's privacy policy.

Note that the `<delivery/>` element SHOULD NOT be used for delivery services that agree to use data only on behalf of the service provider for completion of the delivery.

3.3.6 The RETENTION Element

Each STATEMENT element MUST contain a RETENTION element that indicates the kind of retention policy that applies to the data referenced in that statement.

`<RETENTION>`

The type of retention policy in effect.

The RETENTION element MUST contain one of the following:

- `<no-retention/>`

  Information is not retained for more than a brief period of time necessary to make use of it during the course of a single online interaction. Information MUST be destroyed following this interaction and MUST NOT be logged, archived, or otherwise stored. This type of retention policy would apply, for example, to services that keep no Web server logs, set cookies only for use during a single session, or collect information to perform a search but do not keep logs of searches performed.

- `<stated-purpose/>`

  For the stated purpose: Information is retained to meet the stated purpose. This requires information to be discarded at the earliest time possible. Sites MUST have a retention policy that establishes a destruction time table. The retention policy MUST be included in or linked from the site's human-readable privacy policy.

- `<legal-requirement/>`

  As required by law or liability under applicable law: Information is retained to meet a stated purpose, but the retention period is longer because of a legal requirement or liability. For example, a law may allow consumers to dispute transactions for a certain time period; therefore a business may for liability reasons decide to maintain records of transactions, or a law may affirmatively require a certain business to maintain records for auditing or other soundness purposes. Sites MUST have a retention policy that establishes a destruction time table. The retention policy MUST be included in or linked from the site's human-readable privacy policy.

- `<business-practices/>`
  Determined by service provider's business practice: Information is retained under a service provider's stated business practices. Sites MUST have a retention policy that establishes a destruction time table. The retention policy MUST be included in or linked from the site's human-readable privacy policy.

- `<indefinitely/>`
  Indefinitely: Information is retained for an indeterminate period of time. The absence of a retention policy would be reflected under this option. Where the recipient is a public fora, this is the appropriate retention policy.

```
[42]    retention        =    "<RETENTION>"
                               retentionvalue
                               *extension
                               "</RETENTION>"


[43]    retentionvalue   =    "<no-retention/>"
                               | ; not retained
                               "<stated-purpose/>"
                               | ; for the stated purpose
                               "<legal-requirement/>"
                               | ; stated purpose by law
                               "<indefinitely/>"
                               | ; indeterminate period of time
                               "<business-practices/>"
                               | ; by business practices
```

3.3.7 The DATA-GROUP and DATA Elements

Each STATEMENT element MUST contain at least one DATA-GROUP element that contains one or more DATA elements. DATA elements are used to describe the type of data that a site collects.

```
<DATA-GROUP>
```
Describes the data to be transferred or inferred.

- base
  base URI ([URI]) for URI references present in `ref` attributes. When this attribute is omitted, the default value is the URI of the P3P base data schema (http://www.w3.org/TR/P3P/base). When the attribute appears as an empty string (""), the base is the local document.

```
<DATA>
```
Describes the data to be transferred or inferred.
- `ref` (*mandatory attribute*)

  URI reference ([URI]), where the fragment identifier part denotes the *name of a data element/set*, and the URI part denotes the corresponding *data schema*. In case the URI part is not present, if the `DATA` element is contained within a `DATA-GROUP` element, then the default base URI is assumed to be the URI of the `base` attribute. In the other cases, as usual, the default base URI is a same-document reference ([URI]).

  Remember that *names of data elements and sets are case-sensitive* (so, for example, `user.gender` is different from `USER.GENDER` or `User.Gender`).
- `optional`

  Indicates whether or not the site requires visitors to submit this data element; "no" indicates that the data element is required, while "yes" indicates that the data element is not required. *The default is "no."* The `optional` attribute is used only in policies (not in data schema definitions).

  Note that user agents should be cautious about using the `optional` attribute in automated decision-making. If the `optional` attribute is associated with a data element directly controlled by the user agent (such as the HTTP `Referer` header or cookies), the user agent should make sure that this data is not transmitted to Web sites at which a data element is optional if the site's policy would not match a user's preferences if the data element was required. Likewise, for data elements that users typically type into forms, user agents should alert users when a site's practices about optional data do not match their preferences.

`DATA` elements can contain the actual data (as already seen in the case of the `ENTITY` element), and can contain related category information.

```
[44]    data-group    =    "<DATA-GROUP"
                            [" base=" quoted-URI]
                            ">"
                            1*dataref
                            *extension
                            "</DATA-GROUP>"


[45]    dataref       =    `<DATA" ref="` URI-reference `"`
                            [" optional=" `"` ("yes"|"no") `"`] ">"
                            [categories] ; the categories of the
                                              data element.
                            [PCDATA] ; the eventual value of the
                                           data element
                            "</DATA>"
```

Here, `URI-reference` is defined as in [URI].

For example, to reference the user's home address city, all the elements of the data set `user.business-info` and (optionally) all the elements of the data set `user.home-info.telecom`, the service would send the following references inside a P3P policy:

```
<DATA-GROUP>
<DATA ref="#user.home-info.city"/>
<DATA ref="#user.home-info.telecom" optional="yes"/>
<DATA ref="#user.business-info"/>
</DATA-GROUP>
```

When the actual value of the data is known, it can be expressed inside the DATA element. For example, as seen in the example policies:

```
<ENTITY>
  <DATA-GROUP>
   <DATA ref="#business.name">CatalogExample</DATA>
   <DATA ref="#business.contact-info.postal.street">4000 Lincoln Ave.
     </DATA>
...
```

3.4 Categories and the CATEGORIES Element

Categories are elements inside data elements that provide hints to users and user agents as to the intended uses of the data. Categories are vital to making P3P user agents easier to implement and use. Note that *categories are not data elements*: they just allow users to express more generalized preferences and rules over the exchange of their data.

The following elements are used to denote data categories:

```
[46]   categories  =  "<CATEGORIES>" 1*category "</CATEGORIES>"

[47]   category    =  "<physical/>"     | ; Physical Contact
                                              Information
                      "<online/>"       | ; Online Contact Information
                      "<uniqueid/>"     | ; Unique Identifiers
                      "<purchase/>"     | ; Purchase Information
                      "<financial/>"    | ; Financial Information
                      "<computer/>"     | ; Computer Information
                      "<navigation/>"   | ; Navigation and Click-stream
                                              Data
                      "<interactive/>"  | ; Interactive Data
                      "<demographic/>"  | ; Demographic and
                                              Socioeconomic Data
                      "<content/>"      | ; Content
                      "<state/>"        | ; State Management Mechanisms
                      "<political/>"    | ; Political Information
                      "<health/>"       | ; Health Information
                      "<preference/>"   | ; Preference Data
```

```
        "<location/>"      |  ; Location Data
        "<government/>"     |  ; Government-issued
                                 Identifiers
        "<other-category>" PCDATA "</other-category>"
                           |  ; Other
```

- `<physical/>`
  Physical Contact Information: Information that allows an individual to be contacted or located in the physical world—such as telephone number or address.
- `<online/>`
  Online Contact Information: Information that allows an individual to be contacted or located on the Internet—such as email. Often, this information is independent of the specific computer used to access the network. (See the category "Computer Information")
- `<uniqueid/>`
  Unique Identifiers: Non-financial identifiers, excluding government-issued identifiers, issued for purposes of consistently identifying or recognizing the individual. These include identifiers issued by a Web site or service.
- `<purchase/>`
  Purchase Information: Information actively generated by the purchase of a product or service, including information about the method of payment.
- `<financial/>`
  Financial Information: Information about an individual's finances including account status and activity information such as account balance, payment or overdraft history, and information about an individual's purchase or use of financial instruments including credit or debit card information. Information about a discrete purchase by an individual, as described in "Purchase Information," alone does not come under the definition of "Financial Information."
- `<computer/>`
  Computer Information: Information about the computer system that the individual is using to access the network—such as the IP number, domain name, browser type or operating system.
- `<navigation/>`
  Navigation and Click-stream Data: Data *passively* generated by *browsing* the Web site—such as which pages are visited, and how long users stay on each page.
- `<interactive/>`
  Interactive Data: Data *actively* generated from or reflecting *explicit interactions* with a service provider through its site—such as queries to a search engine, or logs of account activity.
- `<demographic/>`
  Demographic and Socioeconomic Data: Data about an individual's characteristics—such as gender, age, and income.

- `<content/>`
  Content: The words and expressions contained in the body of a communication—such as the text of email, bulletin board postings, or chat room communications.

- `<state/>`
  State Management Mechanisms: Mechanisms for maintaining a stateful session with a user or automatically recognizing users who have visited a particular site or accessed particular content previously—such as HTTP cookies.

- `<political/>`
  Political Information: Membership in or affiliation with groups such as religious organizations, trade unions, professional associations, political parties, etc.

- `<health/>`
  Health Information: Information about an individual's physical or mental health, sexual orientation, use or inquiry into health care services or products, and purchase of health care services or products.

- `<preference/>`
  Preference Data: Data about an individual's likes and dislikes—such as favorite color or musical tastes.

- `<location/>`
  Location Data: Information that can be used to identify an individual's current physical location and track them as their location changes—such as GPS position data.

- `<government/>`
  Government-issued Identifiers: Identifiers issued by a government for purposes of consistently identifying the individual.

- `<other-category>` *string* `</other-category>`
  Other: Other types of data not captured by the above definitions. (A human readable explanation should be provided in these instances, between the `<other-category>` and the `</other-category>` tags.)

The Computer, Navigation, Interactive and Content categories can be distinguished as follows. The Computer category includes information about the user's computer including IP address and software configuration. Navigation data describes actual user behavior related to browsing. When an IP address is stored in a log file with information related to browsing activity, both the Computer category and the Navigation category should be used. Interactive Data is data actively solicited to provide some useful service at a site beyond browsing. Content is information exchanged on a site for the purposes of communication.

The Other category should be used only when data is requested that does not fit into any other category.

P3P uses categories to give users and user agents additional hints as to what type of information is requested from a service. While most data in the base data schema is in a known category (or a set of known categories), some data elements can be in a number of different categories, depending on the situation. The former are called *fixed-category data elements* (or "fixed data elements" for short), the latter *variable-category data elements* ("variable data elements"). Both types of elements are described in Section 5.7.

3.5 Extension Mechanism: The `EXTENSION` Element

P3P provides a flexible and powerful mechanism to extend its syntax and semantics using one element: `EXTENSION`. This element is used to indicate portions of the policy which belong to an extension. The meaning of the data within the `EXTENSION` element is defined by the extension itself.

`<EXTENSION>`
Describes an extension to the syntax.

- `optional`
  This attribute determines if the extension is *mandatory* or *optional*. A *mandatory* extension is indicated by giving the `optional` attribute a value of `no`. A *mandatory* extension to the P3P syntax means that applications that do not understand this extension cannot understand the meaning of the whole policy (or data schema). An *optional* extension, indicated by giving the optional attribute a value of `yes`, means that applications that do not understand this extension can safely ignore the contents of the `EXTENSION` element, and proceed to process the whole policy (or data schema) as usual. The `optional` attribute is not required; its default value is `yes`.

```
[48]   extension  =   "<EXTENSION" [" optional=" `"` ("yes"|"no") `"`]
                       ">" PCDATA "</EXTENSION>"
```

For example, if www.catalog.example.com would like to add to P3P a feature to indicate that a certain set of data elements were only to be collected from users living in the United States, Canada, or Mexico, it could add a mandatory extension like this:

```
<DATA-GROUP>
...
<EXTENSION optional="no">
<COLLECTION-GEOGRAPHY type="include" xmlns=
   "http://www.catalog.example.com/P3P/region">
<USA/><Canada/><Mexico/>
</COLLECTION-GEOGRAPHY>
</EXTENSION>
</DATA-GROUP>
```

On the other hand, if www.catalog.example.com would like to add an extension stating what country the server is in, an optional extension might be more appropriate, such as the following:

```
<POLICY>
<EXTENSION optional="yes">
<ORIGIN xmlns="http://www.catalog.example.com/P3P/
   origin" country="USA"/>
</EXTENSION>
...
</POLICY>
```

The xmlns attribute is significant since it specifies the namespace for interpreting the names of elements and attributes used in the extension. Note that, as specified in [XML-Name], the namespace URI is just intended to be a unique identifier for the XML entities used by the extension. Nevertheless, service providers MAY provide a page with a description of the extension at the corresponding URI.

3.6 Import and Export of User Preferences

User agents MUST document a method by which preferences can be imported and processed, and SHOULD document a method by which preferences can be exported.

4 COMPACT POLICIES

Compact policies are summarized P3P policies that provide hints to user agents to enable the user agent to make quick, synchronous decisions about applying policy. Compact policies are a performance optimization that is OPTIONAL for either user agents or servers. User agents that are unable to obtain enough information from a compact policy to make a decision according to a user's preferences SHOULD fetch the full policy.

In P3Pv1, compact policies contain policy information related to cookies only. The web server is responsible for building a P3P compact policy to represent the cookies referenced in a full policy. The policy specified in a P3P compact policy applies to data stored within all cookies set in the same HTTP response as the compact policy, all cookies set by scripts associated with that HTTP response, and also to data linked to the cookies.

4.1 Referencing Compact Policies

Any document transferred by HTTP MAY include a P3P compact policy through the P3P header. If a site is using P3P headers, it MAY include this on responses for all appropriate request methods, including HEAD and OPTION requests.

To specify a compact policy within the P3P header, a site specifies the compact policy in the P3P header (cf. Section 2.2.2). The P3P compact policy header has a quoted string that may contain one or more delimited tokens (the "compact policy"). Tokens can appear in any order, and the space character (" ") is the only valid delimiter. The syntax for this header is as follows:

```
[49]   compact-policy-field      =   `CP="` compact-policy `"`

[50]   compact-policy            =   compact-token *(" " compact-token)
```

```
[51]    compact-token              =   compact-access           |
                                        compact-disputes         |
                                        compact-remedies         |
                                        compact-non-identifiable |
                                        compact-purpose          |
                                        compact-recipient        |
                                        compact-retention        |
                                        compact-categories       |


                                        compact-test
```

In keeping with the rules for other HTTP headers, the name of the P3P header may be written with any casing. The contents should be specified using the casing precisely as specified in this document.

User agents MAY process the P3P-compact-policy-field.

If an HTTP response includes more than one compact policy, P3P user agents MUST ignore all compact policies after the first one.

4.2 Compact Policy Vocabulary

P3P compact policies use tokens representing the following elements from the P3P vocabulary: ACCESS, CATEGORIES, DISPUTES, NON-INDENTIFIABLE, PURPOSE, RECIPIENT, REMEDIES, RETENTION, TEST.

If a token appears more than once in a single compact policy, the compact policy has *the same semantics* as if that token appeared only once. If an unrecognized token appears in a compact policy, the compact policy has *the same semantics* as if that token was not present.

The P3P compact policy vocabulary is expressed using a developer-readable language to reduce the number of bytes transferred over the wire within a HTTP response header. The syntax of the tokens follows.

4.2.1 Compact ACCESS

Information in the ACCESS element is represented in compact policies using tokens composed by a three letter code:

```
[52]    compact-access             =   "NOI" |  ; for <nonident/>
                                        "ALL" |  ; for <all/>
                                        "CAO" |  ; for <contact-and-other/>
                                        "IDC" |  ; for <ident-contact/>
                                        "OTI" |  ; for <other-ident/>
                                        "NON"    ; for <none/>
```

4.2.2 Compact DISPUTES

If a full P3P policy contains a DISPUTES-GROUP element that contains one or more DISPUTES elements, then the server should signal the user agent by providing a single "DSP" token in the P3P-compact policy field:

```
[53]   compact-disputes           =   "DSP" ; there are some DISPUTES
```

4.2.3 Compact REMEDIES

Information in the REMEDIES element is represented in compact policies as follows:

```
[54]   compact-remedies           =   "COR" | ; for <correct/>
                                       "MON" | ; for <money/>
                                       "LAW"   ; for <law/>
```

4.2.4 Compact NON-IDENTIFIABLE

The presence of the NON-IDENTIFIABLE element in every statement of the policy is signaled by the NID token (note that the NID token MUST NOT be used unless the NON-IDENTIFIABLE element is present in every statement within the policy):

```
[55]   compact-non-identifiable  =   "NID" ; for <NON-IDENTIFIABLE/>
```

4.2.5 Compact PURPOSE

Purposes are expressed in P3P compact policy format using tokens composed by a three letter code plus an optional one letter attribute. Such an optional attribute encodes the value of the "required" attribute in full P3P policies: its value can be "a", "i" and "o", which mean that the "required" attribute in the corresponding P3P policy must be set to "always", "opt-in" and "opt-out" respectively.

If a P3P compact policy needs to specify one or more other-purposes in its full P3P policy, a single OTP flag is used to signal the user agent that other-purposes exist in the full P3P policy.

The corresponding associations among P3P purposes and compact policy codes follow:

```
[56]   compact-purpose   =   "CUR"          | ; for <current/>
                             "ADM" [creq] | ; for <admin/>
                             "DEV" [creq] | ; for <develop/>
                             "TAI" [creq] | ; for <tailoring/>
                             "PSA" [creq] | ; for <pseudo-analysis/>
                             "PSD" [creq] | ; for <pseudo-decision/>
                             "IVA" [creq] | ; for <individual-analysis/>
                             "IVD" [creq] | ; for <individual-decision/>
                             "CON" [creq] | ; for <contact/>
```

```
                            "HIS" [creq] | ; for <historical/>
                            "TEL" [creq] | ; for <telemarketing/>
                            "OTP" [creq]   ; for <other-purpose/>

[57]   creq             =  "a"| ;"always"
                            "i"| ;"opt-in"
                            "o"  ;"opt-out"
```

4.2.6 Compact RECIPIENT

Recipients are expressed in P3P compact policy format using a three letter code plus an optional one letter attribute. Such an optional attribute encodes the value of the "required" attribute in full P3P policies: its value can be "a", "i" and "o", which mean that the "required" attribute in the corresponding P3P policy must be set to "always", "opt-in" and "opt-out" respectively.

The corresponding associations among P3P recipients and compact policy codes follow:

```
[58]   compact-recipient    =  "OUR"        | ; for <ours/>
                                "DEL" [creq] | ; for <delivery/>
                                "SAM" [creq] | ; for <same/>
                                "UNR" [creq] | ; for <unrelated/>
                                "PUB" [creq] | ; for <public/>
                                "OTR" [creq]   ; for <other-recipient/>
```

4.2.7 Compact RETENTION

Information in the RETENTION element is represented in compact policies as follows:

```
[59]   compact-retention    =  "NOR" | ; for <no-retention/>
                                "STP" | ; for <stated-purpose/>
                                "LEG" | ; for <legal-requirement/>
                                "BUS" | ; for <business-practices/>
                                "IND"   ; for <indefinitely/>
```

4.2.8 Compact CATEGORIES

Categories are represented in compact policies as follows:

```
[60]   compact-categories   =  "PHY" | ; for <physical/>
                                "ONL" | ; for <online/>
                                "UNI" | ; for <uniqueID/>
                                "PUR" | ; for <purchase/>
                                "FIN" | ; for <financial/>
                                "COM" | ; for <computer/>
                                "NAV" | ; for <navigation/>
                                "INT" | ; for <interactive/>
                                "DEM" | ; for <demographic/>
```

```
"CNT"  |  ;  for <content/>
"STA"  |  ;  for <state/>
"POL"  |  ;  for <political/>
"HEA"  |  ;  for <health/>
"PRE"  |  ;  for <preference/>
"LOC"  |  ;  for <location/>
"GOV"  |  ;  for <government/>
"OTC"     ;  for <other-category/>
```

Note that if a P3P policy specifies one or more other-category in its full P3P policy, a **single** OTC token is used to signal the user agent that other-categorys exist in the full P3P policy.

4.2.9 Compact TEST

The presence of the TEST element is signaled by the TST token:

```
[61]    compact-test        =       "TST" ; for <TEST/>
```

4.3 Compact Policy Scope

When a P3P compact policy is included in a HTTP response header, it applies to cookies set by the current response. This includes cookies set through the use of a HTTP SET-COOKIE header or cookies set by script.

4.4 Compact Policy Lifetime

To use compact policies, the validity of the full P3P policy must span the lifetime of the cookie. There is no method to indicate that policy is valid beyond the life of the cookie because the value of user agent caching is marginal, since sites would not know when to optimize by not sending the compact policy. When a server sends a compact policy, it is asserting that the compact policy and corresponding full P3P policy will be in effect for at least the lifetime of the cookie to which it applies.

4.5 Transforming a P3P Policy to a Compact Policy

When using P3P compact policies, the web site is responsible for building a compact policy by summarizing the policy referenced by the COOKIE-INCLUDE elements of a P3P policy reference file. If a site's policy reference file uses COOKIE-EXCLUDE elements then the site will need to manage sending the correct P3P compact policies to the user agent given the cookies set in a specific response.

The transformation of a P3P policy to a P3P compact policy may result in a loss of descriptive policy information—the compact policy may not contain all of the policy information specified in the full P3P policy. The information from the full policy that is discarded when building a compact policy includes expiry, data group/data-schema elements, entity elements, consequences elements, and disputes elements are reduced.

Full policies that include mandatory extensions MUST NOT be represented as compact policies.

All of the purposes, recipients, and categories that appear in multiple statements in a full policy MUST be aggregated in a compact policy, as described in section 3.3.1. When performing the aggregation, a web site MUST disclose all relevant tokens (for instance, observe Example 4.1, where multiple retention policies are specified).

In addition, for each fixed category data element appearing in a statement the associated category as defined in the associated schema MUST be included in the compact policy.

Example 4.1:
___

Consider the following P3P policy:

```
<POLICY name="sample"
  discuri="http://www.example.com/cookiepolicy.html"
  opturi="http://www.example.com/opt.html">
  <ENTITY>
    <DATA-GROUP>
      <DATA ref="#business.name">Example, Corp.</DATA>
      <DATA ref="#business.contact-info.online.email">
         privacy@example.com</DATA>
    </DATA-GROUP>
  </ENTITY>
  <ACCESS><none/></ACCESS>
  <DISPUTES-GROUP>
    <DISPUTES resolution-type="service"
     service="http://www.example.com/privacy.html"
     short-description="Please contact our customer service desk with
                    privacy concerns by emailing
                    privacy@example.com"/>
  </DISPUTES-GROUP>
  <STATEMENT>
    <PURPOSE><admin/><develop/><pseudo-decision/></PURPOSE>
    <RECIPIENT><ours/></RECIPIENT>
    <RETENTION><indefinitely/></RETENTION>
    <DATA-GROUP>
      <DATA ref="#dynamic.cookies">
        <CATEGORIES><preference/><navigation/></CATEGORIES>
      </DATA>
    </DATA-GROUP>
  </STATEMENT>
  <STATEMENT>
    <PURPOSE><individual-decision required="opt-out"/></PURPOSE>
    <RECIPIENT><ours/></RECIPIENT>
    <RETENTION><stated-purpose/></RETENTION>
    <DATA-GROUP>
```

```
        <DATA ref="#user.name.given"/>
        <DATA ref="#dynamic.cookies">
          <CATEGORIES><preference/><uniqueid/></CATEGORIES>
        </DATA>
      </DATA-GROUP>
    </STATEMENT>
  </POLICY>
```

The corresponding compact policy is:

```
"NON DSP ADM DEV PSD IVDo OUR IND STP PHY PRE NAV UNI"
```

4.6 Transforming a Compact Policy to a P3P Policy

Some user agents may attempt to generate a full P3P policy from a compact policy, for use in evaluating user preferences. They will not be able to provide values for the ENTITY and DISPUTES elements as well as a number of the attributes. However:

In case there *are not* multiple different values of compact retention, they should be able to generate a policy with an appropriate ACCESS element, and: a single STATEMENT element that contains the appropriate RECIPIENT, RETENTION, and PURPOSE elements, as well as a dynamic.miscdata element with the appropriate CATEGORIES.

In case there *are* multiple different values of compact retention, they should be able to generate a policy with an appropriate ACCESS element, and: multiple STATEMENT elements (as many as the different values of the compact retention) that contain a different corresponding value for the RETENTION element, the appropriate RECIPIENT, and PURPOSE elements, as well as a dynamic.miscdata element with the appropriate CATEGORIES.

Note that, in agreement with the non-ambiguity requirements stated in Section 2.4.1, a site MUST honor a compact policy for a given URI in any case (even when the full policy referenced in the policy reference file for that URI does not correspond, as per Section 4.5, to the compact policy itself).

5 DATA SCHEMAS

A *data schema* is a description of a set of data. P3P includes a way to describe data schemas so that services can communicate to user agents about the data they collect. A data schema is built from a number of *data elements*, which are specific items of data a service might collect.

Data elements in a data schema can have the following properties:

- Data element name. The name of the data element is used when a P3P policy includes this data element in a <DATA> element. This is required on all data elements.
- Descriptive name or short name. A data element's short name provides a short, human-understandable name for the data element. The short name is not required, but it is strongly recommended.

- Long description. The long description of a data element provides a more detailed, human-understandable definition of the data element. Like the short name, the long description is not required, but it is strongly recommended.
- Category or categories. Most data elements have categories assigned to them when they are defined in a data schema. See Categories for more information on categories.

Data elements are organized into a hierarchy. A data element automatically includes all of the data elements below it in the hierarchy. For example, the data element representing "the user's name" includes the data elements representing "the user's given name," "the user's family name," and so on. The hierarchy is based on the data element name. Thus the data elements `user.name.given`, `user.name.family`, and `user.name.nickname` are all children of the data element `user.name`, which is in turn a child of the data element `user`.

P3P has defined a data schema called the *P3P base data schema* that includes a large number of data elements commonly used by services.

Services may declare new data elements by creating and publishing their own data schemas. This is done with the `<DATASCHEMA>` element. These can either be published in standalone XML files, which are then referenced by policies that use them, or they can be embedded in the policies files that reference them. The `<DATASCHEMA>` element is defined as follows:

```
[62]    dataschema    =    "<DATASCHEMA" [` xmlns="http://www.w3.org/2001/
                           09/P3Pv1"`] ">"
                           *(datadef|datastruct|extension)
                           "</DATASCHEMA>"
```

A standalone data schema has the `<DATASCHEMA>` element as the first XML element in the file. It must have the appropriate namespace defined in the `xmlns` attribute to identify it as a P3P data schema, as follows:

```
<DATASCHEMA xmlns="http://www.w3.org/2001/09/P3Pv1">
<DATA-STRUCT ... />
...
<DATA-DEF ... />
</DATASCHEMA>
```

When a data schema is declared inside a policy file, then the `<DATASCHEMA>` element is still used (as described in Section 3.2.1, "The `<POLICIES>` Element"), but no namespace attribute is given.

5.1 Natural Language Support for Data Schemas

Data schemas contain a number of fields in natural language. Services publishing a data schema MAY wish to translate these fields into multiple languages. The data element short and long names MAY be translated, but the data element name MUST NOT be translated—this field needs to stay constant across translations of a data schema.

If a service is going to provide a data schema in multiple natural languages, then it SHOULD examine the `Accept-Language` HTTP request-header on requests for that data schema to pick the best available alternative.

5.2 Data Structures

Data schemas often need to reuse a common group of data elements. P3P data schemas support this through data structures. A data structure is a named, abstract definition of a group of data elements. When a data element is defined, it can be defined as being of an unstructured type, in which case it has no child elements. The data element can also be defined as being of a specific structured type, in which case the data element will be automatically expanded to include as sub-elements all of the elements defined in the data structure. For example, the following structure is used to represent a date and time:

```
<!-- "date" Data Structure -->
<DATA-STRUCT name="date.ymd.year"
    short-description="Year"/>

<DATA-STRUCT name="date.ymd.month"
    short-description="Month"/>

<DATA-STRUCT name="date.ymd.day"
    short-description="Day"/>

<DATA-STRUCT name="date.hms.hour"
    short-description="Hour"/>

<DATA-STRUCT name="date.hms.minute"
    short-description="Minute"/>

<DATA-STRUCT name="date.hms.second"
    short-description="Second"/>
```

Now we shall define a "meeting" data element, which has a time and place for the meeting:

```
<DATA-DEF name="meeting.time"
    short-description="Meeting time"
    structref="#date"/>
<DATA-DEF name="meeting.place"
    short-description="Meeting place/>
```

Since `meeting.place` does not reference a structure, it is of an unstructured type, and has no child elements. The `meeting.time` element uses the `date` structure. By declaring this, the following sub-elements are created:

```
meeting.time.ymd.year
meeting.time.ymd.month
meeting.time.ymd.day
meeting.time.hms.hour
meeting.time.hms.minute
meeting.time.hms.second
```

A P3P policy can now declare that it collects the `meeting` data element, which implies that it collects all of the sub-elements of `meeting`, or it can use data elements lower down the hierarchy—`meeting.time`, for example, or `meeting.time.ymd.day`.

5.3 The `DATA-DEF` and `DATA-STRUCT` Elements

`<DATA-DEF>` and `<DATA-STRUCT>`
Define a data element or a data structure, respectively. Data structures are reusable structured type definitions that can be used to build data elements. Data elements are declared within a `<STATEMENT>` in a P3P policy to describe data covered by that statement.

The following attributes are common to these two elements:

- `name` (*mandatory attribute*)
  Indicates the name of the data element or data structure. Remember that names of data element and data structures are *case-sensitive*, so, for example, `user.gender` is different from `USER.GENDER` or `User.Gender`. Furthermore, in names of data elements and structures no number character can appear immediately following a dot.

- `structref`
  URI reference ([URI]), where the fragment identifier part denotes the *structure*, and the URI part denotes the corresponding data schema where it is defined. The default base URI is a same-document reference ([URI]). Data elements or data structures without a `structref` attribute (and, so, without an associated structure) are called *unstructured*.

- `short-description`
  A string denoting the short display name of the data element or structure, no more than 255 characters.

  The `DATA-DEF` and `DATA-STRUCT` elements can also contain a long description of the data element or structure, using the `LONG-DESCRIPTION` element.

```
[63]  datadef      =   "<DATA-DEF name=" quotedstring
                        [` structref="` URI-reference `"`]
                        [" short-description=" quotedstring]
                        ">"
                        [categories] ; the categories of the
                                        data element.
                        [longdescription] ; the long description of
                                            the data element
                        "</DATA-DEF>"
```

```
[64]  datastruct  =   "<DATA-STRUCT name=" quotedstring
                       [` structref="` URI-reference `"`]
                       [" short-description=" quotedstring]
                       ">"
                       [categories] ; the categories of the
                                        Data Structure.
                       [longdescription] ; the long description of the
                                            Data Structure
                       "</DATA-STRUCT>"
```

Here, `URI-reference` is defined as in [URI].

Data elements can be structured, much like in common programming languages: structures are hierarchical (tree-like) descriptions of data elements: this hierarchical description is performed in the `name` attribute using a full stop (".") character as separator.

P3P provides the *P3P base data schema*, which has built-in definitions of a number of widely used structures and data elements. All P3P implementations are required to understand the P3P base data schema, so the structures and elements it defines are always available to P3P implementers.

5.3.1 Categories in P3P Data Schemas

Categories can be assigned to data structures or data elements. The following rules define how those category definitions are meant to be used:

1. `<DATA-STRUCT>` elements MAY include category definitions with them. If a structure definition includes categories, then all uses of those structures in data definitions and data structures pick up those categories. If a structure contains no categories, then the categories for that structure MAY be defined when it is used in another structure or data element. Otherwise, a data element using this structure is a variable-category element. Any uses of a variable-category data element in a policy require that its categories be listed in the policy.

2. A `<DATA-DEF>` with an unstructured type is a variable-category data element if no categories are defined in the `<DATA-DEF>`, and has exactly those categories listed in the `<DATA-DEF>` if any categories are included.

3. A `<DATA-DEF>` or `<DATA-STRUCT>` with a structured type which has no categories defined on that structure produces a variable-category data element/structure if no categories are defined in the `<DATA-DEF>` or `<DATA-STRUCT>`. If the `<DATA-DEF>` or `<DATA-STRUCT>` does have categories listed, then those categories are applied to that data element, and all of its sub-elements. In other words, categories are pushed down into sub-elements when defining a data element to be of a structured type, and the structured type does not define any categories.

4. A `<DATA-DEF>` using a structured type which has categories defined on that structure picks up all the categories listed on the structure. In addition, categories may be listed in the `<DATA-DEF>`, and these are added to the categories defined in the

structure. These categories are defined only at the level of that data element, and are not "pushed down" to any sub-elements.

5. A `<DATA-STRUCT>` that has no categories assigned to it, and which is using a structured subtype which has categories defined on the subtype picks up all the categories listed on the subtype.

6. A `<DATA-STRUCT>` that has categories assigned to it, and which is using a structured subtype replaces all of the categories listed on the subtype.

7. There is a "bubble-up" rule for categories when referencing data elements: data elements, must at a minimum, include all categories defined by any of its children. This rule applies recursively, so for example, all categories defined by data elements `foo.a.w`, `foo.a.y`, and `foo.b.z` MUST be considered to apply to data element `foo`.

8. A `<DATA-STRUCT>` cannot be defined with some variable-category elements and some fixed-category elements. Either all of the sub-elements of a category must be in the variable category, or else all of them must have one or more assigned categories.

5.3.2 P3P Data Schema Example

Consider the case where the company HyperSpeedExample wishes to describe the features of a vehicle, using a structure called `vehicle`. This structure includes:

- The vehicle's model type (`vehicle.model`),
- The vehicle's color (`vehicle.color`),
- The vehicle's year of manufacture (`vehicle.built.year`), and
- The vehicle's price (`vehicle.price`).

If HyperSpeedExample also wants to include in the definition of a vehicle the location of manufacture, it could add other fields to the structure with all the relevant data like country, street address, postal code, and so on. But, each part of a structure can use other structures as well: *structures can be composed*. In this case, the *P3P base data schema* already provides a structure `postal`, describing all the postal information of a location. So, the final definition of the structure vehicle is

- `vehicle.model` (unstructured)
- `vehicle.color` (unstructured)
- `vehicle.price` (unstructured)
- `vehicle.built.year` (unstructured)
- `vehicle.built.where` (with structure `postal` from the base data schema)

The structure `postal` has fields `postal.street`, `postal.city`, and so on. Since we have applied the structure `postal` to `vehicle.built.where`, it means that we can access the street and city of a vehicle using the descriptions `vehicle.built.where.street` and `vehicle.built.where.city` respectively. So, by applying a structure (in this case, `postal`) we can build very complex descriptions in a modular way.

HyperSpeedExample wants to declare that all of the vehicle information will be in the `<preference/>` category. The `vehicle.model`, `vehicle.color`, `vehicle.price`, and `vehicle.built.year` fields are all unstructured types, so assigning them to the `<preference/>` category accomplishes this for those fields. Since vehicle is a structure definition, assigning the `<preference/>` category to `vehicle.built.where` will override (replace) the categories defined on all of the sub-elements of `vehicle.built.where`, placing all of them in the `<preference/>` category, even though the `postal` structure was originally defined as being in other categories.

As said, structures do not contain data elements; they are just abstract data types. We can use them to rapidly build structured collections of data elements. Going on with the example, HyperSpeedExample needs this abstract description of the features of a vehicle because it wants to actually exchange data about cars and motorcycles. So, it could define two data elements called `car` and `motorcycle`, both with the above structure `vehicle`.

This description of the data elements and data structures is encoded in XML using a data schema. In the HyperSpeedExample case, it would be something like:

```
<DATASCHEMA xmlns="http://www.w3.org/2001/09/P3Pv1">
<DATA-STRUCT name="vehicle.model"
    short-description="Model">
    <CATEGORIES><preference/></CATEGORIES>
</DATA-STRUCT>
<DATA-STRUCT name="vehicle.color"
    short-description="Color">
    <CATEGORIES><preference/></CATEGORIES>
</DATA-STRUCT>
<DATA-STRUCT name="vehicle.built.year"
    short-description="Construction Year">
    <CATEGORIES><preference/></CATEGORIES>
</DATA-STRUCT>
<DATA-STRUCT name="vehicle.built.where"
    structref="http://www.w3.org/TR/P3P/base#postal"
    short-description="Construction Place">
    <CATEGORIES><preference/></CATEGORIES>
</DATA-STRUCT>
<DATA-DEF name="car" structref="#vehicle"/>
<DATA-DEF name="motorcycle" structref="#vehicle"/>
</DATASCHEMA>
```

Continuing with the example, in order to reference a car model and construction year, Hyperspeed *or any other service* could send the following references inside a P3P policy:

```
<DATA-GROUP>
  <!-- First, the "car.model" data element, whose definition is in
  the data schema at http://www.HyperSpeed.example.com/models-schema
  -->
<DATA ref="http://www.HyperSpeed.example.com/models-schema#car.model"/>
```

```
<!-- And second, the "car.built.year" data element, whose definition
  is the data schema at http://www.HyperSpeed.example.com/
  models-schema
  -->
<DATA ref="http://www.HyperSpeed.example.com/
  models-schema#car.built.year"/>
</DATA-GROUP>
```

Using the `base` attribute, the above references can be written in an even more compact way:

```
<DATA-GROUP base="http://www.HyperSpeed.example.com/models-schema">
    <DATA ref="#car.model"/>
    <DATA ref="#car.built.year"/>
</DATA-GROUP>
```

Alternatively, the data schema could be *embedded* directly into a policy file. In this case, the policy file could look like:

```
<POLICIES xmlns="http://www.w3.org/2001/09/P3Pv1">
<!-- Embedded data schema -->
<DATASCHEMA>
<DATA-STRUCT name="vehicle.model"
    short-description="Model">
    <CATEGORIES><preference/></CATEGORIES>
</DATA-STRUCT>
<DATA-STRUCT name="vehicle.color"
    short-description="Color">
    <CATEGORIES><preference/></CATEGORIES>
</DATA-STRUCT>
<DATA-STRUCT name="vehicle.built.year"
    short-description="Construction Year"">
    <CATEGORIES><preference/></CATEGORIES>
</DATA-STRUCT>
<DATA-STRUCT name="vehicle.built.where"
    structref="http://www.w3.org/TR/P3P/base#postal"
    short-description="Construction Place">
    <CATEGORIES><preference/></CATEGORIES>
</DATA-STRUCT>
<DATA-DEF name="car" structref="#vehicle"/>
<DATA-DEF name="motorcycle" structref="#vehicle"/>
</DATASCHEMA>
<!-- end of embedded data schema -->
<POLICY name="policy1" discuri="http://www.example.com/disc1">
...
<DATA-GROUP base="">
<DATA ref="#car.model"/>
<DATA ref="#car.built.year"/>
</DATA-GROUP>
...
```

```
</POLICY>
<POLICY name="policy2" discuri="http://www.example.com/disc2"> .... </
POLICY>
<POLICY name="policy3" discuri="http://www.example.com/disc3"> .... </
POLICY>
</POLICIES>
```

Note that in any case there MUST NOT be more than one data schema per file.

5.3.3 Use of Data Element Names

Note that the data element names specified in the base data schema or in extension data schemas may be used for purposes other than P3P policies. For example, Web sites may use these names to label HTML form fields. By referring to data the same way in P3P policies and forms, automated form-filling tools can be better integrated with P3P user agents.

5.4 Persistence of Data Schemas

An essential requirement on data schemas is the *persistence of data schemas*: data schemas that can be fetched at a certain URI can only be changed by extending the data schema in a *backward-compatible* way (that is to say, changing the data schema does not change the meaning of any policy using that schema). This way, the URI of a policy acts in a sense like a unique identifier for the data elements and structures contained therein: any data schema that is not backward-compatible *must therefore use a new different URI*.

Note that a useful application of the persistence of data schema is given for example in the case of multi-lingual sites: multiple language versions (translations) of the same data schema can be offered by the server, using the HTTP "Content-Language" response header field to properly indicate that a particular language has been used for the data schema.

5.5 Basic Data Structures

The Basic Data Structures are structures used by the P3P base data schema (and possibly, due to their basic nature, they should be reused as much as possible by other different data schemas). All P3P-compliant user agent implementations MUST be aware of the Basic Data Structures. Each table below specifies the elements of a basic data structure, the categories associated, their structures, and the display names shown to users. More than one category may be associated with a fixed data element. However, each base data element is assigned to only one category whenever possible. Data schema designers are recommended to do the same.

5.5.1 Dates

The date structure specifies a date. Since date information can be used in different ways, depending on the context, all date information is tagged as being of "variable" category (see Section 5.7.2). For example, schema definitions can explicitly set the corresponding category in the element referencing this data structure, where soliciting the birthday of a user might be "Demographic and Socioeconomic Data," while the expiration date of a credit card might belong to the "Purchase Information" category.

date	Category	Structure	Short Display Name
ymd.year	*(variable-category)*	*unstructured*	Year
ymd.month	*(variable-category)*	*unstructured*	Month
ymd.day	*(variable-category)*	*unstructured*	Day
hms.hour	*(variable-category)*	*unstructured*	Hour
hms.minute	*(variable-category)*	*unstructured*	Minute
hms.second	*(variable-category)*	*unstructured*	Second
fractionsecond	*(variable-category)*	*unstructured*	Fraction of Second
timezone	*(variable-category)*	*unstructured*	Time Zone

The "time zone" information is for example described in the time standard [ISO8601]. Note that "date.ymd" and "date.hms" can be used to fast reference the year/month/day and hour/minutes/seconds blocks respectively.

5.5.2 Names

The personname structure specifies information about the naming of a person.

personname	Category	Structure	Short Display Name
prefix	Demographic and Socioeconomic Data	*unstructured*	Name Prefix
given	Physical Contact Information	*unstructured*	Given Name (First Name)
family	Physical Contact Information	*unstructured*	Family Name (Last Name)
middle	Physical Contact Information	*unstructured*	Middle Name
suffix	Demographic and Socioeconomic Data	*unstructured*	Name Suffix
nickname	Demographic and Socioeconomic Data	*unstructured*	Nickname

5.5.3 Logins

The login structure specifies information (IDs and passwords) for computer systems and Web sites which require authentication. Note that this data element should not be used for computer systems or Web sites which use digital certificates for authentication: in those cases, the *certificate* structure should be used.

login	Category	Structure	Short Display Name
id	Unique Identifiers	*unstructured*	Login ID
password	Unique Identifiers	*unstructured*	Login Password

The "id" field represents the ID portion of the login information for a computer system. Often, user IDs are made public, while passwords are kept secret. This does not include any type of biometric authentication mechanisms.

The "password" field represents the password portion of the login information for a computer system. This is a secret data value, usually a character string, that is used in authenticating a user. Passwords are typically kept secret, and are generally considered to be sensitive information

5.5.4 Certificates

The certificate structure is used to specify identity certificates (like, for example, X.509).

certificate	Category	Structure	Short Display Name
key	Unique Identifiers	*unstructured*	Certificate Key
format	Unique Identifiers	*unstructured*	Certificate Format

The "format" field is used to represent the information of an IANA registered public key or authentication certificate format, while the "key" field is used to represent the corresponding certificate key.

5.5.5 Telephones

The telephonenum structure specifies the characteristics of a telephone number.

telephonenum	Category	Structure	Short Display Name
intcode	Physical Contact Information	*unstructured*	International Telephone Code
loccode	Physical Contact Information	*unstructured*	Local Telephone Area Code
number	Physical Contact Information	*unstructured*	Telephone Number
ext	Physical Contact Information	*unstructured*	Telephone Extension
comment	Physical Contact Information	*unstructured*	Telephone Optional Comments

5.5.6 Contact Information

The contact structure is used to specify contact information. Services can specify precisely which set of data they need, postal, telecommunication, or online address information.

contact	Category	Structure	Short Display Name
postal	Physical Contact Information, Demographic and Socioeconomic Data	postal	Postal Address Information
telecom	Physical Contact Information	telecom	Telecommunications Information
online	Online Contact Information	online	Online Address Information

5.5.6.1 Postal

The postal structure specifies a postal mailing address.

postal	Category	Structure	Short Display Name
name	Physical Contact Information, Demographic and Socioeconomic Data	personname	Name
street	Physical Contact Information	*unstructured*	Street Address
city	Demographic and Socioeconomic Data	*unstructured*	City
stateprov	Demographic and Socioeconomic Data	*unstructured*	State or Province
postalcode	Demographic and Socioeconomic Data	*unstructured*	Postal Code
country	Demographic and Socioeconomic Data	*unstructured*	Country Name
organization	Demographic and Socioeconomic Data	*unstructured*	Organization Name

The "country" field represents the information of the name of the country (for example, one among the countries listed in [ISO3166]).

5.5.6.2 Telecommunication

The telecom structure specifies telecommunication information about a person.

telecom	Category	Structure	Short Display Name
telephone	Physical Contact Information	telephonenum	Telephone Number
fax	Physical Contact Information	telephonenum	Fax Number
mobile	Physical Contact Information	telephonenum	Mobile Telephone Number
pager	Physical Contact Information	telephonenum	Pager Number

5.5.6.3 Online

The online structure specifies online information about a person.

`online`	Category	Structure	Short Display Name
email	Online Contact Information	*unstructured*	Email Address
uri	Online Contact Information	*unstructured*	Home Page Address

5.5.7 Access Logs and Internet Addresses

Two structures used for representing forms of Internet addresses are provided. The `uri` structure covers Universal Resource Identifiers (URI), which are defined in more detail in [URI]. The `ipaddr` structure represents IP addresses and Domain Name System (DNS) hostnames.

5.5.7.1 URI

`uri`	Category	Structure	Short Display Name
authority	*(variable-category)*	*unstructured*	URI Authority
stem	*(variable-category)*	*unstructured*	URI Stem
querystring	*(variable-category)*	*unstructured*	Query-string Portion of URI

The authority of a URI is defined as the `authority` component in [URI]. The stem of a URI is defined as the information contained in the portion of the URI after the authority and up to (and including) the first "?" character in the URI, and the querystring is the information contained in the portion of the URI after the first "?" character. For URIs which do not contain a "?"character, the stem is the entire URI, and the querystring is empty.

Since URI information can be used in different ways, depending on the context, all the fields in the `uri` structure are tagged as being of "variable" category. Schema definitions MUST explicitly set the corresponding category in the element referencing this data structure.

5.5.7.2 ipaddr

The `ipaddr` structure represents the hostname and IP address of a system.

`ipaddr`	Category	Structure	Short Display Name
hostname	Computer Information	*unstructured*	Complete Host and Domain Name
partialhostname	Demographic	*unstructured*	Partial Hostname
fullip	Computer Information	*unstructured*	Full IP Address
partialip	Demographic	*unstructured*	Partial IP Address

The `hostname` element is used to represent collection of either the simple hostname of a system, or the full hostname including domain name. The `partialhostname` element represents the information of a fully-qualified hostname which has had *at least* the host portion removed from the hostname. In other words, everything up to the first "." in the fully-qualified hostname MUST be removed for an address to qualify as a "partial hostname."

The `fullip` element represents the information of a full IP version 4 or IP version 6 address. The `partialip` element represents an IP version 4 address (only—not a version 6 address) which has had *at least* the last 7 bits of information removed. This removal MUST be done by replacing those bits with a fixed pattern for all visitors (for example, all 0s or all 1s).

Certain Web sites are known to make use not of the visitor's entire IP address or hostname, but rather make use of a reduced form of that information. By collecting only a subset of the address information, the site visitor is given some measure of anonymity. It is certainly not the intent of this specification to claim that these "stripped" IP addresses or hostnames are impossible to associate with an individual user, but rather that it is significantly more difficult to do so. Sites which perform this data reduction MAY wish to declare this practice in order to more-accurately reflect their practices.

5.5.7.3 Access Log Information

The `loginfo` structure is used to represent information typically stored in Web-server access logs.

loginfo	Category	Structure	Short Display Name
uri	Navigation and click-stream data	uri	URI of Requested Resource
timestamp	Navigation and click-stream data	date	Request Timestamp
clientip	Computer Information, Demographic and Socioeconomic Data	ipaddr	Client's IP Address or Hostname
other.httpmethod	Navigation and click-stream data	*unstructured*	HTTP Request Method
other.bytes	Navigation and click-stream data	*unstructured*	Data Bytes in Response
other.statuscode	Navigation and click-stream data	*unstructured*	Response Status Code

The resource in the HTTP request is captured by the `uri` field. The time at which the server processes the request is represented by the `timestamp` field. Server implementations are free to define this field as the time the request was received, the time that the server began sending the response, the time that sending the response was complete, or some other convenient representation of the time the request was processed. The IP address of the client system making the request is given by the `clientip` field.

The `other` data fields represent other information commonly stored in Web server access logs. `other.httpmethod` is the HTTP method (such as GET, POST, etc.) in the client's request. `other.bytes` indicates the number of bytes in the response-body sent by the server. `other.statuscode` is the HTTP status code on the request, such as 200, 302, or 404 (see section 6.1.1 of [HTTP1.1] for details).

5.5.7.4 Other HTTP Protocol Information

The `httpinfo` structure represents information carried by the HTTP protocol which is not covered by the loginfo structure.

`httpinfo`	Category	Structure	Short Display Name
referer	Navigation and click-stream data	uri	Last URI Requested by the User
useragent	Computer Information	*unstructured*	User Agent Information

The `useragent` field represents the information in the HTTP `User-Agent` header (which gives information about the type and version of the user's Web browser), and/or the HTTP `accept*` headers.

The `referer` field represents the information in the HTTP `Referer` header, which gives information about the previous page visited by the user. Note that this field is mis-spelled in exactly the same way as the corresponding HTTP header.

5.6 The Base Data Schema

All P3P-compliant user agent implementations MUST be aware of the data elements in the P3P base data schema. The P3P base data schema includes the definition of the basic data structures, and four data element sets: **user, thirdparty, business** and **dynamic**. The user, thirdparty and business sets include elements that users and/or businesses might provide values for, while the `dynamic` set includes elements that are dynamically generated in the course of a user's browsing session. User agents may support a variety of mechanisms that allow users to provide values for the elements in the user set and store them in a data repository, including mechanisms that support multiple personae. Users may choose not to provide values for these data elements.

The formal XML definition of the P3P base data schema is given in Appendix 3. In the following sections, the base data elements and sets are explained one by one. In the future there will be in all likelihood *demand for the creation of other data sets and elements*. Obvious applications include catalogue, payment, and agent/system attribute schemas (an extensive set of system elements is provided for example in http://www.w3.org/TR/NOTE-agent-attributes).

Each table below specifies a set, the elements within the set, the category associated with the element, its structure, and the display name shown to users. More than one category may be associated with a fixed data element. However, each base data element is assigned to only one category whenever possible. It is recommended that data schema designers do the same.

5.6.1 User Data

The `user` data set includes general information about the user.

user	Category	Structure	Short Display Name
name	Physical Contact Information, Demographic and Socioeconomic Data	personname	User's Name
bdate	Demographic and Socioeconomic Data	date	User's Birth Date
login	Unique Identifiers	login	User's Login Information
cert	Unique Identifiers	certificate	User's Identity Certificate
gender	Demographic and Socioeconomic Data	*unstructured*	User's Gender (Male or Female)
employer	Demographic and Socioeconomic Data	*unstructured*	User's Employer
department	Demographic and Socioeconomic Data	*unstructured*	Department or Division of Organization Where User is Employed
jobtitle	Demographic and Socioeconomic Data	*unstructured*	User's Job Title
home-info	Physical Contact Information, Online Contact Information, Demographic and Socioeconomic Data	contact	User's Home Contact Information
business-info	Physical Contact Information, Online Contact Information, Demographic and Socioeconomic Data	contact	User's Business Contact Information

Note that this data set includes elements that are actually sets of data themselves. These sets are defined in the Data Structures subsection of this document. The short display name for an individual element contained within a data set is defined as the concatenation of the short display names that have been defined for the set and the element, separated by a separator appropriate for the language/script in question, e.g., a comma for English. For example, the short display name for `user.home-info.postal.postalcode` could be "User's Home Contact Information, Postal Address Information, Postal code." User agent implementations may prefer to develop their own short display names rather than using the concatenated names when displaying information for the user.

5.6.2 Third Party Data

The `thirdparty` data set allows users and businesses to provide values for a related third party. This can be useful whenever third party information needs to be exchanged, for example when ordering a present online that should be sent to another person, or when providing information about one's spouse or business partner. Such information could be stored in a user repository alongside the user data set. User agents may offer to store multiple such `thirdparty` data sets and allow users to select the appropriate values from a list when necessary.

The `thirdparty` data set is identical with the user data set. See section 5.6.1 User Data for details.

5.6.3 Business Data

The `business` data set features a subset of `user` data relevant for organizations. In P3P1.0, this data set is primarily used for declaring the policy entity, though it should also be applicable to business-to-business interactions.

business	Category	Structure	Short Display Name
name	Demographic and Socioeconomic Data	*unstructured*	Organization Name
department	Demographic and Socioeconomic Data	*unstructured*	Department or Division of Organization
cert	Unique Identifiers	certificate	Organization Identity Certificate
contact-info	Physical Contact Information, Online Contact Information, Demographic and Socioeconomic Data	contact	Contact Information for the Organization

5.6.4 Dynamic Data

In some cases, there is a need to specify data elements that do not have fixed values that a user might type in or store in a repository. In the P3P base data schema, all such elements are grouped under the `dynamic` data set. Sites may refer to the types of data they collect using the dynamic data set only, rather than enumerating all of the specific data elements.

dynamic	Category	Structure	Short Display Name
clickstream	Navigation and Click-stream Data, Computer Information	loginfo	Click-stream Information
http	Navigation and Click-stream Data, Computer Information	httpinfo	HTTP Protocol Information
clientevents	Navigation and Click-stream Data	*unstructured*	User's Interaction with a Resource
cookies	*(variable-category)*	*unstructured*	Use of HTTP Cookies
miscdata	*(variable-category)*	*unstructured*	Miscellaneous Non-base Data Schema Information

searchtext	Interactive Data	*unstructured*	Search Terms
interactionrecord	Interactive Data	*unstructured*	Server Stores the Transaction History

These elements are often implicit in navigation or Web interactions. They should be used with categories to describe the type of information collected through these methods. A brief description of each element follows.

clickstream

The clickstream element is expected to apply to practically all Web sites. It represents the combination of information typically found in Web server access logs: the IP address or hostname of the user's computer, the URI of the resource requested, the time the request was made, the HTTP method used in the request, the size of the response, and the HTTP status code in the response. Web sites that collect standard server access logs as well as sites which do URI path analysis can use this data element to describe how that data will be used. Web sites that collect only some of the data elements listed for the clickstream element MAY choose to list those specific elements rather than the entire dynamic.clickstream element. This allows sites with more limited data-collection practices to accurately present those practices to their visitors.

http

The http element contains additional information contained in the HTTP protocol. See the definition of the httpinfo structure for descriptions of specific elements. Sites MAY use the dynamic.http field as a shorthand to cover all the elements in the httpinfo structure if they wish, or they MAY reference the specific elements in the httpinfo structure.

clientevents

The clientevents element represents data about how the user interacts with their Web browser while interacting with a resource. For example, an application may wish to collect information about whether the user moved their mouse over a certain image on a page, or whether the user ever brought up the help window in a Java applet. This kind of information is represented by the dynamic.clientevents data element. Much of this interaction record is represented by the events and data defined by the Document Object Model (DOM) Level 2 Events [DOM2-Events]. The clientevents data element also covers any other data regarding the user's interaction with their browser while the browser is displaying a resource. The exception is events which are covered by other elements in the base data schema. For example, requesting a page by clicking on a link is part of the user's interaction with their browser while viewing a page, but merely collecting the URL the user has clicked on does not require declaring this data element; clickstream covers that event. However, the DOM event DOMFocusIn (representing the user moving their mouse over an object on a page) is not covered by any other existing element, so if a site is collecting the occurrence of this event, then it needs to state that it collects the dynamic.clientevents element. Items covered by this data element are typically collected by client-side scripting languages, such as JavaScript, or by client-side applets, such as ActiveX or Java applets. Note that while the previous discussion has been in terms of a user viewing a resource, this data element also applies to Web applications which do not display resources visually—for example, audio-based Web browsers.

cookies

The `cookies` element should be used whenever HTTP cookies are set or retrieved by a site. Please note that `cookies` is a *variable data element* and requires the explicit declaration of usage categories in a policy.

miscdata

The `miscdata` element references information collected by the service that the service does not reference using a specific data element. Categories have to be used to better describe these data: sites MUST reference a separate `miscdata` element in their policies for each category of miscellaneous data they collect.

searchtext

The `searchtext` element references a specific type of solicitation used for searching and indexing sites. For example, if the only fields on a search engine page are search fields, the site only needs to disclose that data element.

interactionrecord

The `interactionrecord` element should be used if the server is keeping track of the interaction it has with the user (i.e., information other than clickstream data, for example account transactions, etc.).

5.7 Categories and Data Elements/Structures

5.7.1 Fixed-Category Data Elements/Structures

Most of the elements in the base data schema are so-called "*fixed*" data elements: they belong to one or at most two category classes. By assigning a category invariably to elements or structures in the base data schema, services and users are able to refer to entire groups of elements simply by referencing the corresponding category. For example, using [APPEL], the privacy preferences exchange language, users can write rules that warn them when they visit a site that collects any data element in a certain category.

When creating data schemas for fixed data elements, schema creators have to explicitly enumerate the categories that these elements belong to. For example:

```
<DATA-STRUCT name="postal.street"     structref="#text"
           short-description="Street Address">
<CATEGORIES><physical/></CATEGORIES>
</DATA-STRUCT>
```

If an element or structure belongs to multiple categories, multiple elements referencing the appropriate categories can be used. For example, the following piece of XML can be used to declare that the data elements in user.name have both category "physical" and "demographic":

```
<DATA-STRUCT name="user.name"       structref="#personname"
             short-description="User's Name">
<CATEGORIES><physical/><demographic/></CATEGORIES>
</DATA-STRUCT>
```

Please note that the category classes of fixed data elements/structures can **not** be overridden, for example by writing rules or policies that assign a different category to a known fixed base data element. User Agents MUST ignore such categories and instead use the original category (or set of categories) listed in the schema definition. User Agents MAY preferably alert the user that a fixed data element is used together with a non-standard category class.

5.7.2 Variable-Category Data Elements/Structures

Not all data elements/structures in the base data schema belong to a pre-determined category class. Some can contain information from a range of categories, depending on a particular situation. Such elements/structures are called *variable-category data elements/ structures* (or "variable data element/structure" for short). Although most variable data elements in the P3P base data schema are combined in the dynamic element set, they can appear in any data set, even mixed with *fixed-category data elements.*

When creating a schema definition for such elements and/or structures, schema authors MUST NOT list an explicit category attribute, otherwise the element/structure becomes *fixed.* For example when specifying the "Year" *Data Structure,* which can take various categories depending on the situation (e.g., when used for a credit card expiration date vs. for a birth date), the following schema definition can be used:

```
<DATA-STRUCT name="date.ymd.year"
             short-description="Year"/>
<!-- Variable Data Structure-->
```

This allows new schema extensions that reference such variable-category *Data Structures* to assign a specific category to derived elements, depending on their usage in that extension. For example, an e-commerce schema extension could thus define a credit card expiration date as follows:

```
<DATA-STRUCT name="Card.ExpDate"          structref="#date.ymd"
             short-description="Card Expiration Date">
<CATEGORIES><purchase/></CATEGORIES>
</DATA-STRUCT>
```

Under these conditions, the variable Data Structure date is assigned a fixed category "Purchase Information" when being used for specifying a credit card expiration date.

Note that while user preferences can list such variable data elements without any additional category information (effectively expressing preferences over *any* usage of this element), services MUST always explicitly specify the categories that apply to the usage of a variable data element in their particular policy. This information has to appear as a category element in the corresponding DATA element listed in the policy, for example as in:

```
<POLICY ... >
   ...
   <DATA
ref="#dynamic.cookies"><CATEGORIES><uniqueid/></CATEGORIES></DATA>
   ...
</POLICY>
```

where a service declares that cookies are used to recognize the user at this site (i.e., category Unique Identifiers).

If a service wants to declare a data element that is in multiple categories, it simply declares the corresponding categories (as shown in the above section):

```
<POLICY ... >
   ...
   <DATA
ref="#dynamic.cookies"><CATEGORIES><uniqueid/><preference/>
   </CATEGORIES></DATA>
   ...
</POLICY>
```

With the above declaration a service announces that it uses cookies both to recognize the user at this site *and* for storing user preference data. Note that for the purpose of P3P there is no difference whether this information is stored in two separate cookies or in a single one.

Finally, note that categories can be inherited as well: *Categories inherit downward when a field is structured, but only into fields which have no predefined category.* Therefore, we suggest to schema authors that they do their best to insure that all applicable categories are applied to new data elements they create.

5.8 Using Data Elements

P3P offers Web sites a great deal of flexibility in how they describe the types of data they collect.

- Sites may describe data generally using the `dynamic.miscdata` element and the appropriate categories.
- Sites may describe data specifically using the data elements defined in the base data schema.
- Sites may describe data specifically using data elements defined in new data schemas.

Any of these three methods may be combined within a single policy.

By using the `dynamic.miscdata` element, sites can specify the types of data they collect without having to enumerate every individual data element. This may be convenient for sites that collect a lot of data or sites belonging to large organizations that want to offer a single P3P policy covering the entire organization. However, the disadvantage of this approach is that user agents will have to assume that the site might collect any data element belonging to the categories referenced by the site. So, for example, if a site's policy states that it collects `dynamic.miscdata` of the physical contact information category, but the only physical contact information it collects is business address, user agents will nonetheless assume that the site might also collect telephone numbers. If the site wishes to be clear that it does not collect telephone numbers or any other physical contact information other than business address, than it should disclose that it collects `user.business-info.contact.postal`. Furthermore, as user agents are developed with automatic form-filling capabilities, it is likely that sites that enumerate the data they collect will be able to better integrate with these tools.

By defining new data schemas, sites can precisely specify the data they collect beyond the base data set. However, if user agents are unfamiliar with the elements defined in these schemas, they will be able to provide only minimal information to the user about these new elements. The information they provide will be based on the category and display names specified for each element.

Regardless of whether a site wishes to make general or specific data disclosures, there are additional advantages to disclosing specific elements from the `dynamic` data set. For example, by disclosing `dynamic.cookies` a site can indicate that it uses cookies and explain the purpose of this use. User agent implementations that offer users cookie control interfaces based on this information are encouraged. Likewise, user agents that by default do not send the HTTP_REFERER header, might look for the `dynamic.http.referer` element in P3P policies and send the header if it will be used for a purpose the user finds acceptable.

6 APPENDICES

Appendix 1: References (Normative)
[CHARMODEL]
M. Dürst, F. Yergeau (Eds.), "Character Model for the World Wide Web," World Wide Web Consortium Working Draft. 29 November 1999.
[DOM2-Events]
T. Pixley (Ed.), "Document Object Model (DOM) Level 2 Events Specification," World Wide Web Consortium, Proposed Recommendation. 27 September 2000.
[HTTP1.0]
T. Berners-Lee, R. Fielding, H. Frystyk, "RFC1945—Hypertext Transfer Protocol—HTTP/1.0," May 1996.
[HTTP1.1]
R. Fielding, J. Gettys, J. Mogul, H. Frystyk, L. Masinter, P. Leach, T. Berners-Lee, "RFC2616—Hypertext Transfer Protocol—HTTP/1.1," June 1999. [Updates RFC2068]

[KEY]

S. Bradner. "RFC2119—Key words for use in RFCs to Indicate Requirement Levels." March 1997.

[P3P-HEADER]

R. Lotenberg, M. Marchiori (Eds.), "The HTTP header for the Platform for Privacy Preferences 1.0 (P3P1.0)" (also available in HTML and XML formats), IETF Internet Draft, August 2001.

[STATE]

Kristol, D., Montulli, L., "RFC2965—HTTP State Management Mechanism." October, 2000 [Obsoletes RFC2109]

[URI]

T. Berners-Lee, R. Fielding, and L. Masinter. "RFC 2396—Uniform Resource Identifiers (URI): Generic Syntax and Semantics." August 1998. [Updates RFC1738]

[UTF-8]

F. Yergeau. "RFC2279—UTF-8, a transformation format of ISO 10646." January 1998.

[XML]

T. Bray, J. Paoli, C. M. Sperberg-McQueen (Eds.). "Extensible Markup Language (XML) 1.0 Specification." World Wide Web Consortium, Recommendation. 10 February 1998.

[XML-Name]

T. Bray, D. Hollander, A. Layman (Eds.). "Namespaces in XML." World Wide Web Consortium, Recommendation. 14 January 1999.

[XML-Schema1]

H. Thompson, D. Beech, M. Maloney, and N. Mendelsohn (Eds.). "XML Schema Part 1: Structures" World Wide Web Consortium Recommendation. 2 May 2001.

[XML-Schema2]

P. Biron, A. Malhotra (Eds.). "XML Schema Part 2: Datatypes" World Wide Web Consortium Recommendation. 2 May 2001.

Appendix 2: References (Non-Normative)

[ABNF]

D. Crocker, P. Overel. "RFC2234—Augmented BNF for Syntax Specifications: ABNF," Internet Mail Consortium, Demon Internet Ltd., November 1997.

[APPEL]

M. Langheinrich (Ed.). "A P3P Preference Exchange Language (APPEL)" World Wide Web Consortium Working Draft.

[COOKIES]

"Persistent Client State—HTTP Cookies," Preliminary Specification, Netscape, 1999.

[HTML]

D. Raggett, A. Le Hors, and I. Jacobs (Eds.). "HTML 4.01 Specification" World Wide Web Consortium.

[ISO3166]

"ISO3166: Codes for The Representation of Names of Countries." International Organization for Standardization.

[ISO8601]

"ISO8601: Data elements and interchange formats—Information interchange—Representation of dates and times." International Organization for Standardization.

[RDF]

O. Lassila and R. Swick (Eds.). "Resource Description Framework (RDF) Model and Syntax Specification." World Wide Web Consortium, Recommendation. 22 February 1999.

[UNICODE]

Unicode Consortium. "The Unicode Standard."

Appendix 3: The P3P Base Data Schema Definition (Normative)

The data schema corresponding to the P3P base data schema follows for easy reference. The schema is also present as a separate file at the URI *http://www.w3.org/TR/P3P/base.*

```
<DATASCHEMA xmlns="http://www.w3.org/2001/09/P3Pv1">
<!-- ********** Base Data Structures ********** -->

<!-- "date" Data Structure -->
<DATA-STRUCT name="date.ymd.year"
    short-description="Year"/>

<DATA-STRUCT name="date.ymd.month"
    short-description="Month"/>

<DATA-STRUCT name="date.ymd.day"
    short-description="Day"/>

<DATA-STRUCT name="date.hms.hour"
    short-description="Hour"/>

<DATA-STRUCT name="date.hms.minute"
    short-description="Minute"/>

<DATA-STRUCT name="date.hms.second"
    short-description="Second"/>

<DATA-STRUCT name="date.fractionsecond"
    short-description="Fraction of Second"/>

<DATA-STRUCT name="date.timezone"
    short-description="Time Zone"/>

<!-- "login" Data Structure -->
<DATA-STRUCT name="login.id"
    short-description="Login ID">
    <CATEGORIES><uniqueid/></CATEGORIES>
</DATA-STRUCT>
```

```
<DATA-STRUCT name="login.password"
    short-description="Login Password">
    <CATEGORIES><uniqueid/></CATEGORIES>
</DATA-STRUCT>

<!-- "personname" Data Structure -->
<DATA-STRUCT name="personname.prefix"
    short-description="Name Prefix">
    <CATEGORIES><demographic/></CATEGORIES>
</DATA-STRUCT>

<DATA-STRUCT name="personname.given"
    short-description="Given Name (First Name)">
    <CATEGORIES><physical/></CATEGORIES>
</DATA-STRUCT>

<DATA-STRUCT name="personname.middle"
    short-description="Middle Name">
    <CATEGORIES><physical/></CATEGORIES>
</DATA-STRUCT>

<DATA-STRUCT name="personname.family"
    short-description="Family Name (Last Name)">
    <CATEGORIES><physical/></CATEGORIES>
</DATA-STRUCT>

<DATA-STRUCT name="personname.suffix"
    short-description="Name Suffix">
    <CATEGORIES><demographic/></CATEGORIES>
</DATA-STRUCT>

<DATA-STRUCT name="personname.nickname"
    short-description="Nickname">
    <CATEGORIES><demographic/></CATEGORIES>
</DATA-STRUCT>

<!-- "certificate" Data Structure -->
<DATA-STRUCT name="certificate.key"
    short-description="Certificate key">
    <CATEGORIES><uniqueid/></CATEGORIES>
</DATA-STRUCT>

<DATA-STRUCT name="certificate.format"
    short-description="Certificate format">
    <CATEGORIES><uniqueid/></CATEGORIES>
</DATA-STRUCT>

<!-- "telephonenum" Data Structure -->
<DATA-STRUCT name="telephonenum.intcode"
    short-description="International Telephone Code">
    <CATEGORIES><physical/></CATEGORIES>
</DATA-STRUCT>
```

```
<DATA-STRUCT name="telephonenum.loccode"
    short-description="Local Telephone Area Code">
    <CATEGORIES><physical/></CATEGORIES>
</DATA-STRUCT>

<DATA-STRUCT name="telephonenum.number"
    short-description="Telephone Number">
    <CATEGORIES><physical/></CATEGORIES>
</DATA-STRUCT>

<DATA-STRUCT name="telephonenum.ext"
    short-description="Telephone Extension">
    <CATEGORIES><physical/></CATEGORIES>
</DATA-STRUCT>

<DATA-STRUCT name="telephonenum.comment"
    short-description="Telephone Optional Comments">
    <CATEGORIES><physical/></CATEGORIES>
</DATA-STRUCT>

<!-- "postal" Data Structure -->
<DATA-STRUCT name="postal.name" structref="#personname">
</DATA-STRUCT>

<DATA-STRUCT name="postal.street"
    short-description="Street Address">
    <CATEGORIES><physical/></CATEGORIES>
</DATA-STRUCT>

<DATA-STRUCT name="postal.city"
    short-description="City">
    <CATEGORIES><demographic/></CATEGORIES>
</DATA-STRUCT>

<DATA-STRUCT name="postal.stateprov"
    short-description="State or Province">
    <CATEGORIES><demographic/></CATEGORIES>
</DATA-STRUCT>

<DATA-STRUCT name="postal.postalcode"
    short-description="Postal Code">
    <CATEGORIES><demographic/></CATEGORIES>
</DATA-STRUCT>

<DATA-STRUCT name="postal.organization"
    short-description="Organization Name">
    <CATEGORIES><demographic/></CATEGORIES>
</DATA-STRUCT>
```

```
<DATA-STRUCT name="postal.country"
    short-description="Country Name">
    <CATEGORIES><demographic/></CATEGORIES>
</DATA-STRUCT>

<!-- "telecom" Data Structure -->
<DATA-STRUCT name="telecom.telephone"
    short-description="Telephone Number"
    structref="#telephonenum">
    <CATEGORIES><physical/></CATEGORIES>
</DATA-STRUCT>

<DATA-STRUCT name="telecom.fax"
    short-description="Fax Number"
    structref="#telephonenum">
    <CATEGORIES><physical/></CATEGORIES>
</DATA-STRUCT>

<DATA-STRUCT name="telecom.mobile"
    short-description="Mobile Telephone Number"
    structref="#telephonenum">
    <CATEGORIES><physical/></CATEGORIES>
</DATA-STRUCT>

<DATA-STRUCT name="telecom.pager"
    short-description="Pager Number"
    structref="#telephonenum">
    <CATEGORIES><physical/></CATEGORIES>
</DATA-STRUCT>

<!-- "online" Data Structure -->
<DATA-STRUCT name="online.email"
    short-description="Email Address">
    <CATEGORIES><online/></CATEGORIES>
</DATA-STRUCT>

<DATA-STRUCT name="online.uri"
    short-description="Home Page Address">
    <CATEGORIES><online/></CATEGORIES>
</DATA-STRUCT>

<!-- "contact" Data Structure -->
<DATA-STRUCT name="contact.postal"
    short-description="Postal Address Information"
    structref="#postal">
</DATA-STRUCT>

<DATA-STRUCT name="contact.telecom"
    short-description="Telecommunications Information"
    structref="#telecom">
    <CATEGORIES><physical/></CATEGORIES>
</DATA-STRUCT>
```

```
<DATA-STRUCT name="contact.online"
     short-description="Online Address Information"
     structref="#online">
     <CATEGORIES><online/></CATEGORIES>
</DATA-STRUCT>

<!-- "uri" Data Structure -->
<DATA-STRUCT name="uri.authority"
     short-description="URI Authority"/>

<DATA-STRUCT name="uri.stem"
     short-description="URI Stem"/>

<DATA-STRUCT name="uri.querystring"
     short-description="Query-string Portion of URI"/>

<!-- "ipaddr" Data Structure -->
<DATA-STRUCT name="ipaddr.hostname"
     short-description="Complete Host and Domain Name">
     <CATEGORIES><computer/></CATEGORIES>
</DATA-STRUCT>

<DATA-STRUCT name="ipaddr.partialhostname"
     short-description="Partial Hostname">
     <CATEGORIES><demographic/></CATEGORIES>
</DATA-STRUCT>

<DATA-STRUCT name="ipaddr.fullip"
     short-description="Full IP Address">
     <CATEGORIES><computer/></CATEGORIES>
</DATA-STRUCT>

<DATA-STRUCT name="ipaddr.partialip"
     short-description="Partial IP Address">
     <CATEGORIES><demographic/></CATEGORIES>
</DATA-STRUCT>

<!-- "loginfo" Data Structure -->
<DATA-STRUCT name="loginfo.uri"
     short-description="URI of Requested Resource"
     structref="#uri">
     <CATEGORIES><navigation/></CATEGORIES>
</DATA-STRUCT>

<DATA-STRUCT name="loginfo.timestamp"
     short-description="Request Timestamp"
     structref="#date">
     <CATEGORIES><navigation/></CATEGORIES>
</DATA-STRUCT>
```

```
<DATA-STRUCT name="loginfo.clientip"
    short-description="Client's IP Address or Hostname"
    structref="#ipaddr">
</DATA-STRUCT>

<DATA-STRUCT name="loginfo.other.httpmethod"
    short-description="HTTP Request Method">
    <CATEGORIES><navigation/></CATEGORIES>
</DATA-STRUCT>

<DATA-STRUCT name="loginfo.other.bytes"
    short-description="Data Bytes in Response">
    <CATEGORIES><navigation/></CATEGORIES>
</DATA-STRUCT>

<DATA-STRUCT name="loginfo.other.statuscode"
    short-description="Response Status Code">
    <CATEGORIES><navigation/></CATEGORIES>
</DATA-STRUCT>

<!-- "httpinfo" Data Structure -->
<DATA-STRUCT name="httpinfo.referer"
    short-description="Last URI Requested by the User"
    structref="#uri">
    <CATEGORIES><navigation/></CATEGORIES>
</DATA-STRUCT>

<DATA-STRUCT name="httpinfo.useragent"
    short-description="User Agent Information">
    <CATEGORIES><computer/></CATEGORIES>
</DATA-STRUCT>

<!-- ********** Base Data Schemas ********** -->

<!-- "dynamic" Data Schema -->
<DATA-DEF name="dynamic.clickstream"
    short-description="Click-stream Information"
    structref="#loginfo">
    <CATEGORIES><navigation/><computer/><demographic/></CATEGORIES>
</DATA-DEF>

<DATA-DEF name="dynamic.http"
    short-description="HTTP Protocol Information"
    structref="#httpinfo">
    <CATEGORIES><navigation/><computer/></CATEGORIES>
</DATA-DEF>

<DATA-DEF name="dynamic.clientevents"
    short-description="User's Interaction with a Resource">
    <CATEGORIES><navigation/></CATEGORIES>
</DATA-DEF>
```

```
<DATA-DEF name="dynamic.cookies"
    short-description="Use of HTTP Cookies"/>

<DATA-DEF name="dynamic.searchtext"
    short-description="Search Terms">
    <CATEGORIES><interactive/></CATEGORIES>
</DATA-DEF>

<DATA-DEF name="dynamic.interactionrecord"
    short-description="Server Stores the Transaction History">
    <CATEGORIES><interactive/></CATEGORIES>
</DATA-DEF>

<DATA-DEF name="dynamic.miscdata"
    short-description="Miscellaneous Non-base Data Schema =
information"/>

<!-- "user" Data Schema -->
<DATA-DEF name="user.name"
    short-description="User's Name"
    structref="#personname">
    <CATEGORIES><physical/><demographic/></CATEGORIES>
</DATA-DEF>

<DATA-DEF name="user.bdate"
    short-description="User's Birth Date"
    structref="#date">
    <CATEGORIES><demographic/></CATEGORIES>
</DATA-DEF>

<DATA-DEF name="user.login"
    short-description="User's Login Information"
    structref="#login">
    <CATEGORIES><uniqueid/></CATEGORIES>
</DATA-DEF>

<DATA-DEF name="user.cert"
    short-description="User's Identity Certificate"
    structref="#certificate">
    <CATEGORIES><uniqueid/></CATEGORIES>
</DATA-DEF>

<DATA-DEF name="user.gender"
    short-description="User's Gender">
    <CATEGORIES><demographic/></CATEGORIES>
</DATA-DEF>

<DATA-DEF name="user.jobtitle"
    short-description="User's Job Title">
    <CATEGORIES><demographic/></CATEGORIES>
</DATA-DEF>
```

```
<DATA-DEF name="user.home-info"
    short-description="User's Home Contact Information"
    structref="#contact">
    <CATEGORIES><physical/><online/><demographic/></CATEGORIES>
</DATA-DEF>

<DATA-DEF name="user.business-info"
    short-description="User's Business Contact Information"
    structref="#contact">
    <CATEGORIES><physical/><online/><demographic/></CATEGORIES>
</DATA-DEF>

<DATA-DEF name="user.employer"
    short-description="Name of User's Employer">
    <CATEGORIES><demographic/></CATEGORIES>
</DATA-DEF>

<DATA-DEF name="user.department"
    short-description="Department or Division of Organization where
    User is Employed">
    <CATEGORIES><demographic/></CATEGORIES>
</DATA-DEF>

<!-- "thirdparty" Data Schema -->
<DATA-DEF name="thirdparty.name"
    short-description="Third Party's Name"
    structref="#personname">
    <CATEGORIES><physical/><demographic/></CATEGORIES>
</DATA-DEF>

<DATA-DEF name="thirdparty.bdate"
    short-description="Third Party's Birth Date"
    structref="#date">
    <CATEGORIES><demographic/></CATEGORIES>
</DATA-DEF>

<DATA-DEF name="thirdparty.login"
    short-description="Third Party's Login Information"
    structref="#login">
    <CATEGORIES><uniqueid/></CATEGORIES>
</DATA-DEF>

<DATA-DEF name="thirdparty.cert"
    short-description="Third Party's Identity Certificate"
    structref="#certificate">
    <CATEGORIES><uniqueid/></CATEGORIES>
</DATA-DEF>

<DATA-DEF name="thirdparty.gender"
    short-description="Third Party's Gender">
    <CATEGORIES><demographic/></CATEGORIES>
</DATA-DEF>
```

```
<DATA-DEF name="thirdparty.jobtitle"
    short-description="Third Party's Job Title">
    <CATEGORIES><demographic/></CATEGORIES>
</DATA-DEF>

<DATA-DEF name="thirdparty.home-info"
    short-description="Third Party's Home Contact Information"
    structref="#contact">
    <CATEGORIES><physical/><online/><demographic/></CATEGORIES>
</DATA-DEF>

<DATA-DEF name="thirdparty.business-info"
    short-description="Third Party's Business Contact Information"
    structref="#contact">
    <CATEGORIES><physical/><online/><demographic/></CATEGORIES>
</DATA-DEF>

<DATA-DEF name="thirdparty.employer"
    short-description="Name of Third Party's Employer">
    <CATEGORIES><demographic/></CATEGORIES>
</DATA-DEF>

<DATA-DEF name="thirdparty.department"
    short-description="Department or Division of Organization where
    Third Party is Employed">
    <CATEGORIES><demographic/></CATEGORIES>
</DATA-DEF>

<!-- "business" Data Schema -->
<DATA-DEF name="business.name"
    short-description="Organization Name">
    <CATEGORIES><demographic/></CATEGORIES>
</DATA-DEF>

<DATA-DEF name="business.department"
    short-description="Department or Division of Organization">
    <CATEGORIES><demographic/></CATEGORIES>
</DATA-DEF>

<DATA-DEF name="business.cert"
    short-description="Organization Identity certificate"
    structref="#certificate">
    <CATEGORIES><uniqueid/></CATEGORIES>
</DATA-DEF>

<DATA-DEF name="business.contact-info"
    short-description="Contact Information for the Organization"
    structref="#contact">
    <CATEGORIES><physical/><online/><demographic/></CATEGORIES>
</DATA-DEF>

</DATASCHEMA>
```

Appendix 4: XML Schema Definition (Normative)

This appendix contains the XML schema, both for P3P policy reference files, for P3P policy documents, and for P3P data schema documents. An XML schema may be used to validate the structure and datastruct values used in an instance of the schema given as an XML document. P3P policy and data schema documents are XML documents that MUST conform to this schema. Note that this schema is based on the XML Schema specification [XML-Schema1][XML-Schema2]. The schema is also present as a separate file at the URI *http://www.w3.org/2001/09/P3Pv1.xsd*.

```
<?xml version='1.0' encoding='UTF-8'?>
<schema
  xmlns='http://www.w3.org/2001/XMLSchema'
  xmlns:p3p='http://www.w3.org/2001/09/P3Pv1'
  targetNamespace='http://www.w3.org/2001/09/P3Pv1'
  elementFormDefault='qualified'>

<!-- Basic P3P Data Type -->
 <simpleType name='yes_no'>
  <restriction base='string'>
   <enumeration value='yes'/>
   <enumeration value='no'/>
  </restriction>
 </simpleType>


<!-- *********** Policy Reference *********** -->
<!-- ************** META ************** -->
 <element name='META'>
  <complexType mixed='true'>
   <sequence>
    <element ref='p3p:POLICY-REFERENCES'/>
    <element ref='p3p:POLICIES' minOccurs='0'/>
   </sequence>
  </complexType>
 </element>

<!-- ******* POLICY-REFERENCES ******** -->
 <element name='POLICY-REFERENCES'>
  <complexType>
   <sequence>
    <element ref='p3p:EXPIRY' minOccurs='0'/>
    <element ref='p3p:POLICY-REF' minOccurs='0' maxOccurs='unbounded'/>
    <element ref='p3p:HINT' minOccurs='0' maxOccurs='unbounded'/>
   </sequence>
  </complexType>
 </element>

 <element name='POLICY-REF'>
  <complexType>
```

```xml
  <sequence>
   <element name='INCLUDE'
            minOccurs='0' maxOccurs='unbounded' type='anyURI'/>
   <element name='EXCLUDE'
            minOccurs='0' maxOccurs='unbounded' type='anyURI'/>
   <element name='COOKIE-INCLUDE'
            minOccurs='0' maxOccurs='unbounded'
            type='p3p:cookie-element'/>
   <element name='COOKIE-EXCLUDE'
            minOccurs='0' maxOccurs='unbounded'
            type='p3p:cookie-element'/>
   <element name='METHOD'
            minOccurs='0' maxOccurs='unbounded' type='anyURI'/>
  </sequence>
  <attribute name='about' type='anyURI' use='required'/>
 </complexType>
</element>

<complexType name='cookie-element'>
 <attribute name='name' type='string' use='optional'/>
 <attribute name='value' type='string' use='optional'/>
 <attribute name='domain' type='string' use='optional'/>
 <attribute name='path' type='string' use='optional'/>
</complexType>

<!-- ************* HINT ************* -->
<element name='HINT'>
 <complexType>
  <attribute name='domain' type='string' use='required'/>
  <attribute name='path' type='string' use='required'/>
 </complexType>
</element>

<!-- ************* EXPIRY ************* -->
<element name='EXPIRY'>
 <complexType>
  <attribute name='max-age' type='nonNegativeInteger' use='optional'/>
  <attribute name='date' type='string' use='optional'/>
 </complexType>
</element>

<!-- *********** POLICIES ************ -->
<element name='POLICIES'>
 <complexType>
  <sequence>
   <element ref='p3p:EXPIRY' minOccurs='0'/>
   <element ref='p3p:DATASCHEMA' minOccurs='0'/>
   <element ref='p3p:POLICY' minOccurs='0' maxOccurs='unbounded'/>
  </sequence>
 </complexType>
</element>
```

```
<!-- *************** Policy *************** -->
<!-- ************* POLICY ************* -->
 <element name='POLICY'>
  <complexType>
   <sequence>
    <element ref='p3p:EXTENSION' minOccurs='0' maxOccurs='unbounded'/>
    <element ref='p3p:TEST' minOccurs='0'/>
    <element ref='p3p:ENTITY'/>
    <element ref='p3p:ACCESS'/>
    <element ref='p3p:DISPUTES-GROUP' minOccurs='0'/>
    <element ref='p3p:STATEMENT' minOccurs='0' maxOccurs='unbounded'/>
    <element ref='p3p:EXTENSION' minOccurs='0' maxOccurs='unbounded'/>
   </sequence>
   <attribute name='discuri' type='anyURI' use='required'/>
   <attribute name='opturi' type='anyURI' use='optional'/>
   <attribute name='name' type='ID' use='required'/>
  </complexType>
 </element>

<!-- ************* TEST ************* -->
 <element name='TEST'>
  <complexType/>
 </element>

<!-- ************* ENTITY ************* -->
 <element name='ENTITY'>
  <complexType>
   <sequence>
    <element ref='p3p:EXTENSION' minOccurs='0' maxOccurs='unbounded'/>
    <element ref='p3p:DATA-GROUP'/>
    <element ref='p3p:EXTENSION' minOccurs='0' maxOccurs='unbounded'/>
   </sequence>
  </complexType>
 </element>

<!-- ************* ACCESS ************* -->
 <element name='ACCESS'>
  <complexType>
   <sequence>
    <choice>
     <element name='nonident' type='p3p:access-value'/>
     <element name='ident-contact' type='p3p:access-value'/>
     <element name='other-ident' type='p3p:access-value'/>
     <element name='contact-and-other' type='p3p:access-value'/>
     <element name='all' type='p3p:access-value'/>
     <element name='none' type='p3p:access-value'/>
    </choice>
    <element ref='p3p:EXTENSION' minOccurs='0' maxOccurs='unbounded'/>
   </sequence>
  </complexType>
 </element>
```

```
<complexType name='access-value'/>

<!-- *********** DISPUTES ************ -->
 <element name='DISPUTES-GROUP'>
  <complexType>
   <sequence>
    <element ref='p3p:DISPUTES' maxOccurs='unbounded'/>
    <element ref='p3p:EXTENSION' minOccurs='0' maxOccurs='unbounded'/>
   </sequence>
  </complexType>
 </element>

 <element name='DISPUTES'>
  <complexType>
   <sequence>
    <element ref='p3p:EXTENSION' minOccurs='0' maxOccurs='unbounded'/>
    <choice minOccurs='0'>
     <sequence>
      <element ref='p3p:LONG-DESCRIPTION'/>
      <element ref='p3p:IMG' minOccurs='0'/>
      <element ref='p3p:REMEDIES' minOccurs='0'/>
      <element ref='p3p:EXTENSION' minOccurs='0' maxOccurs='unbounded'/>
     </sequence>
     <sequence>
      <element ref='p3p:IMG'/>
      <element ref='p3p:REMEDIES' minOccurs='0'/>
      <element ref='p3p:EXTENSION' minOccurs='0' maxOccurs='unbounded'/>
     </sequence>
     <sequence>
      <element ref='p3p:REMEDIES'/>
      <element ref='p3p:EXTENSION' minOccurs='0' maxOccurs='unbounded'/>
     </sequence>
    </choice>
   </sequence>
   <attribute name='resolution-type' use='required'>
    <simpleType>
     <restriction base='string'>
      <enumeration value='service'/>
      <enumeration value='independent'/>
      <enumeration value='court'/>
      <enumeration value='law'/>
     </restriction>
    </simpleType>
   </attribute>
   <attribute name='service' type='anyURI' use='required'/>
   <attribute name='verification' type='string' use='optional'/>
   <attribute name='short-description' type='string' use='optional'/>
  </complexType>
 </element>
```

```
<!-- ******** LONG-DESCRIPTION ******** -->
 <element name='LONG-DESCRIPTION'>
  <simpleType>
   <restriction base='string'/>
  </simpleType>
 </element>

<!-- ************* IMG ************* -->
 <element name='IMG'>
  <complexType>
   <attribute name='src' type='anyURI' use='required'/>
   <attribute name='width' type='nonNegativeInteger' use='optional'/>
   <attribute name='height' type='nonNegativeInteger' use='optional'/>
   <attribute name='alt' type='string' use='required'/>
  </complexType>
 </element>

<!-- *********** REMEDIES *********** -->
 <element name='REMEDIES'>
  <complexType>
   <sequence>
    <choice maxOccurs='unbounded'>
     <element name='correct' type='p3p:remedies-value'/>
     <element name='money' type='p3p:remedies-value'/>
     <element name='law' type='p3p:remedies-value'/>
    </choice>
    <element ref='p3p:EXTENSION' minOccurs='0' maxOccurs='unbounded'/>
   </sequence>
  </complexType>
 </element>

 <complexType name='remedies-value'/>

<!-- ********** STATEMENT *********** -->
 <element name='STATEMENT'>
  <complexType>
   <sequence>
    <element ref='p3p:EXTENSION' minOccurs='0' maxOccurs='unbounded'/>
    <element name='CONSEQUENCE' minOccurs='0' type='string'/>
    <element name='NON-IDENTIFIABLE' minOccurs='0'>
     <complexType/>
    </element>
    <element ref='p3p:PURPOSE'/>
    <element ref='p3p:RECIPIENT'/>
    <element ref='p3p:RETENTION'/>
    <element ref='p3p:DATA-GROUP' maxOccurs='unbounded'/>
    <element ref='p3p:EXTENSION' minOccurs='0' maxOccurs='unbounded'/>
   </sequence>
  </complexType>
 </element>
```

```
 <complexType name='non-identifiable'/>

<!-- ************ PURPOSE ************* -->
 <element name='PURPOSE'>
  <complexType>
   <sequence>
    <choice maxOccurs='unbounded'>
     <element name='current' type='p3p:purpose-value'/>
     <element name='admin' type='p3p:purpose-value'/>
     <element name='develop' type='p3p:purpose-value'/>
     <element name='tailoring' type='p3p:purpose-value'/>
     <element name='pseudo-analysis' type='p3p:purpose-value'/>
     <element name='pseudo-decision' type='p3p:purpose-value'/>
     <element name='individual-analysis' type='p3p:purpose-value'/>
     <element name='individual-decision' type='p3p:purpose-value'/>
     <element name='contact' type='p3p:purpose-value'/>
     <element name='historical' type='p3p:purpose-value'/>
     <element name='telemarketing' type='p3p:purpose-value'/>
     <element name='other-purpose'>
      <complexType mixed='true'>
       <attribute name='required' use='optional'
        type='p3p:required-value'/>
      </complexType>
     </element>
    </choice>
    <element ref='p3p:EXTENSION' minOccurs='0' maxOccurs='unbounded'/>
   </sequence>
  </complexType>
 </element>

 <simpleType name='required-value'>
  <restriction base='string'>
   <enumeration value='always'/>
   <enumeration value='opt-in'/>
   <enumeration value='opt-out'/>
  </restriction>
 </simpleType>

 <complexType name='purpose-value'>
  <attribute name='required' use='optional' type='p3p:required-value'/>
 </complexType>

<!-- *********** RECIPIENT ************ -->
 <element name='RECIPIENT'>
  <complexType>
   <sequence>
    <choice maxOccurs='unbounded'>
     <element name='ours'>
      <complexType>
       <sequence>
```

```
        <element ref='p3p:recipient-description' minOccurs='0'
         maxOccurs='unbounded'/>
       </sequence>
      </complexType>
     </element>
     <element name='same' type='p3p:recipient-value'/>
     <element name='other-recipient' type='p3p:recipient-value'/>
     <element name='delivery' type='p3p:recipient-value'/>
     <element name='public' type='p3p:recipient-value'/>
     <element name='unrelated' type='p3p:recipient-value'/>
    </choice>
    <element ref='p3p:EXTENSION' minOccurs='0' maxOccurs='unbounded'/>
   </sequence>
  </complexType>
 </element>

 <complexType name='recipient-value'>
  <sequence>
   <element ref='p3p:recipient-description' minOccurs='0'
    maxOccurs='unbounded'/>
  </sequence>
  <attribute name='required' use='optional' type='p3p:required-value'/>
 </complexType>

 <element name='recipient-description'>
  <complexType mixed='true'/>
 </element>

<!-- ********** RETENTION ************ -->
 <element name='RETENTION'>
  <complexType>
   <sequence>
    <choice>
     <element name='no-retention' type='p3p:retention-value'/>
     <element name='stated-purpose' type='p3p:retention-value'/>
     <element name='legal-requirement' type='p3p:retention-value'/>
     <element name='indefinitely' type='p3p:retention-value'/>
     <element name='business-practices' type='p3p:retention-value'/>
    </choice>
    <element ref='p3p:EXTENSION' minOccurs='0' maxOccurs='unbounded'/>
   </sequence>
  </complexType>
 </element>

 <complexType name='retention-value'/>

<!-- ************** DATA ************** -->
 <element name='DATA-GROUP'>
  <complexType>
   <sequence>
    <element ref='p3p:DATA' maxOccurs='unbounded'/>
```

```
    <element ref='p3p:EXTENSION' minOccurs='0' maxOccurs='unbounded'/>
   </sequence>
   <attribute name='base' type='anyURI'
              use='optional' default='http://www.w3.org/TR/P3P/base'/>
  </complexType>
 </element>

 <element name='DATA'>
  <complexType mixed='true'>
   <sequence minOccurs='0' maxOccurs='unbounded'>
    <element ref='p3p:CATEGORIES'/>
   </sequence>
   <attribute name='ref' type='anyURI' use='required'/>
   <attribute name='optional' use='optional'
    default='no' type='p3p:yes_no'/>
  </complexType>
 </element>


<!-- ************** Data Schema ************* -->
<!-- *********** DATASCHEMA *********** -->
 <element name='DATASCHEMA'>
  <complexType>
   <choice minOccurs='0' maxOccurs='unbounded'>
    <element ref='p3p:DATA-DEF'/>
    <element ref='p3p:DATA-STRUCT'/>
    <element ref='p3p:EXTENSION'/>
   </choice>
  </complexType>
 </element>

 <element name='DATA-DEF' type='p3p:data-def'/>
 <element name='DATA-STRUCT' type='p3p:data-def'/>

 <complexType name='data-def'>
  <sequence>
   <element ref='p3p:CATEGORIES' minOccurs='0'/>
   <element ref='p3p:LONG-DESCRIPTION' minOccurs='0'/>
  </sequence>
  <attribute name='name' type='ID' use='required'/>
  <attribute name='structref' type='anyURI' use='optional'/>
  <attribute name='short-description' type='string' use='optional'/>
 </complexType>

<!-- *********** CATEGORIES *********** -->
 <element name='CATEGORIES'>
  <complexType>
   <choice maxOccurs='unbounded'>
    <element name='physical' type='p3p:categories-value'/>
    <element name='online' type='p3p:categories-value'/>
    <element name='uniqueid' type='p3p:categories-value'/>
```

```
   <element name='purchase' type='p3p:categories-value'/>
   <element name='financial' type='p3p:categories-value'/>
   <element name='computer' type='p3p:categories-value'/>
   <element name='navigation' type='p3p:categories-value'/>
   <element name='interactive' type='p3p:categories-value'/>
   <element name='demographic' type='p3p:categories-value'/>
   <element name='content' type='p3p:categories-value'/>
   <element name='state' type='p3p:categories-value'/>
   <element name='political' type='p3p:categories-value'/>
   <element name='health' type='p3p:categories-value'/>
   <element name='preference' type='p3p:categories-value'/>
   <element name='location' type='p3p:categories-value'/>
   <element name='government' type='p3p:categories-value'/>
   <element name='other-category' type='string'/>
  </choice>
 </complexType>
</element>

<complexType name='categories-value'/>

<!-- ********** EXTENSION *********** -->
<element name='EXTENSION'>
 <complexType mixed='true'>
  <choice minOccurs='0' maxOccurs='unbounded'>
   <any minOccurs='0' maxOccurs='unbounded' processContents='skip'/>
  </choice>
  <attribute name='optional' use='optional' default='yes'
   type='p3p:yes_no'/>
 </complexType>
</element>

</schema>
```

Appendix 5: XML DTD Definition (Non-Normative)

This appendix contains the DTD for policy documents and for data schemas. The DTD is also present as a separate file at the URI *http://www.w3.org/2001/09/P3Pv1.dtd*.

```
<!-- ************** Entities ************** -->
<!ENTITY % URI "CDATA">
<!ENTITY % NUMBER "CDATA">

<!-- ********** Policy Reference ********** -->

<!-- ************* META ************* -->
<!ELEMENT META (#PCDATA | POLICY-REFERENCES | POLICIES)*>

<!-- ******* POLICY-REFERENCES ******** -->
<!ELEMENT POLICY-REFERENCES (EXPIRY?, POLICY-REF*, HINT*)>
```

```
<!-- ********** POLICY-REF ********** -->
<!ELEMENT POLICY-REF (INCLUDE*,
    EXCLUDE*,
    METHOD*)>
<!ATTLIST POLICY-REF
    about %URI; #REQUIRED >

<!-- ************* HINT ************* -->
<!ELEMENT HINT EMPTY>
<!ATTLIST HINT
    domain CDATA  #IMPLIED
    path   CDATA  #IMPLIED >

<!-- ************ EXPIRY ************ -->
<!ELEMENT EXPIRY EMPTY>
<!ATTLIST EXPIRY
    max-age %NUMBER; #IMPLIED
    date    CDATA    #IMPLIED >

<!-- ********** POLICIES ********** -->
<!ELEMENT POLICIES (EXPIRY?, DATASCHEMA?,
    POLICY*)>

<!-- ***** INCLUDE/EXCLUDE/METHOD ***** -->
<!ELEMENT INCLUDE         (#PCDATA)>
<!ELEMENT EXCLUDE         (#PCDATA)>
<!ELEMENT COOKIE-INCLUDE   EMPTY>
<!ATTLIST COOKIE-INCLUDE
    name   CDATA  #IMPLIED
    value  CDATA  #IMPLIED
    domain CDATA  #IMPLIED
    path   CDATA  #IMPLIED>
<!ELEMENT COOKIE-EXCLUDE   EMPTY>
<!ATTLIST COOKIE-EXCLUDE
    name   CDATA  #IMPLIED
    value  CDATA  #IMPLIED
    domain CDATA  #IMPLIED
    path   CDATA  #IMPLIED>
<!ELEMENT METHOD          (#PCDATA)>

<!-- *************** Policy *************** -->

<!-- ************ POLICY ************ -->
<!ELEMENT POLICY (EXTENSION*,
    TEST,
    ENTITY,
    ACCESS,
    DISPUTES-GROUP?,
    STATEMENT*,
    EXTENSION*)>
```

```
<!ATTLIST POLICY
    name    ID   #REQUIRED
    discuri %URI; #REQUIRED
    opturi  %URI; #IMPLIED>

<!-- ******** TEST ******** -->
<!ELEMENT TEST EMPTY>

<!-- ************* ENTITY ************* -->
<!ELEMENT ENTITY (EXTENSION*, DATA-GROUP, EXTENSION*)>

<!-- ************* ACCESS ************* -->
<!ELEMENT ACCESS ((nonident
    | all
    | contact-and-other
    | ident-contact
    | other-ident
    | none),
    EXTENSION*)>
<!ELEMENT nonident          EMPTY>
<!ELEMENT all               EMPTY>
<!ELEMENT contact-and-other EMPTY>
<!ELEMENT ident-contact     EMPTY>
<!ELEMENT other-ident       EMPTY>
<!ELEMENT none              EMPTY>

<!-- ************ DISPUTES ************ -->
<!ELEMENT DISPUTES-GROUP (DISPUTES+, EXTENSION*)>
<!ELEMENT DISPUTES (EXTENSION*,
    ( (LONG-DESCRIPTION, IMG?, REMEDIES?, EXTENSION*)
      | (IMG, REMEDIES?, EXTENSION*)
      | (REMEDIES, EXTENSION*) )?)>
<!ATTLIST DISPUTES
    resolution-type  (service | independent | court | law) #REQUIRED
    service          %URI;                                 #REQUIRED
    verification     CDATA                                 #IMPLIED
    short-description CDATA                                #IMPLIED >

<!-- ******** LONG-DESCRIPTION ******** -->
<!ELEMENT LONG-DESCRIPTION (#PCDATA)>

<!-- ************* IMG ************** -->
<!ELEMENT IMG EMPTY>
<!ATTLIST IMG
    src    %URI;    #REQUIRED
    width  %NUMBER; #IMPLIED
    height %NUMBER; #IMPLIED
    alt    CDATA    #REQUIRED >

<!-- ************ REMEDIES ************ -->
<!ELEMENT REMEDIES ((correct | money | law)+, EXTENSION*)>
```

```
<!ELEMENT correct EMPTY>
<!ELEMENT money   EMPTY>
<!ELEMENT law     EMPTY>

<!-- ********** STATEMENT *********** -->
<!ELEMENT STATEMENT (EXTENSION*,
    NON-IDENTIFIABLE?,
    CONSEQUENCE?,
    PURPOSE,
    RECIPIENT,
    RETENTION,
    DATA-GROUP+,
    EXTENSION*)>

<!-- ********** CONSEQUENCE *********** -->
<!ELEMENT CONSEQUENCE (#PCDATA)>

<!-- ******** NON-IDENTIFIABLE ******** -->
<!ELEMENT NON-IDENTIFIABLE (EMPTY)>

<!-- ************ PURPOSE ************ -->
<!ELEMENT PURPOSE ((current
    | admin
    | develop
    | customization
    | tailoring
    | pseudo-analysis
    | pseudo-decision
    | individual-analysis
    | individual-decision
    | contact
    | historical
    | telemarketing
    | other-purpose)+,
    EXTENSION*)>

<!ENTITY % pur_att
        "required (always | opt-in | opt-out) #IMPLIED">
<!ELEMENT current            EMPTY>
<!ATTLIST current            %pur_att;>
<!ELEMENT admin              EMPTY>
<!ATTLIST admin              %pur_att;>
<!ELEMENT develop            EMPTY>
<!ATTLIST develop            %pur_att;>
<!ELEMENT customization      EMPTY>
<!ATTLIST customization      %pur_att;>
<!ELEMENT tailoring          EMPTY>
<!ATTLIST tailoring          %pur_att;>
<!ELEMENT pseudo-analysis    EMPTY>
<!ATTLIST pseudo-analysis    %pur_att;>
<!ELEMENT pseudo-decision    EMPTY>
```

```
<!ATTLIST pseudo-decision       %pur_att;>
<!ELEMENT individual-analysis EMPTY>
<!ATTLIST individual-analysis %pur_att;>
<!ELEMENT individual-decision EMPTY>
<!ATTLIST individual-decision %pur_att;>
<!ELEMENT contact             EMPTY>
<!ATTLIST contact             %pur_att;>
<!ELEMENT profiling           EMPTY>
<!ATTLIST profiling           %pur_att;>
<!ELEMENT historical          EMPTY>
<!ATTLIST historical          %pur_att;>
<!ELEMENT telemarketing       EMPTY>
<!ATTLIST telemarketing       %pur_att;>
<!ELEMENT other-purpose       (#PCDATA)>
<!ATTLIST other-purpose       %pur_att;>

<!-- ********** RECIPIENT *********** -->
<!ELEMENT RECIPIENT ((ours
     | same
     | other-recipient
     | delivery
     | public
     | unrelated)+,
     EXTENSION*)>
<!ELEMENT ours                 (recipient-description*)>
<!ELEMENT same                 (recipient-description*)>
<!ATTLIST same                 %pur_att;>
<!ELEMENT other-recipient      (recipient-description*)>
<!ATTLIST other-recipient      %pur_att;>
<!ELEMENT delivery             (recipient-description*)>
<!ATTLIST delivery             %pur_att;>
<!ELEMENT public               (recipient-description*)>
<!ATTLIST public               %pur_att;>
<!ELEMENT unrelated            (recipient-description*)>
<!ATTLIST unrelated            %pur_att;>
<!ELEMENT recipient-description (#PCDATA)>

<!-- ********** RETENTION *********** -->
<!ELEMENT RETENTION ((no-retention
     | stated-purpose
     | legal-requirement
     | indefinitely
     | business-practices),
     EXTENSION*)>
<!ELEMENT no-retention        EMPTY>
<!ELEMENT stated-purpose      EMPTY>
<!ELEMENT legal-requirement   EMPTY>
<!ELEMENT indefinitely        EMPTY>
<!ELEMENT business-practices EMPTY>
```

```
<!-- ************* DATA ************* -->
<!ELEMENT DATA-GROUP (DATA+, EXTENSION*)>
<!ATTLIST DATA-GROUP
    base      %URI;        "http://www.w3.org/TR/P3P/base" >
<!ELEMENT DATA (#PCDATA | CATEGORIES)*>
<!ATTLIST DATA
    ref       %URI;        #REQUIRED
    optional (yes | no) "no" >


<!-- *********** DATA SCHEMA *********** -->
<!ELEMENT DATASCHEMA (DATA-DEF | DATA-STRUCT | EXTENSION)*>

<!ELEMENT DATA-DEF    (CATEGORIES?, LONG-DESCRIPTION?)>
<!ATTLIST DATA-DEF
    name              ID    #REQUIRED
    structref         %URI; #IMPLIED
    short-description CDATA #IMPLIED  >

<!ELEMENT DATA-STRUCT (CATEGORIES?, LONG-DESCRIPTION?)>
<!ATTLIST DATA-STRUCT
    name              ID    #REQUIRED
    structref         %URI; #IMPLIED
    short-description CDATA #IMPLIED  >

<!-- *********** CATEGORIES *********** -->
<!ELEMENT CATEGORIES (physical
   | online
   | uniqueid
   | purchase
   | financial
   | computer
   | navigation
   | interactive
   | demographic
   | content
   | state
   | political
   | health
   | preference
   | location
   | government
   | other-category)+>
<!ELEMENT physical    EMPTY>
<!ELEMENT online      EMPTY>
<!ELEMENT uniqueid    EMPTY>
<!ELEMENT purchase    EMPTY>
<!ELEMENT financial   EMPTY>
<!ELEMENT computer    EMPTY>
<!ELEMENT navigation  EMPTY>
<!ELEMENT interactive EMPTY>
```

```
<!ELEMENT demographic EMPTY>
<!ELEMENT content     EMPTY>
<!ELEMENT state       EMPTY>
<!ELEMENT political   EMPTY>
<!ELEMENT health      EMPTY>
<!ELEMENT preference  EMPTY>
<!ELEMENT location    EMPTY>
<!ELEMENT government  EMPTY>
<!ELEMENT other       EMPTY>


<!-- *********** EXTENSION *********** -->
<!ELEMENT EXTENSION (#PCDATA)>
<!ATTLIST EXTENSION
    optional (yes | no) "yes" >
```

Appendix 6: ABNF Notation (Non-Normative)

The formal grammar of P3P is given in this specification using a slight modification of [ABNF]. The following is a simple description of the ABNF.

```
name = (elements)
```

where `<name>` is the name of the rule, `<elements>` is one or more rule names or terminals combined through the operands provided below. Rule names are case-insensitive.

```
(element1 element2)
```

elements enclosed in parentheses are treated as a single element, whose contents are strictly ordered.

```
<a>*<b>element
```

at least `<a>` and at most `<b>` occurrences of the element.

($1*4$`<element>` means *one to four elements.*)

```
<a>element
```

exactly `<a>` occurrences of the element.

(4`<element>` means *exactly 4 elements.*)

```
<a>*element
```

`<a>` or more elements

($4*$`<element>` means *4 or more elements.*)

```
*<b>element
```

0 to `<b>` elements.

($*5$`<element>` means *0 to 5 elements.*)

```
*element
```

0 or more elements.

($*$`<element>` means *0 to infinite elements.*)

```
[element]
```
optional element, equivalent to *1(element).
(`[element]` means *0 or 1 element*.)

`"string"` or `'string'`
matches the literal string given inside double quotes.
Other notations used in the productions are:
`;` or `/* ... */`
comment.

Appendix 7: P3P Guiding Principles (Non-Normative)

This appendix describes the intent of P3P development and recommends guidelines regarding the responsible use of P3P technology. An earlier version was published in the W3C Note "P3P Guiding Principles."

The Platform for Privacy Preferences Project (P3P) has been designed to be flexible and support a diverse set of user preferences, public policies, service provider polices, and applications. This flexibility will provide opportunities for using P3P in a wide variety of innovative ways that its designers had not imagined. The P3P Guiding Principles were created in order to: express the intentions of the members of the P3P working groups when designing this technology and suggest how P3P can be used most effectively in order to maximize privacy and user confidence and trust on the Web. In keeping with our goal of flexibility, this document does not place requirements upon any party. Rather, it makes recommendations about 1) what *should* be done to be consistent with the intentions of the P3P designers and 2) how to maximize user confidence in P3P implementations and Web services. P3P was intended to help protect privacy on the Web. We encourage the organizations, individuals, policy-makers and companies who use P3P to embrace the guiding principles in order to reach this goal.

Information Privacy

P3P has been designed to promote privacy and trust on the Web by enabling service providers to disclose their information practices, and enabling individuals to make informed decisions about the collection and use of their personal information. P3P user agents work on behalf of individuals to reach agreements with service providers about the collection and use of personal information. Trust is built upon the mutual understanding that each party will respect the agreement reached.

Service providers should preserve trust and protect privacy by applying relevant laws and principles of data protection and privacy to their information practices. The following is a list of privacy principles and guidelines that helped inform the development of P3P and may be useful to those who use P3P:

- CMA Code of Ethics & Standards of Practice: Protection of Personal Privacy
- 1981 Council of Europe Convention For the Protection of Individuals with Regard to Automatic Processing of Personal Data
- CSA—Q830-96 Model Code for the Protection of Personal Information

- Directive 95/46/EC of the European Parliament and of the Council of 24 October 1995 on the protection of individuals with regard to the processing of personal data and on the free movement of such data
- The DMA's Marketing Online Privacy Principles and Guidance and The DMA Guidelines for Ethical Business Practice
- OECD Guidelines on the Protection of Privacy and Transborder Flows of Personal Data
- Online Privacy Alliance Guidelines for Online Privacy Policies

In addition, service providers and P3P implementers should recognize and address the special concerns surrounding children's privacy.

Notice and Communication

Service providers should provide timely and effective notices of their information practices, and user agents should provide effective tools for users to access these notices and make decisions based on them.

Service providers should:

- Communicate explicitly about data collection and use, expressing the purpose for which personal information is collected and the extent to which it may be shared.
- Use P3P privacy policies to communicate about all information they propose to collect through a Web interaction.
- Prominently post clear, human-readable privacy policies.

User agents should:

- Provide mechanisms for displaying a service's information practices to users.
- Provide users an option that allows them to easily preview and agree to or reject each transfer of personal information that the user agent facilitates.
- Not be configured by default to transfer personal information to a service provider without the user's consent.
- Inform users about the privacy-related options offered by the user agent.

Choice and Control

Users should be given the ability to make meaningful choices about the collection, use, and disclosure of personal information. Users should retain control over their personal information and decide the conditions under which they will share it.

Service providers should:

- Limit their requests to information necessary for fulfilling the level of service desired by the user. This will reduce user frustration, increase trust, and enable relationships with many users, including those who may wish to have an anonymous, pseudonymous, customized, or personalized relationship with the service.
- Obtain informed consent prior to the collection and use of personal information.

- Provide information about the ability to review and if appropriate correct personal information.

User agents should:

- Include configuration tools that allow users to customize their preferences.
- Allow users to import and customize P3P preferences from trusted parties.
- Present configuration options to users in a way that is neutral or biased towards privacy.
- Be usable without requiring the user to store user personal information as part of the installation or configuration process.

Fairness and Integrity

Service providers should treat users and their personal information with fairness and integrity. This is essential for protecting privacy and promoting trust.

Service providers should:

- Accurately represent their information practices in a clear and unambiguous manner—never with the intention of misleading users.
- Use information only for the stated purpose and retain it only as long as necessary.
- Ensure that information is accurate, complete, and up-to-date.
- Disclose accountability and means for recourse.
- For as long as information is retained, continue to treat information according to the policy in effect when the information was collected, unless users give their informed consent to a new policy.

User agents should:

- Act only on behalf of the user according to the preferences specified by the user.
- Accurately represent the practices of the service provider.

Security

While P3P itself does not include security mechanisms, it is intended to be used in conjunction with security tools. Users' personal information should always be protected with reasonable security safeguards in keeping with the sensitivity of the information.

Service providers should:

- Provide mechanisms for protecting any personal information they collect.
- Use appropriate trusted protocols for the secure transmission of data.

User agents should:

- Provide mechanisms for protecting the personal information that users store in any data repositories maintained by the agent.
- Use appropriate trusted protocols for the secure transmission of data.
- Warn users when an insecure transport mechanism is being used.

Appendix 8: Working Group Contributors (Non-Normative)

This specification was produced by the P3P Specification Working Group. The following individuals participated in the P3P Specification Working Group, chaired by Lorrie Cranor (AT&T): Mark Ackerman (University of California, Irvine), Margareta Björksten (Nokia), Eric Brunner (Engage), Joe Coco (Microsoft), Brooks Dobbs (DoubleClick), Rajeev Dujari (Microsoft), Matthias Enzmann (GMD), Patrick Feng (RPI), Aaron Goldfeder (Microsoft), Dan Jaye (Engage), Marit Koehntopp (Privacy Commission of Land Schleswig-Holstein, Germany), Yuichi Koike (NEC/W3C), Yusuke Koizumi (ENC), Daniel LaLiberte (Crystaliz), Marc Langheinrich (NEC/ETH Zurich), Daniel Lim (PrivacyBank), Ran Lotenberg (IDcide), Massimo Marchiori (W3C/MIT/UNIVE), Christine McKenna (Phone.com, Inc.), Mark Nottingham (Akamai), Paul Perry (Microsoft), Jules Polonetsky (DoubleClick), Martin Presler-Marshall (IBM), Joel Reidenberg (Fordham Law School), Dave Remy (Geotrust), Ari Schwartz (CDT), Noboru Shimizu (ENC), Rob Smibert (Jotter Technologies Inc.), Tri Tran (AvenueA), Mark Uhrmacher (DoubleClick), Danny Weitzner (W3C), Michael Wallent (Microsoft), Rigo Wenning (W3C), Betty Whitaker (NCR), Allen Wyke (Engage), Kevin Yen (Netscape), Sam Yen (Citigroup), Alan Zausner (American Express).

The P3P Specification Working Group inherited a large part of the specification from previous P3P Working Groups. The Working Group would like to acknowledge the contributions of the members of these previous groups (affiliations shown are the members' affiliations at the time of their participation in each Working Group).

The P3P Implementation and Deployment Working Group, chaired by Rolf Nelson (W3C) and Marc Langheinrich (NEC/ETH Zurich): Mark Ackerman (University of California, Irvine), Rob Barrett (IBM), Joe Coco (Microsoft), Lorrie Cranor (AT&T), Massimo Marchiori (W3C/MIT), Gabe Montero (IBM), Stephen Morse (Netscape), Paul Perry (Microsoft), Ari Schwartz (CDT), Gabriel Speyer (Citibank), Betty Whitaker (NCR).

The P3P Syntax Working Group, chaired by Steve Lucas (Matchlogic): Lorrie Cranor (AT&T), Melissa Dunn (Microsoft), Daniel Jaye (Engage Technologies), Massimo Marchiori (W3C/MIT), Maclen Marvit (Narrowline), Max Metral (Firefly), Paul Perry (Firefly), Martin Presler-Marshall (IBM), Drummond Reed (Intermind), Joseph Reagle (W3C).

The P3P Vocabulary Harmonization Working Group, chaired by Joseph Reagle (W3C): Liz Blumenfeld (America Online), Ann Cavoukian (Information and Privacy Commission/Ontario), Scott Chalfant (Matchlogic), Lorrie Cranor (AT&T), Jim Crowe (Direct Marketing Association), Josef Dietl (W3C), David Duncan (Information and Privacy Commission/Ontario), Melissa Dunn (Microsoft), Patricica Faley (Direct Marketing Association), Marit Köhntopp (Privacy Commissioner of Schleswig-Holstein, Germany), Tony Lam (Hong Kong Privacy Commissioner's Office), Tara Lemmey (Narrowline), Jill Lesser (America Online), Steve Lucas (Matchlogic), Deirdre Mulligan (Center for Democracy and Technology), Nick Platten (Data Protection Consultant, formerly of DG XV, European Commission), Ari Schwartz (Center for Democracy and Technology), Jonathan Stark (TRUSTe).

The P3P Protocols and Data Transport Working Group, chaired by Yves Leroux (Digital): Lorrie Cranor (AT&T), Philip DesAutels (Matchlogic), Melissa Dunn (Microsoft), Peter Heymann (Intermind), Tatsuo Itabashi (Sony), Dan Jaye (Engage), Steve Lucas

(Matchlogic), Jim Miller (W3C), Michael Myers (VeriSign), Paul Perry (FireFly), Martin Presler-Marshall (IBM), Joseph Reagle (W3C), Drummond Reed (Intermind), Craig Vodnik (Pencom Web Worlds).

The P3P Vocabulary Working Group, chaired by Lorrie Cranor (AT&T): Mark Ackerman (W3C), Philip DesAutels (W3C), Melissa Dunn (Microsoft), Joseph Reagle (W3C), Upendra Shardanand (Firefly).

The P3P Architecture Working Group, chaired by Martin Presler-Marshall (IBM): Mark Ackerman (W3C), Lorrie Cranor (AT&T), Philip DesAutels (W3C), Melissa Dunn (Microsoft), Joseph Reagle (W3C).

Finally, Appendix 7 is drawn from the W3C Note "P3P Guiding Principles," whose signatories are: Azer Bestavros (Bowne Internet Solutions), Ann Cavoukian (Information and Privacy Commission Ontario Canada), Lorrie Faith Cranor (AT&T Labs-Research), Josef Dietl (W3C), Daniel Jaye (Engage Technologies), Marit Köhntopp (Land Schleswig-Holstein), Tara Lemmey (Narrowline; TRUSTe), Steven Lucas (MatchLogic), Massimo Marchiori (W3C/MIT), Dave Marvit (Fujitsu Labs), Maclen Marvit (Narrowline Inc.), Yossi Matias (Tel Aviv University), James S. Miller (MIT), Deirdre Mulligan (Center for Democracy and Technology), Joseph Reagle (W3C), Drummond Reed (Intermind), Lawrence C. Stewart (Open Market, Inc.).

Change log from the *15 December 2000 Candidate Recommendation*:

- Fixed errors in examples and typos throughout
- Made many minor (non-substantive) wording changes
- Added 2.4.7 Absence of Policy Reference File
- Added paragraph to intro to 3.2 to explain requirements for handling policies containing errors
- Changed categories of some of the elements of the `postal` structure and the `dynamic.clickstream.clientip` element in the base data schema
- Changed 3.3.3 to clarify how the `NON-IDENTIFIABLE` element is aggregated across statements
- Added language to 4.2 to clarify user agent behavior when encountering duplicate or unknown tokens in compact policies
- Added section 2.4.8 Asynchronous Evaluation
- Change other-purpose token from `OPT` to `OTP` in 4.2 Compact Policies
- Added language to 2.4.1 to clarify the precedence of multiple policy reference files and make section more clear generally
- Added definitions of data structure and data schema to 1.3 Terminology
- Added new data element user.login to base data schema
- Many wording changes throughout related to the term "identifiable"
- Changed language in 2.3.2.5 to state that policy reference files are applied relative to the DNS hosts that reference them
- Changed the intro to 4 to clarify compact policy scope
- Changed definition of the current purpose in 3.3.4 and removed the `required` attribute from the current purpose

- Removed the `customization` purpose in 3.3.4
- Changed the definition of the `tailoring` purpose in 3.3.4
- The use of HTTP headers for determining policy and policy reference file expiry was eliminated
- `EXPIRY` element moved to child of `POLICIES` element (instead of child of `POLICY` element)
- 2.3.2.7 was changed to clarify expiry rules for cookie policies
- Syntax of `COOKIE-INCLUDE` and `COOKIE-EXCLUDE` changed
- `POLICY` elements must now be contained within a `POLICIES` element and must have a `name` attribute
- Changes to 2.4.3 to clarify the meaning of the safezone
- New section 2.2.4 HTTP ports and other protocols
- `EMBEDDED-INCLUDE` removed
- New section 2.3.2.6 Policy Reference Hints
- Added language to 2.2.2 and 2.2.3 to clarify that when the `policyref` attribute is a relative URI, that URI is interpreted relative to the request URI
- Added language to 2.3.2.1.1 to clarify the significance of order in policy reference files
- Added language to 2.3.2.5 and 2.3.2.8 to clarify the use of the `METHOD` element
- An embedded `DATASCHEMA` is now child of `POLICIES` rather than of `POLICY`

Internet Resources

APPLICATION SERVERS

http://www.bea.com/products/weblogic/server/papers.shtml

ASSOCIATION

http://www.ipv6forum.org/
http://www.mda-mobiledata.org/
http://www.ipdr.org/
http://www.etis.org/
http://www.cibernet.com/
http://www.waaglobal.org/

BILLING

http://www.billing.org/
http://www.portal.com

CLIENT

http://www.blaupunkt.de
http://www.pioneer-eur.com/products/car/carelec.htm

COMMERCE

http://www.mobiletransaction.org
http://www.mastercardintl.com/newtechnology/mcommerce/whatis/glossary.html

DEVELOPER TOOLS

Aeris.net's MicroBurst Developer Program: *http://www.aeris.net/sitev2/devnet_login.htm*

Aether Developer Zone: *http://devzone.aethersystems.com/*

AT&T Wireless Data Developer Program: *http://www.attws.com/bus/lcorp/wireless_ip/developers/index.jhtml*

Broadbeam's Developer Program: *http://www.nettechrf.com/*

CDMA Online: *http://www.cdmaonline.com/*

CDPD technical specification: *http://www.wirelessdata.org/develop/cdpdspec/index.asp*

Cingular Developer Program: *http://alliance.cingularinteractive.com/dev/cda/home*

Ericsson Developers Zone: *http://www.ericsson.com/developerszone*

Geoworks Developer Program: *http://www.geoworks.com/partners/devzone/index.html*

GSM World: *http://www.gsmworld.com/*

GSM Data Window: *http://www.gsmworld.com/gsmdata/*

Logica Application Provider Program: *http://www.logica.com/telecoms/products/app/*

Motorola's Developer Program: *http://developers.motorola.com/developers/*

Newcom Technologies Pty. Ltd.: *http://www.newcom.com.au/*

Nextel Developer Program: *http://developer.nextel.com/*

Nokia Developer Program: *http://forum.nokia.com/main.html*

Novatel Wireless Developer Program: *http://www.novatelwireless.com/support/developer_program*

Open Mobile Alliance: *http://www.openmobilealliance.org/*

Openwave Developer Programs: *http://www.openwave.com/alliances/developer/*

OracleMobile Online Studio: *http://otn.oracle.com/hosted_dev/oracle_mobile/content.html*

PacketVideo Developer Program: *https://www.pvinfonet.com/pvdevnet/*

RIM Developer Zone: *http://developers.rim.net/*

Sierra Wireless Developers Toolkit: *http://www.sierrawireless.com/developers/*

WAP Forum: *http://www.wapforum.org/*

WaveLink Developer Program: *http://www.wavelink.com/devconnect/*

E112

http://europa.eu.int/information_society/topics/telecoms/regulatory/userinfo/99comrev/index_en.htm

E911

http://www.fcc.gov/e911/

EMERGENCY SERVICES

http://www.fcc.gov/e911/

http://www.comcare.org/

E-OTD

http://www.fcc.gov/e911/aerial.pdf
http://www.cursor-system.com/sitefiles/cursor/cursor_howmain.htm

ERICSSON MOBILE POSITIONING SYSTEM

http://www.ericsson.com/mps/

GENERAL GIS

http://www.esri.com/library/whitepapers/pdfs/

GEOCODING

http://www.sgsi.com/geoc0198.htm
http://www.na.teleatlas.com/pdfs/geocoding.pdf

GEOCODING/SOUNDEX

http://www.nara.gov/genealogy/soundex/soundex.html

GPS

http://www.trimble.com/gps
http://www.crosland.co.uk/aboutgps.htm

GML

http://www.jlocationservices.com/company/galdos/articles/introduction_to_gml.htm

JAVA

http://developer.java.sun.com/developer/products/wireless/getstart/articles/whyjava/

JAVA LOCATION SERVICES

http://www.jlocationservices.com

LOCALIZATION

http://www.mozilla.org
http://www.globalization.com

LOCALIZATION ASSOCIATION

http://www.lisa.org/

LOCATION PATTERN MATCHING

http://www.911dispatch.com/911_file/lpm.html
http://www.uswcorp.com/USWCMainPages/our.htm

MAPPING

http://www.nima.mil
http://www.noaa.gov
http://www.usgs.gov
http://terraserver.homeadvisor.msn.com/

MAP PROJECTION

http://everest.hunter.cuny.edu/mp/
http://www.nationalgeographic.com/features/2000/exploration/projections/index.html

MICROSOFT .NET

http://www.microsoft.com/net/whitepapers.asp

NOKIA GATEWAY MOBILE LOCATION CENTER

http://www.nokia.com/pc_files/GMLC_datasheet.pdf

PERSONALIZATION/PRIVACY

http://www.w3.org/P3P/

POSITION AUGMENTATION/DEAD RECKONING

http://www.gpsworld.com/0601/0601wood.html

POSITIONING

http://bestgsm.virtualave.net/other/mobposit/mobposit.html
http://www.gpsworld.com
http://www.mobilepositioning.com/

POSITIONING AND PRIVACY

http://www.ftc.gov/bcp/workshops/wireless/agenda.htm

PRIVACY

http://www.cdt.org/privacy/issues/location/
http://www.cdt.org/privacy/issues/location/001122ctia.pdf
http://www.ftc.gov/bcp/workshops/wireless/index.html
http://www.privacytimes.com/

PRIVACY/CRM

http://www.crmcommunity.com/

PROVISIONING

http://www.wapforum.org/what/technical.htm

ROAD PRICING

http://www.minvenw.nl/rws/projects/nvvp/
http://www.Transport-Pricing.Net
http://www.ettm.com
http://www.vtpi.org/tdm/tdm35.htm
http://www.roadpricing.nl

SECURITY

http://www.umtsworld.com/technology/security.htm

SECURITY AND LDAP

http://www.ietf.org/rfc/rfc2820.txt

SERVER SIZING

http://www.dell.com/us/en/slg/topics/products_size_pedge_sizing.htm
http://www.sun.com/solutions/third-party/global/oracle/pdf/nca-app.pdf
http://activeanswers.compaq.com/ActiveAnswers/Render/1,1027,4815-6-100-225-1,00.htm

STANDARDS

http://www.magicservicesforum.org

TRAFFIC/RDS–TMC

http://www.rds.org.uk/episode/rdstmcbrochure.htm

WAP

http://www.wapforum.org/what/technical.htm
http://www.wapforum.org/what/whitepapers.htm

WIRELESS

http://www.gprsworld.com/
http://www.fiercewireless.com
http://www.wirelessinternetdaily.com/

WIRELESS ASSOCIATION

http://www.gsmworld.com/
http://www.umts-forum.org/

WIRELESS SECURITY

http://www.cnp-wireless.com/wsp.html
http://www.tivoli.com/news/press/analyst/idc_security_ebus_enabler.pdf
http://www.tivoli.com/products/solutions/security/secureway_wpnds.html#reports
http://www.neohapsis.com/
http://www.brookson.com/gsm/contents.htm
http://www.mobileinfo.com/Security/
http://www.networkcomputing.com/1202/1202f1d1.html
http://www1.wapforum.org/tech/documents/WAP-187-TransportE2ESec-20010628-a.pdf
http://www1.wapforum.org/tech/documents/WAP-261-WTLS-20010406-a.pdf

References

3GPP TS 33.102, *Technical Specification Group Services and System Aspects, 3G Security, Security Architecture (Release 4),* 3rd Generation Partnership Project, 2001.

Ben-Natan, R., and Sasson, O., *IBM Websphere Starter Kit*, McGraw Hill, 2000.

Bray, T., Paoli, J., and Sperberg-McQueen, C.M., Eds., "Extensible Markup Language (XML) 1.0," 2nd edition, October 2000, W3C Recommendation, *http://www.w3.org/TR/2000/ REC-xml.*

Brookson, C., "GSM (and PCN) Security and Encryption," GSM Security, 1994.

Burrough, P., and McDonnell, R., *Principles of Geographical Information Systems,* 2nd edition, Oxford University Press, 1998.

Chatterjee, A., Kaas, H., Kumaresh, T. V., and Wojcik, P., "A Roadmap for Telematics," *McKinsey Quarterly,* 2002, Number 2.

Cockcroft, A., and Pettit, R., *Sun Performance and Tuning: Java and the Internet*, 2nd edition, Prentice Hall, 1999.

de La Beujardiére, J., Ed., "Web Map Service Implementation Specification," Open GIS Consortium Inc., June 2001.

Demers, M., *Fundamentals of Geographic Information Systems*, Wiley, 2000.

Djuknic, G. M., and Richton, R.E., "Geolocation and Assisted GPS," *Computer*, February 2001, pp. 123–125.

Dorling, D., and Fairbairn, D., *Mapping: Ways of Representing the World*, Addison-Wesley, 1997.

Dornan, A., *The Essential Guide to Wireless Communications Applications*, Prentice Hall, 2001.

Drane, C., Macnaughton, M., and Scott, C., "Positioning GSM Telephones," *IEEE Comm Magazine*, April 1998, pp. 46–59.

Ferraiolo, J., Ed., "Scalable Vector Graphics (SVG) 1.0 Specification," W3C Recommendation, September 2001.

Fry, D., "Localization Industry Primer," Location Industry Standards Association, May 2001.

Goldfarb, C., and Prescod, P., *Charles F. Goldfarb's XML Handbook*, 4th edition, Prentice Hall, 2001.

Green, J., Betti, D., and Davison, J., "Mobile Location Services: Market Strategies," White Paper, Ovum, December 2000.

GSM 03.72: "Digital Cellular Telecommunications System (Phase 2+); Location Services (LCS); (Functional Description) – Stage 2." 3GPP Organizational Partners, France, 2001.

Hall, M., and Brown, L., *Core Web Programming*, Prentice Hall, 2001.

Harvey, L., "Personify's Profit Platform," Patricia Seybold Group, October 2000.

Hayes, I., "Wireless Location Based Services: Finding Your Way Out of the Box," *Executive Update*, Cutter Consortium, 2001.

HeyAnita, Inc., "Voice Portal Technologies," VoiceXML Forum Presentation, 2001.

Hjelm, J., *Designing Wireless Information Services*, Wiley, 2000.

Hunter, J., and Crawford, W., *Java Servlet Programming*, 2nd edition, O'Reilly & Associates, 2001.

Jones, C., *Geographical Information Systems and Computer Cartography*, Addison-Wesley, 1997.

Kanemitsu, H., and Kamada, T., "POIX: Point of Interest eXchange Language Specification," W3C Note, June 1999.

King, C., Dalton, C., and Osmanoglu, T. E., *Security Architecture*, RSA Press, 2001.

Kitazawa, K., Konishi, Y., and Shibasaki, R., "A Method of Map Matching For Personal Positioning Systems," GIS Development, ACRS, 2000.

Korkea-aho, M., Tang, H., Racz, D., Polk, J., and Takahashi, K., "A Common Spatial Location Dataset," IETF Draft (draft-korkea-aho-spatial-dataset-01.txt), Work in progress, May 2001.

Korkea-aho, M., and Tang, H., "Spatial Location Payload," IETF Draft (draft-korkea-aho-spatial-location-payload-00.txt), Work in progress, May 2001.

Leventhal, M., Lewis, D., Fuchs, M., and Fuchs, M., *Designing XML Internet Applications*, Prentice Hall, 1998.

Mahy, R., "A Simple Text Format for the Spatial Location Protocol (SLoP)," IETF Draft, Work in progress, July 2000.

Marchand, R., *Building a Carrier Grade Service*, VoiceGenie Technologies, Inc., 2001.

Mathews, B., Lee, D., Dister, B., Bowler, J., Cooperstein, H., Jindal, A., Nguyen, T., Wu, P., and Sandal, T., "Vector Markup Language (VML)," W3C Note, May 1998.

Mobile Electronic Transactions Initiative, *MeT Account Based Payment*, Version A, 2001.

Mobile Electronic Transactions Initiative, *MeT Consistent User Experience*, Version 1.1, 2001.

Mobile Electronic Transactions Initiative, *The MeT Core Specification*, Version 1.1, 2001.

Mobile Electronic Transactions Initiative, *MeT Event Ticketing*, Version A, 2001.

Mobile Electronic Transactions Initiative, *The MeT Initiative: Enabling Mobile E-commerce*, Version 2.0, White paper, 2001.

Mobile Electronic Transactions Initiative, *MeT Ticketing Framework*, Version 1.0, 2001.

Muchow, J., *Core J2ME Technology*, Prentice Hall, 2001.

Nilsson, N., *Principles of Artificial Intelligence*, Tioga Publishing, 1980.

Nokia Mobile Phones, *Mobile Location Services*, White paper, 2001.

Nokia Mobile Phones, *Nokia Terminals and Applications for the 3G Market*, White paper, 2001.

Nouvier, J., "Road Safety and Telematics: How Far Have We Got?," CERTU, 2000.

Openwave Systems, Inc., *WML 1.3 Developer's Guide*, Openwave Systems, August 2001.

Pieper, I. R., "Mobimiles: Introducing an Integrated Concept for Variable Road Pricing Based on Time, Location and Vehicle," report prepared for Project Organisation KMH Road Pricing System, Ministry of Transport, Public Works and Water Management (Dutch Government), July 2001.

Plewe, B., *GIS Online: Information Retrieval, Mapping and the Internet*, OnWord Press, 1997.

Portal Software, Inc., *Mobile Commerce: Deploying Billing Infrastructure to Support Mobile Commerce*, White paper, 2001.

Rischpater, R., *Wireless Web Development*, Apress, 2000.

Samaha, J., Kayssi, A., Achkar, R., and Azar, M., "G3 Integrates Three System Technologies," *Computer*, October 2000, pp. 107–110.

Savander, N., "Concrete Building Blocks for Mobile Services," Nokia Networks, 2001.

Sekiguchi, M., Takayama, K., Naito, H., Maeda, Y., Horai, H., and Toriumi, M., "NaVigation Markup Language (NVML)," W3C Note, August 1999.

Sielemann, M., "What's the Approach to Telematics From the Mobile Operator's Perspective?," Vodafone, EyeforAuto Frankfurt Presentation, September 2001.

Snell, J., Tidwell, D., and Kulchenko, P., *Programming Web Services With SOAP*, O'Reilly & Associates, 2001.

Srinivasan, S., and Brown, E., "Is Speech Recognition Becoming Mainstream?," *Computer*, April 2002, pp. 38–41.

Sun Microsystems, *MIDP APIs for Wireless Applications,* White paper, February 2001.

Takahashi, K., and Tang, H., "Location-Based Service Scenarios for Privacy Analysis," IETF Draft (draft-takahashi-spatial-privacy-scenario-00.txt), Work in progress, July 2001.

Thomas, A., *Selecting Enterprise JavaBeans Technology*, Patricia Seybold Group, 1998.

Ueno, K., Artus, D., Brown, L., Clark, L., Gerken, C., Hambrick, G., Iyenger, A., Joines, S., Kapadia, S., Ramdani, M., Roca, J., Son, S., VanOosten, J., and Zhang, C., *Websphere V3.5 Handbook*, Prentice Hall, 2001.

UMTS Forum Report No. 11, "Enabling UMTS/Third Generation Services and Applications," October 2000.

U.S. Army Corps of Engineers 1110-1-1003: "NAVSTAR Global Positioning Systems Surveying," August 1996.

VanderMeer, J., "Will Wireless Location-Based Services Pay Off?," *Business Geographics,* Adams Business Media, 2000.

VeriSign, Inc., *Enabling the Promise of Wireless Technology,* White paper, 2001.

WAP-187-TransportE2Esec-200110628-a, "WAP Transport Layer End-to-End Security," Wireless Application Protocol Forum Ltd., June 2001.

WAP-261-WTLS-20010406-a, "Wireless Transport Layer Security," Wireless Application Protocol Forum Ltd., April 2001.

Watts, D., McKnight, G., Mitura, P., Neophytou, C., and Gulver, M., *Tuning Netfinity Servers for Performance: Getting the Most Out of Windows 2000 and Windows NT 4.0*, Prentice Hall, 2001.

Yi-Bing, L., and Chlamtac, I., *Wireless and Mobile Network Architectures,* Wiley, 2001.

Zhao, Y., *Vehicle Location and Navigation Systems*, Artech House, 1997.

Index

A

A* algorithm, 45, 46–48
A-GPS, *See* Assisted GPS (A-GPS)
Abbreviations, 141–44
Acxiom, 60
Adeptra, 129
ALERT-C, 123
Amazon.com, 82
American Automobile Association, 94
Amplitude modulation (AM), 11
Angle of arrival (AOA), 69, 71
Application developers, new opportunities
       and challenges for, 7–8
Application server, 17–28
   architecture, 22–28
   business logic layer, 18, 22, 23–25
   data access layer, 18, 22, 25
   defined, 17, 18
   importance of, 18–20
   intellectual property protection, 19
   J2EE application server, 20–22
   manageability, 19
   network communications security, 19
   performance, 19–20
   presentation layer, 18, 22, 23

   reuse, 19
   scalability, 21
   systems management and reporting inter-
      face, 21, 22
Assisted GPS (A-GPS), 67, 72, 84, 93
ATM Locator Service (Visa), 103
ATX Technologies, 94
   case study, 95
Authentication center (AuC), 12
Auto manufacturers, 4–5

B

Base station controller (BSC), 13
Base station subsystem (BSS), 13
Base transceiver stations (BTSs), 13
BEA Weblogic Server, 25
"Best of breed" maps, 131–36
   example merge, 136–39
   map database conflation, 132–36
   need for, 131–32
Billing, 91–96
   call detail record (CDR), 93
   location-based, 130
   roaming, 93
   technology and business models, 92–93

Binary Runtime Environment for Wireless (BREW) (Qualcomm), 8, 116–17
Business logic layer, 18, 22, 23–25

C

CA Unicenter, 21
Cambridge Positioning, 68
Cascading style sheets (CSS), 110
CDMA, 13
Cell global identity (CGI), 66
Cell-ID, 65–66
Cell of origin (COO), 65–66
Cellpoint, 66
Cellular digital packet data (CDPD), 103
Cellular Telecommunications & Internet Association (CTIA), 89
cHTML, 101
Client platforms, 101–18
    Microsoft Windows CE, 103–5
    onboard systems, 106–9
    Palm OS, 102–3
    screen details, 102
    Symbian OS, 105–6
Client protocols/languages, 109–17
    Binary Runtime Environment for Wireless (BREW) (Qualcomm), 116–17
    Extensible Markup Language (XML), 109–10
    Java and J2ME, 115–16
    Simple Object Access Protocol (SOAP), 110–11
    voice and VoiceXML, 113–14
    Wireless Application Protocol (WAP), 112–13
    Wireless Markup Language (WML), 112–13
Code Division Multiple Access (CDMA), 13
Combined NavTech/GDT solution (Kivera), 136–39
    "empties," 138
    inconsistent source data, 138
    inspection problems, 138–39
    loops and control nodes, 138
    problem, 136–37
    sewing together of data sets, 137

ComCARE Alliance, 124, 127
Commercial application servers, 25
Common Client Interface (CCI), 25
CommuterLink system, 128
Concierge and travel services, 127–29
Conformal projections, 38
Connected Device Configuration (CDC), 115
Connected Limited Device Configuration (CLDC), 115
Consumer expectations, roadmaps, 50–52
    Europe, 51–52
    North America, 50–51
Context, 82
Context-based services, 2
Cross Country, 94
Customer relationship management (CRM), 84

D

Data access layer, 18, 22, 25
    commercial application servers, 25
    J2EE Connector Architecture (JCA), 25
Data-capable mobile phones, 1–2
Dead reckoning, augmenting GPS with, 68
Differential GPS (D-GPS), 67
Digital map databases, 131–39
    "best of breed" maps, 131–36
Digital maps, 32–34
    collecting data from the field, 34
    creation/maintenance process, 33–34
    defined, 32–33
    source information, gathering, 34
    source map data conversion, 34
DoCoMo, 5, 91
DoubleClick, 84
Dun & Bradstreet, 60

E

EJBs, *See* Enterprise JavaBeans (EJBs)
Ellipsoid, 37
Emergency assistance, 123–27
    Roadside Telematics Corporation, 124–27
Emergency Services application (Road-Medic), 103, 124, 125–27
Enhanced-observed time difference (E-OTD), 18, 68, 93

Enterprise JavaBeans (EJBs), 23–25
  entity beans, 24–25
  sample EJB environment, 24
  session beans, 24–25
Equal area projections, 38
Equidistant projections, 38
Ericsson, 66, 68, 112
European roadmaps, consumer expectations, 51–52
Experian, 60
Extensible Markup Language (XML), 8, 101
Extensible Stylesheet Language (XSL), 110

F

Floating car data (FCD), 120
Frequency modulation (FM), 11

G

Gateway GPRS support node (GGSN), 13
Gateway mobile location center (GMLC), 64
Geocoding, 39–44
  address cleanup, 42–43
  address input, 40
  address standardization, 41
    differences in, 43
  defined, 39
  how it works, 40–42
  importance to mobile location services, 43–44
  reverse, 44
  rural delivery and post office boxes, 43
  site address and billing address, 43
  soundex, 41–43
  static map database and dynamic communities, 43
  technical definition of, 39–40
Geographic data structures, computer storage of, 36
Geographic information systems (GIS), 2
Geographic literacy, roadmaps, 52
Geographical data types, 34–36
  arcs, 35
  computer storage of geographic data structures, 36
  lines, 35
  linked attributes, 36
  nodes, 35
  points, 34–35
  polygons, 36
Geography Markup Language (GML), 18, 23, 145–218
Global positioning system (GPS), 8, 66–68
  assisted GPS (A-GPS), 67, 72, 84, 93
  augmenting with dead reckoning, 68
  differential GPS (D-GPS), 67
  how it works, 66–67
Global System for Mobile Communications (GSM), 8, 13
  positioning server architecture in, 64
  and security, 77
GML, *See* Geography Markup Language (GML)
Government regulation, 6–7
  E911/E112, 6
  road pricing, 6–7
GPRS (General Packet Radio Service), 13–14, 91
GPS, *See* Global positioning system (GPS)
gpsOne (Qualcomm), 84
GSM, *See* Global System for Mobile Communications (GSM)
Graphics Interchange Format (GIF), 50

H

Hagel, John, 83
Hand-based positioning, 93
Handset-based positioning, 84
HDML, 101
Home location register (HLR), 12
HotSync, 103
HP Openview, 21
Hybrid positioning, 84
Hypertext Markup Language (HTML), 23, 101
Hypertext Transfer Protocol (HTTP), 23

I

IBM Tivoli, 21
IBM Websphere Everyplace Server, 25
In-vehicle mobile location services, 4–5
  value chain, 15
Inductive loops, 120

InfoUsa, 60
Internationalization, 117–18

J

J2EE application server, 20–22, *See also*
    Application server
  defined, 21–22
  integrated management environment, 21
  security, 22
  transaction semantics, 22
J2EE Connector Architecture (JCA), 25
J2EE web application network architecture,
    25–26
  advantages, 27
  application server cluster, 26
  database server cluster, 26
  disadvantages, 27
  load balancer cluster, 26
J2ME, 115–16
  configurations/profiles, 116
  when to use, 115
Jakarta Tomcat (Apache), 25
Java, 115–16
Java 2 MicroEdition (J2ME), 8
Java 2 Standard Edition (J2SE), 115
Java Database Connectivity (JDBC) driver,
    25
Java server pages (JSPs), 23
Java servlets, 23

K

King, Stephen, 82
Kivera, 57–59
  combined NavTech/GDT solution, 136–39
    "empties," 138
    inconsistent source data, 138
    inspection problems, 138–39
    loops and control nodes, 138
    problem, 136–37
    sewing together of data sets, 137
  extraction program, 57
  geocoding engine, 59
  Location Server API Guide, 59
  Merge/Delete Engine, 60
  POI database, 57–59

L

lastminute.com, 128
Latitude, 37
Launched combined DVD/navigation sys-
    tems, *See* Onboard systems
LBS client devices, 106–9
  types of, 106
Localization, 117–18
  map data challenge, 118
  voice system challenge (Europe), 118
Localization Industry Standard Association,
    118
Location-based advertising and marketing, 129
Location-based applications, 3
Location-based billing, 130
Location measurement units (LMUs), 64
Location pattern matching, 72
Location server, 14–15
Longitude, 37

M

m-commerce (mobile commerce), 97–100
  applications, 97–98
  defined, 97–99
  local information in, 97–98
  Mobile Electronic Transactions Standard
    (MeT), 99–100
  mobility management, 98
  real-time ticket-based applications, 98–99
Map data challenge, 118
Map database conflation, 132–36
  cross-over points, run-time determination
    of, 133–34
  data conflation and compilation, 132–33
  database merging, 134–35
  grid-based merging, 135
  overlay strategy, 133
  precompiled cross-over reference points,
    134
  selective area merging, 136
Map database quality/coverage, 17
Map image generation, 50–56
  consumer expectations, roadmaps, 50–52
  decision making and maps, 50
  roadmaps, geographic literacy, 52
  visual orientation of people, 50

Map matching, 72
MapQuest, 2
Market drivers, 3, 4–7
  auto manufacturers and mobile operators, 4–5
  government regulation, 6–7
MeT, *See* Mobile Electronic Transactions Standard (MeT)
MetroOne, 128
Microsoft Mobile Information Server, 25
Microsoft .NET, 8, 87–88
Microsoft Passport, 87–88
Microsoft Windows CE, 103–5
MIDlets, 115
Minc, Alain, 3
Mobile Electronic Transactions Standard (MeT), 8, 99–100
  core functions, 99–100
  defined, 99
Mobile equipment (ME), 12
Mobile Information Device Profile (MIDP), 115
Mobile Location Protocol (LIF), 110, 219–312
Mobile location service applications, 119–30
  concierge and travel services, 127–29
  emergency assistance, 123–27
  location-based advertising and marketing, 129
  location-based billing, 130
  navigation, 119–20
  positioning accuracy and speed requirements for, 73
  traffic, 120–23
Mobile location services, 15
  application opportunities, 3
  applications, 119–30
  categories of, 4
  defined, 1–8
  deployment environment, 9
  deployment of, 3
  development of term, 3
  in-vehicle, 4–5
  industry, 14–15
  infrastructure design, 17–18
  market drivers, 3, 4–7

solution, building, 9–16
solution components, 15
solution infrastructure, 16
subscribers, 94
Mobile Marketing Association (MMA), 89–90
Mobile network, 14
Mobile phones, 1–2
Mobile positioning, 63–73
  angle of arrival (AOA), 69, 71
  cell of origin (COO), 65–66
  enhanced-observed time difference (E-OTD), 68
  global positioning system (GPS), 66–68
  location pattern matching, 72
  map matching, 72
  time difference of arrival (TDOA), 69–71
Mobile station (MS), 12
Mobile Streams, 73
Mobile switching center (MSC), 12
Mobility management, 5
MobiMiles, 6
Motorola, 112
Mozilla Project, 117
Multimodal directions, 5
Multiple access techniques, 13

N

Net Worth (Hagel/Singer), 83
Network switching subsystem (NSS), 12–13
Neverlost (Magellan), 119
Nokia, 66, 112
Nora, Simon, 3
North American roadmaps, consumer expectations, 51–52
NTT DoCoMo, 5, 91

O

Onboard systems, 106–9
  Pioneer systems, 108–9
  TravelPilot DX-N system (Blaupunkt), 106–7
  TravelPilot DX-R70 system (Blaupunkt), 107–8
OnStar onboard system (General Motors), 4, 14, 94, 109
  emergency assistance, 123–24

Open GIS Consortium, 18
Openwave, 112
Orange (France), 60

P

P3P, 8, 84–87
    defined, 85
    how it works, 85–87
    specification, 313–434
Palm OS, 102–3, 112
Palm.net network, 103
Personal mobile location services, 5
Personalization, 81–90
    defined, 81–82
    privacy issues, 88–90
        industry self-regulation efforts, 89–90
    transactive content, 82
    user profiling, 83
Personalization and filing system, 83–88
    customer relationship management, inte-
        gration with, 84
    generation and control of location infor-
        mation, 84
    Microsoft .NET and Microsoft Passport,
        87–88
    P3P, 84–87
Personify, 84
Pixels, 52
Platform for Privacy Preferences (P3P), *See*
    P3P
POI databases, 57–60
    data extraction, 57–58
    geocode POIs, 59
    geocoded data, 59
    index creation, 60
    merge/delete POIs, 60
    POI sources, 60
    routing with POIs, 60
    unique identifiers, 57–58
POI searches, 56–60
    category search, 57
    name search, 56
    phone search, 57
    POI databases, 57–60
Portal Software, 84, 93, 130
    case study, 95

PQA format, 103
Presentation layer, 18, 22, 23
Priceline, 128
Projection process, 38
Projections and coordinate systems, 37–38
    conformal projections, 38
    equal area projections, 38
    equidistant projections, 38
    latitude, 37
    longitude, 37
    projection process, 38
    true directions projections, 38
Public safety answering points (PSAP), 123
Public switched data network (PSDN), 13

Q

Qualcomm, 8, 116-17

R

Radio Data Services-Traffic Message Chan-
    nel (RDS-TMC), 123, 128
Raster maps, 52–55
    advantages, 54–55
    disadvantages, 54–55
RDS-TMC, *See* Radio Data Services-Traffic
    Message Channel (RDS-TMC)
Real-time attribute editing, 61
Real-time traffic, and updating of attributes in
    real time, 61
Reebok, 129
Remote Monitoring (RMON), 21
Reverse geocoding, 44
Road pricing, 6–7
Road sensors, 120
Roadmaps:
    consumer expectations, 50–52
        Europe, 51–52
        North America, 50–51
    geographic literacy, 52
    linking to maps, 122–23
RoadMedic Emergency Services application,
    103, 124, 125–27
Routing, 45–50
    multiple depot–multiple vehicle node rout-
        ing, 49
    multiple traveling salesman problem, 49

Routing *(cont.)*,
    segment attributes, 49–50
    shortest path problem, 45–48
    single depot–multiple vehicle node rout-
        ing, 49
    traveling salesman problem, 48–48

S

Security, 75–80
    and GSM, 77
    infrastructure, 75
    in wireless networks, special concerns,
        76–79
    Wireless Transport Layer Security
        (WTLS), 77–79
Serving GPRS support node (SGSN), 13
Serving mobile location center (SMLC), 64
Short Message Service (SMS), 5
Siebel, 84, 93
Signalsoft, 130
Simple Network Management Protocol
        (SNMP), 21
Simple Object Access Protocol (SOAP), 8, 18
Singer, Marc, 83
Soundex:
    applying scoring rules, 42
    coding guide, 41–42
    encoding examples, 42
    mismatches, 43
    performing a search, 41
Spatial analysis, 29–62
    digital maps, 32–34
    geocoding, 39–44
    geographical data types, 34–36
    hybrid database model, 31
    map image generation, 50–56
    object-oriented database model, 31
    POI searches, 56–60
    projections and coordinate systems, 37–38
    raster maps, 52–55
    real-time attribute editing, 61
    relational database model, 31
    routing, 45–50
    sample server, 29–31
    vector maps, 55–56
Spatial analysis technologies, 2

Spread spectrum, 13
Standard Location Immediate Report DTD,
        110
String of pearls, 72
Subscriber identity module (SIM), 12
Sun Microsystems, 115
Symbian OS, 105–6

T

T-Motion (Germany/Austria), 60
TCP/IP, *See* Transmission Control Protocol/
        Internet Protocol (TCP/IP)
TDMA, 13
Tegaron (DaimlerChrysler), 4, 14, 94, 109
Telematics, 94
    use of term, 3
TellMe Network Inc., 128
Tessellation, 32–33
Third-party POI databases, 57
3rd Generation Partnership Project, 71
Time difference of arrival (TDOA), 69–71
Time Division Multiple Access (TDMA), 13
Timing advance (TA) information, 66
TomTom CityMaps, 105–6
Traffic, 120–23
    historical and congestion-based traffic,
        122
    incident-based traffic, 121
    linking to maps, 122–23
    product integration and use in LBS, 122–
        23
Transactive content, 82
Transmission Control Protocol/Internet
        Protocol (TCP/IP), 13, 112
Travel services, 127–29
Travelocity, 128
True directions projections, 38
Trueposition, 71

U

UDP protocol, 112
Universal Transverse Mercator (UTM), 38
U.S. Wireless Communications and Public
        Safety Act of 1999, 89
U.S. Wireless Corporation, 72
User profiling, 83

V

Vector maps:
    advantages, 55
    disadvantages, 55–56
Vector Markup Language (VML), 50
ViaMichelin, 60
Visitor location register (VLR), 12
VML, *See* Vector Markup Language (VML)
Voice system challenge (Europe), 118
VoiceXML (VXML), 8, 23, 101, 113–14
    sample document, 114
VoiceXML (VXML) Forum, 114

W

WAP, *See* Wireless Application Protocol
        (WAP)
Web clipping format, 103
Websphere Everyplace Server (IBM), 93
Williams, Larry, 124, 125
Windows CE for Automotive (WCEfA), 104
Wingcast (Ford Motor Company), 4–5
Wireless Application Protocol (WAP), 8, 91,
        112–13
    security for, 77–79
Wireless data, 13–14
Wireless data networks, 2
Wireless Markup Language Script (WMLS),
        115
Wireless Markup Language (WML), 18, 23,
        101, 112–13
Wireless networks, 9–14
    amplitude, 9
    analog and digital radio systems, 11
        attenuation, 9
        cellular networks, 11–13
        electromagnetic spectrum, 10
        electromagnetic waves, generating, 10
        frequency, 9
        modulation, 11
        phase modulation, 11
        radio basics, 9–10
        radio signals, 9–11
        transverse waves, 9–10
        wavebands, 11
        wavelength, 9
        wireless data, 13–14
Wireless personal digital assistants (PDAs), 1
Wireless Transaction Protocol (WTP), 112
Wireless Transport Layer Security (WTLS),
        77–79, 112
WML, *See* Wireless Markup Language
        (WML)
WMLS, *See* Wireless Markup Language
        Script (WMLS)
WTLS, *See* Wireless Transport Layer Secu-
        rity (WTLS)
WTP, *See* Wireless Transaction Protocol
        (WTP)

X

XML (Extensible Markup Language), 8, 101
XSL Transformations (XSLT), 110

Z

ZagMe, 129

http://www.phptr.com/

Prentice Hall PTR   InformIT   InformIT Online Books   Financial Times Prentice Hall   ft.com   PTG Interactive   Reuters

TOMORROW'S SOLUTIONS FOR TODAY'S PROFESSIONALS

Prentice Hall Professional Technical Reference

Browse | Book Series | What's New | User Groups | Alliances | Special Sales | Contact Us

Search | Help | Home

Quick Search

PTR Favorites

Find a Bookstore

Book Series

Special Interests

Newsletters

Press Room

International

Best Sellers

Solutions Beyond the Book

Shopping Bag

Keep Up to Date with
PH PTR Online

We strive to stay on the cutting edge of what's happening in professional computer science and engineering. Here's a bit of what you'll find when you stop by **www.phptr.com**:

What's new at PHPTR? We don't just publish books for the professional community, we're a part of it. Check out our convention schedule, keep up with your favorite authors, and get the latest reviews and press releases on topics of interest to you.

Special interest areas offering our latest books, book series, features of the month, related links, and other useful information to help you get the job done.

User Groups Prentice Hall Professional Technical Reference's User Group Program helps volunteer, not-for-profit user groups provide their members with training and information about cutting-edge technology.

Companion Websites Our Companion Websites provide valuable solutions beyond the book. Here you can download the source code, get updates and corrections, chat with other users and the author about the book, or discover links to other websites on this topic.

Need to find a bookstore? Chances are, there's a bookseller near you that carries a broad selection of PTR titles. Locate a Magnet bookstore near you at www.phptr.com.

Subscribe today! Join PHPTR's monthly email newsletter! Want to be kept up-to-date on your area of interest? Choose a targeted category on our website, and we'll keep you informed of the latest PHPTR products, author events, reviews and conferences in your interest area.

Visit our mailroom to subscribe today! **http://www.phptr.com/mail_lists**

LICENSE AGREEMENT AND LIMITED WARRANTY

Company's only obligation under these limited warranties is, at the Company's option, return of the warranted item for a refund of any amounts paid by you or replacement of the item. Any replacement of SOFTWARE or media under the warranties shall not extend the original warranty period. The limited warranty set forth above shall not apply to any SOFTWARE which the Company determines in good faith has been subject to misuse, neglect, improper installation, repair, alteration, or damage by you. EXCEPT FOR THE EXPRESSED WARRANTIES SET FORTH ABOVE, THE COMPANY DISCLAIMS ALL WARRANTIES, EXPRESS OR IMPLIED, INCLUDING WITHOUT LIMITATION, THE IMPLIED WARRANTIES OF MERCHANTABILITY AND FITNESS FOR A PARTICULAR PURPOSE. EXCEPT FOR THE EXPRESS WARRANTY SET FORTH ABOVE, THE COMPANY DOES NOT WARRANT, GUARANTEE, OR MAKE ANY REPRESENTATION REGARDING THE USE OR THE RESULTS OF THE USE OF THE SOFTWARE IN TERMS OF ITS CORRECTNESS, ACCURACY, RELIABILITY, CURRENTNESS, OR OTHERWISE.

IN NO EVENT, SHALL THE COMPANY OR ITS EMPLOYEES, AGENTS, SUPPLIERS, OR CONTRACTORS BE LIABLE FOR ANY INCIDENTAL, INDIRECT, SPECIAL, OR CONSEQUENTIAL DAMAGES ARISING OUT OF OR IN CONNECTION WITH THE LICENSE GRANTED UNDER THIS AGREEMENT, OR FOR LOSS OF USE, LOSS OF DATA, LOSS OF INCOME OR PROFIT, OR OTHER LOSSES, SUSTAINED AS A RESULT OF INJURY TO ANY PERSON, OR LOSS OF OR DAMAGE TO PROPERTY, OR CLAIMS OF THIRD PARTIES, EVEN IF THE COMPANY OR AN AUTHORIZED REPRESENTATIVE OF THE COMPANY HAS BEEN ADVISED OF THE POSSIBILITY OF SUCH DAMAGES. IN NO EVENT SHALL LIABILITY OF THE COMPANY FOR DAMAGES WITH RESPECT TO THE SOFTWARE EXCEED THE AMOUNTS ACTUALLY PAID BY YOU, IF ANY, FOR THE SOFTWARE.

SOME JURISDICTIONS DO NOT ALLOW THE LIMITATION OF IMPLIED WARRANTIES OR LIABILITY FOR INCIDENTAL, INDIRECT, SPECIAL, OR CONSEQUENTIAL DAMAGES, SO THE ABOVE LIMITATIONS MAY NOT ALWAYS APPLY. THE WARRANTIES IN THIS AGREEMENT GIVE YOU SPECIFIC LEGAL RIGHTS AND YOU MAY ALSO HAVE OTHER RIGHTS WHICH VARY IN ACCORDANCE WITH LOCAL LAW.

ACKNOWLEDGMENT

YOU ACKNOWLEDGE THAT YOU HAVE READ THIS AGREEMENT, UNDERSTAND IT, AND AGREE TO BE BOUND BY ITS TERMS AND CONDITIONS. YOU ALSO AGREE THAT THIS AGREEMENT IS THE COMPLETE AND EXCLUSIVE STATEMENT OF THE AGREEMENT BETWEEN YOU AND THE COMPANY AND SUPERSEDES ALL PROPOSALS OR PRIOR AGREEMENTS, ORAL, OR WRITTEN, AND ANY OTHER COMMUNICATIONS BETWEEN YOU AND THE COMPANY OR ANY REPRESENTATIVE OF THE COMPANY RELATING TO THE SUBJECT MATTER OF THIS AGREEMENT.

Should you have any questions concerning this Agreement or if you wish to contact the Company for any reason, please contact in writing at the address below.

Robin Short
Prentice Hall PTR
One Lake Street
Upper Saddle River, New Jersey 07458

About the CD-ROM

The CD-ROM included with *Mobile Location Services: The Definitive Guide* contains a number of resources that you may find valuable in developing mobile location services. Located in the /resources directory of the CD-ROM is a list of technical and development related Internet Web sites. The /software directory contains several software products that provide a hands-on, code-level perspective of the concepts discussed in this book or serve as tools for proof of concept development.

Internet Resources (in the /resources directory)

A list of Internet Web sites related to mobile location services.

Software (in the /software directory)

- **JaGo 0.6.5 (http://katla.giub.uni-bonn.de/jago/)**
  JaGo is a Java framework for geospatial solutions. It is based on common GI standards (ISO, OGC) and allows building applications with spatially referenced content.
  Software Requirements: Java 2 Standard Edition Software Development Kit

- **MapServer 3.5 (http://mapserver.gis.umn.edu/)**
  MapServer is an OpenSource development environment for building spatially enabled Internet applications. The software builds upon other popular OpenSource or freeware systems like Shapelib, FreeType, Proj.4, libTIFF, Perl, and others. MapServer will run where most commercial systems won't or can't, on Linux/Apache platforms.
  Software Requirements: HTTP Web Server (Apache or Microsoft IIS)

- **OpenMap 4.5 (http://openmap.bbn.com/)**
  OpenMap™ is a JavaBeans™-based toolkit for building applications and applets needing geographic information. Using OpenMap components, you can access data from legacy applications, in-place, in a distributed setting. At its core, OpenMap is a set of Swing components that understand geographic coordinates. These components help you show map data, and help you handle user input events to manipulate that data.
  Software Requirements: Application Server (such as Apache TomCat). Java 2 Standard Edition Software Development Kit.

- **MapIt! 1.0.2.1 (http://www.mapit.de/index.en.html)**
  MapIt! is a serverside Web application for raster maps. Navigation and points of interest are easily configured.
  Software Requirements: No additional software required for Windows. Linux requires Python and the Python Imaging Library.

Technical Support

Prentice Hall does not offer technical support for any of the programs on the CD-ROM. However, if the CD-ROM is damaged, you may obtain a replacement copy by sending an e-mail that describes the problem to: disc_exchange@prenhall.com.